Digital Signal Processing
and Applications with the
TMS320C6713 and
TMS320C6416 DSK

TOPICS IN DIGITAL SIGNAL PROCESSING

Digital Signal Processing and Applications with the TMS320C6713 and TMS320C6416 DSK

SECOND EDITION

Rulph Chassaing

Worcester Polytechnic Institute

Donald Reay

Heriot-Watt University

WILEY-INTERSCIENCE

A JOHN WILEY & SONS, INC., PUBLICATION

Published by John Wiley & Sons, Inc., Hoboken, New Jersey
Published simultaneously in Canada

For general information on our other products and services or for technical support, please contact
our Customer Care Department within the United States at 877-762-2974, outside the United States
at 317-572-3993 or fax 317-572-4002.

Wiley also publishes its books in a variety of electronic formats. Some content that appears in print
may not be available in electronic formats. For more information about Wiley products, visit our
web site at www.wiley.com.

Library of Congress Cataloging-in-Publication Data:

Chassaing, Rulph.
  Digital signal processing and applications with the TMS320C6713 and TMS320C6416 DSK /
Rulph Chassaing, Donald Reay.—2nd ed.
    p. cm.
  ISBN 978-0-470-13866-3 (cloth/CD)
  1. Signal processing–Digital techniques. 2. Texas Instruments TMS320
series microprocessors–Programming. I. Reay, Donald. II. Title.
  TK5102.9.C47422 2008
  621.382′2—dc22

                                                                        20070290065

Printed in the United States of America

10 9 8 7 6 5 4 3 2 1

To Reiko
And to the memory of Rulph and of Jay

Contents

Preface

Since the publication of the first edition of this book in 2005, Texas Instruments has released a new version of Code Composer Studio (CCS). Consequently, although nearly all of the program examples presented in the first edition will work with the DSK, some of the detailed instructions for using CCS described in the first edition are no longer accurate. Every effort has been made to ensure that this edition is compatible with Version 3 of Code Composer Studio. Slight changes have been made to the program examples to the extent that the examples provided with this edition should not be mixed with the earlier versions.

Sadly, Rulph Chassaing passed away in 2005. I had the privilege and the pleasure of being able to work with Rulph after attending his workshop at the TI developer conference in 1999. We corresponded regularly while he was writing his book on the C6711 DSK and Rulph kindly included some of the program examples I had developed. I helped Rulph to present a workshop at the TI developer conference in 2002, and we maintained contact while he wrote the first edition of this book. I have used Rulph's books, on the C31, C6711, and C6713 processors, for teaching both at Heriot-Watt (UK) and at Zhejiang (PRC) universities.

Rulph's books are an extensive and valuable resource for teaching DSP hands-on in a laboratory setting. They contain a wealth of practical examples—programs that run on TI DSKs (nearly all in real-time) and illustrate vividly many key concepts in digital signal processing. It would have been a great shame if the continued use of this text had been compromised by incompatibilities with the latest version of CCS.

While thoroughly checking the first edition and attempting to ensure the compatibility (with CCS) and integrity of the example programs, I have taken the opportunity to develop and to add more (particularly in Chapters 2, 5, 6, and 9) and to evolve a slightly more narrative structure (particularly in Chapters 2, 4, 5, and 6).

A small amount of material from the first edition has been dropped. Due to their natures, Chapters 3, 8, and 10 have been left very much unchanged.

While it contains a degree of introductory DSP theory, some details of the architecture of the C6713 and C6416 processors, an introduction to assembly language programming for those processors, and no little instruction on the use of Code Composer Studio, the emphasis of this book is on illustrating DSP concepts hands-on in a laboratory environment using real audio frequency signals.

The strength of this book lies, I believe, in the number (and utility) of program examples. I hope that professors and instructors will be able to pick material from the book in order to hold their own hands-on laboratory classes.

I am thankful to Robert Owen of the Texas Instruments University Program in Europe for support of the DSP teaching facilities at Heriot-Watt University and to Cathy Wicks of the Texas Instruments University Program in North America for the initial suggestion of updating the book and for her continued support. Walter J. Gomes III (Jay) and I mapped out the update to this book before he passed away last year. The thought of his enthusiasm for the project has been a constant motivation.

I thank my colleague at Heriot-Watt University, Dr. Keith Brown, for his help in testing program examples and for his suggestions. But above all, I thank Rulph for inspiring me to get involved in teaching hands-on DSP.

DONALD REAY

Heriot-Watt University
Edinburgh, United Kingdom
January 2008

Preface to the First Edition

Digital signal processors, such as the TMS320 family of processors, are used in a wide range of applications, such as in communications, controls, speech processing, and so on. They are used in cellular phones, digital cameras, high-definition television (HDTV), radio, fax transmission, modems, and other devices. These devices have also found their way into the university classroom, where they provide an economical way to introduce real-time digital signal processing (DSP) to the student.

Texas Instruments introduced the TM320C6x processor, based on the very-long-instruction-word (VLIW) architecture. This new architecture supports features that facilitate the development of efficient high-level language compilers. Throughout the book we refer to the C/C++ language simply as C. Although TMS320C6x/assembly language can produce fast code, problems with documentation and maintenance may exist. With the available C compiler, the programmer must "let the tools do the work." After that, if the programmer is not satisfied, Chapters 3 and 8 and the last few examples in Chapter 4 can be very useful.

This book is intended primarily for senior undergraduate and first-year graduate students in electrical and computer engineering and as a tutorial for the practicing engineer. It is written with the conviction that the principles of DSP can best be learned through interaction in a laboratory setting, where students can appreciate the concepts of DSP through real-time implementation of experiments and projects. The background assumed is a course in linear systems and some knowledge of C.

Most chapters begin with a theoretical discussion, followed by representative examples that provide the necessary background to perform the concluding experiments. There are a total of 105 programming examples, most using C code, with a few in assembly and linear assembly code. A list of these examples appears on page xvii. A total of 22 students' projects are also discussed. These projects cover a wide range of applications in filtering, spectrum analysis, modulation techniques, speech processing, and so on.

Programming examples are included throughout the text. This can be useful to the reader who is familiar with both DSP and C programming but who is not necessarily an expert in both. Many assignments are included at the end of Chapters 1–6.

This book can be used in the following ways:

1. For a DSP course with a laboratory component, using parts of Chapters 1–9. If needed, the book can be supplemented with some additional theoretical materials, since its emphasis is on the practical aspects of DSP. It is possible to cover Chapter 7 on adaptive filtering following Chapter 4 on finite impulse response (FIR) filtering (since there is only one example in Chapter 7 that uses materials from Chapter 5). It is my conviction that adaptive filtering should be incorporated into an undergraduate course in DSP.

2. For a laboratory course using many of the examples and experiments from Chapters 1–7 and Chapter 9. The beginning of the semester can be devoted to short programming examples and experiments and the remainder of the semester for a final project. The wide range of sample projects (for both undergraduate and graduate students) discussed in Chapter 10 can be very valuable.

3. For a senior undergraduate or first-year graduate design project course using selected materials from Chapters 1–10.

4. For the practicing engineer as a tutorial and reference, and for workshops and seminars, using selected materials throughout the book.

In Chapter 1 we introduce the tools through three programming examples. These tools include the powerful Code Composer Studio (CCS) provided with the TMS320C6713 DSP starter kit (DSK). It is essential to perform these examples before proceeding to subsequent chapters. They illustrate the capabilities of CCS for debugging, plotting in both the time and frequency domains, and other matters. Appendix H contains several programming examples using the TMS320C6416 DSK.

In Chapter 2 we illustrate input and output (I/O) with the AIC23 stereo codec on the DSK board through many programming examples. Chapter 3 covers the architecture and the instructions available for the TMS320C6x processor. Special instructions and assembler directives that are useful in DSP are discussed. Programming examples using both assembly and linear assembly are included in this chapter.

In Chapter 4 we introduce the z-transform and discuss FIR filters and the effect of window functions on these filters. Chapter 5 covers infinite impulse response (IIR) filters. Programming examples to implement real-time FIR and IIR filters are included. Appendix D illustrates MATLAB for the design of FIR and IIR filters.

Chapter 6 covers the development of the fast Fourier transform (FFT). Programming examples on FFT are included using both radix-2 and radix-4 FFT. In Chapter 7 we demonstrate the usefulness of the adaptive filter for a number of applications with least mean squares (LMS). Programming examples are included to illustrate the gradual cancellation of noise or system identification. Students have been very receptive to applications in adaptive filtering. Chapter 8 illustrates techniques for code optimization.

In Chapter 9 we introduce DSP/BIOS and discuss a number of schemes (Visual C++, MATLAB, etc.) for real-time data transfer (RTDX) and communication between the PC and the DSK.

Chapter 10 discusses a total of 22 projects implemented by undergraduate and graduate students. They cover a wide range of DSP applications in filtering, spectrum analysis, modulation schemes, speech processing, and so on.

A CD is included with this book and contains all the programs discussed. See page xxi for a list of the folders that contain the support files for the examples and projects.

Over the last 10 years, faculty members from over 200 institutions have taken my workshops on "DSP and Applications." Many of these workshops were supported by grants from the National Science Foundation (NSF) and, subsequently, by Texas Instruments. I am thankful to NSF, Texas Instruments, and the participating faculty members for their encouragement and feedback. I am grateful to Dr. Donald Reay of Heriot-Watt University, who contributed several examples during his review of my previous book based on the TMS320C6711 DSK. I appreciate the many suggestions made by Dr. Mounir Boukadoum of the University of Quebec, Dr. Subramaniam Ganesan from Oakland University, and Dr. David Kozel from Purdue University at Calumet. I also thank Dr. Darrell Horning of the University of New Haven, with whom I coauthored my first book, *Digital Signal Processing with the TMS320C25*, for introducing me to "book writing." I thank all the students at Roger Williams University, the University of Massachusetts at Dartmouth, and Worcester Polytechnic Institute (WPI) who have taken my real-time DSP and senior design project courses, based on the TMS320 processors, over the last 20 years. The contribution of Aghogho Obi, from WPI, is very much appreciated.

The continued support of many people from Texas Instruments is also very much appreciated: Cathy Wicks and Christina Peterson, in particular, have been very supportive of this book.

Special appreciation: The laboratory assistance of Walter J. Gomes III in several workshops and during the development of many examples has been invaluable. His contribution is appreciated.

RULPH CHASSAING

List of Examples

Programs/Files on Accompanying CD

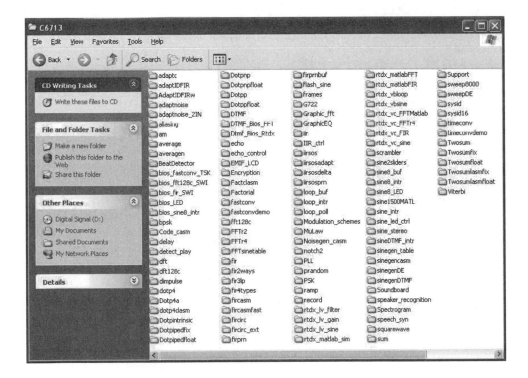

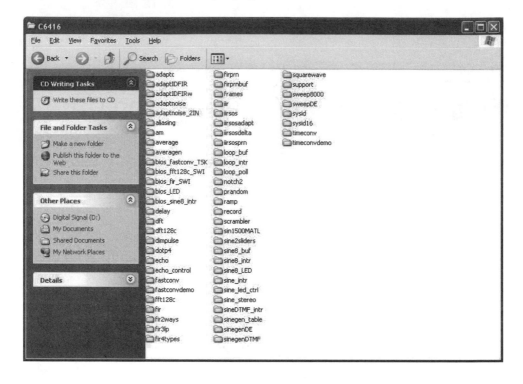

1

DSP Development System

- Installing and testing Code Composer Studio Version 3.1
- Use of the TMS320C6713 or TMS320C6416 DSK
- Programming examples

This chapter describes how to install and test Texas Instruments' integrated development environment (IDE), Code Composer Studio (CCS), for either the TMS320C6713 or the TMS320C6416 Digital Signal Processing Starter Kit (DSK). Three example programs that demonstrate hardware and software features of the DSK and CCS are presented. It is recommended strongly that you review these examples before proceeding to subsequent chapters. The detailed instructions contained in this chapter are specific to CCS Version 3.1.

1.1 INTRODUCTION

The Texas Instruments TMS320C6713 and TMS320C6416 Digital Signal Processing Starter Kits are low cost development platforms for real-time digital signal processing applications. Each comprises a small circuit board containing either a TMS320C6713 floating-point digital signal processor or a TMS320C6416 fixed-point digital signal processor and a TLV320AIC23 analog interface circuit (codec) and connects to a host PC via a USB port. PC software in the form of Code Composer Studio (CCS) is provided in order to enable software written in C or assembly

Digital Signal Processing and Applications with the TMS320C6713 and TMS320C6416 DSK,
Second Edition   By Rulph Chassaing and Donald Reay
Copyright © 2008 John Wiley & Sons, Inc.

language to be compiled and/or assembled, linked, and downloaded to run on the DSK. Details of the TMS320C6713, TMS320C6416, TLV320AIC23, DSK, and CCS can be found in their associated datasheets [36–38]. The purpose of this chapter is to introduce the installation and use of either DSK.

A digital signal processor (DSP) is a specialized form of microprocessor. The architecture and instruction set of a DSP are optimized for real-time digital signal processing. Typical optimizations include hardware multiply-accumulate (MAC) provision, hardware circular and bit-reversed addressing capabilities (for efficient implementation of data buffers and fast Fourier transform computation), and Harvard architecture (independent program and data memory systems). In many cases, DSPs resemble microcontrollers insofar as they provide single chip computer solutions incorporating onboard volatile and nonvolatile memory and a range of peripheral interfaces and have a small footprint, making them ideal for embedded applications. In addition, DSPs tend to have low power consumption requirements. This attribute has been extremely important in establishing the use of DSPs in cellular handsets. As may be apparent from the foregoing, the distinctions between DSPs and other, more general purpose, microprocessors are blurred. No strict definition of a DSP exists. Semiconductor manufacturers bestow the name DSP on products exhibiting some, but not necessarily all, of the above characteristics as they see fit.

The C6x notation is used to designate a member of the Texas Instruments (TI) TMS320C6000 family of digital signal processors. The architecture of the C6x digital signal processor is very well suited to numerically intensive calculations. Based on a very-long-instruction-word (VLIW) architecture, the C6x is considered to be TI's most powerful processor family.

Digital signal processors are used for a wide range of applications, from communications and control to speech and image processing. They are found in cellular phones, fax/modems, disk drives, radios, printers, hearing aids, MP3 players, HDTV, digital cameras, and so on. Specialized (particularly in terms of their onboard peripherals) DSPs are used in electric motor drives and a range of associated automotive and industrial applications. Overall, DSPs are concerned primarily with real-time signal processing. Real-time processing means that the processing must keep pace with some external event; whereas nonreal-time processing has no such timing constraint. The external event to keep pace with is usually the analog input. While analog-based systems with discrete electronic components including resistors and capacitors are sensitive to temperature changes, DSP-based systems are less affected by environmental conditions such as temperature. DSPs enjoy the major advantages of microprocessors. They are easy to use, flexible, and economical.

A number of books and articles have been published that address the importance of digital signal processors for a number of applications [1–22]. Various technologies have been used for real-time processing, from fiber optics for very high frequency applications to DSPs suitable for the audio frequency range. Common applications using these processors have been for frequencies from 0 to 96 kHz. It is standard within telecommunications systems to sample speech at 8 kHz (one sample every 0.125 ms). Audio systems commonly use sample rates of 44.1 kHz (compact disk) or 48 kHz. Analog/digital (A/D)-based data-logging boards in the megahertz sampling rate range are currently available.

1.2 DSK SUPPORT TOOLS

Most of the work presented in this book involves the development and testing of short programs to demonstrate DSP concepts. To perform the experiments described in the book, the following tools are used:

1. *A Texas Instruments DSP starter kit (DSK).* The DSK package includes:
   (a) Code Composer Studio (CCS), which provides the necessary software support tools. CCS provides an integrated development environment (IDE), bringing together the C compiler, assembler, linker, debugger, and so on.
   (b) A circuit board (the TMS320C6713 DSK is shown in Figure 1.1) containing a digital signal processor and a 16-bit stereo codec for analog signal input and output.
   (c) A universal synchronous bus (USB) cable that connects the DSK board to a PC.
   (d) A +5 V universal power supply for the DSK board.
2. *A PC.* The DSK board connects to the USB port of the PC through the USB cable included with the DSK package.
3. *An oscilloscope, spectrum analyzer, signal generator, headphones, microphone, and speakers.* The experiments presented in subsequent chapters of this book are intended to demonstrate digital signal processing concepts in real-time, using audio frequency analog input and output signals. In order to appreciate those concepts and to get the greatest benefit from the experiments, some forms of signal source and sink are required. As a bare minimum, a microphone and either headphones or speakers are required. A far greater benefit will be acquired if a signal generator is used to generate sinusoidal, and other, test signals and an oscilloscope and spectrum analyzer are used to display, measure, and analyze input and output signals. Many modern digital oscilloscopes incorporate FFT functions, allowing the frequency content of signals to be displayed. Alternatively, a number of software packages that use a PC equipped with a soundcard to implement virtual instruments are available.

All the files and programs listed and discussed in this book (apart from some of the student project files in Chapter 10) are included on the accompanying CD. A list of all the examples is given on pages xxi–xxvi.

1.2.1 C6713 and C6416 DSK Boards

The DSK packages are powerful, yet relatively inexpensive, with the necessary hardware and software support tools for real-time signal processing [23–43]. They are complete DSP systems. The DSK boards, which measure approximately 5 × 8 inches, include either a 225-MHz C6713 floating-point digital signal processor or a 1-GHz C6416 fixed-point digital signal processor and a 16-bit stereo codec TLV320AIC23 (AIC23) for analog input and output.

The onboard codec AIC23 [38] uses sigma–delta technology that provides analog-to-digital conversion (ADC) and digital-to-analog conversion (DAC) functions. It uses a 12-MHz system clock and its sampling rate can be selected from a range of alternative settings from 8 to 96 kHz.

A daughter card expansion facility is also provided on the DSK boards. Two 80-pin connectors provide for external peripheral and external memory interfaces.

The DSK boards each include 16 MB (megabytes) of synchronous dynamic RAM (SDRAM) and 512 kB (kilobytes) of flash memory. Four connectors on the boards provide analog input and output: MIC IN for microphone input, LINE IN for line input, LINE OUT for line output, and HEADPHONE for a headphone output (multiplexed with line output). The status of four user DIP switches on the DSK board can be read from within a program running on the DSP and provide the user with a feedback control interface. The states of four LEDs on the DSK board can be controlled from within a program running on the DSP. Also onboard the DSKs are voltage regulators that provide 1.26 V for the DSP cores and 3.3 V for their memory and peripherals.

1.2.2 TMS320C6713 Digital Signal Processor

The TMS320C6713 (C6713) is based on the very-long-instruction-word (VLIW) architecture, which is very well suited for numerically intensive algorithms. The internal program memory is structured so that a total of eight instructions can be fetched every cycle. For example, with a clock rate of 225 MHz, the C6713 is capable of fetching eight 32-bit instructions every 1/(225 MHz) or 4.44 ns.

Features of the C6713 include 264 kB of internal memory (8 kB as L1P and L1D Cache and 256 kB as L2 memory shared between program and data space), eight functional or execution units composed of six ALUs and two multiplier units, a 32-bit address bus to address 4 GB (gigabytes), and two sets of 32-bit general-purpose registers.

The C67xx processors (such as the C6701, C6711, and C6713) belong to the family of the C6x floating-point processors; whereas the C62xx and C64xx belong to the family of the C6x fixed-point processors. The C6713 is capable of both fixed- and floating-point processing. The architecture and instruction set of the C6713 are discussed in Chapter 3.

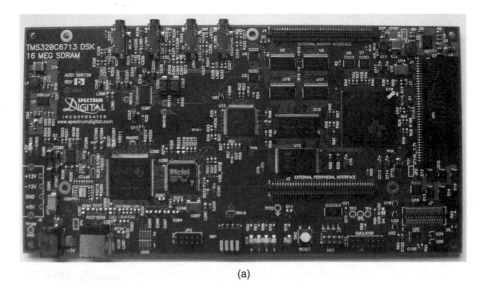

(a)

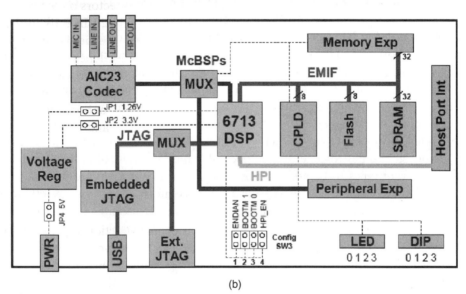

(b)

FIGURE 1.1. TMS3206713-based DSK board: (a) board and (b) block diagram. (Courtesy of Texas Instruments.)

1.2.3 TMS320C6416 Digital Signal Processor

The TMS320C6416 (C6416) is based on the VELOCITI advanced very-long-instruction-word (VLIW) architecture, which is very well suited for numerically intensive algorithms. The internal program memory is structured so that a total of eight instructions can be fetched every cycle. For example, with a clock rate of 1 GHz, the C6416 is capable of fetching eight 32-bit instructions every 1/(1 GHz) or 1.0 ns.

Features of the C6416 include 1056 kB of internal memory (32 kB as L1P and L1D cache and 1024 kB as L2 memory shared between program and data space), eight functional or execution units composed of six ALUs and two multiplier units, a 32-bit address bus to address 4 GB (gigabytes), and two sets of 32-bit general-purpose registers.

1.3 CODE COMPOSER STUDIO

Code Composer Studio (CCS) provides an integrated development environment (IDE) for real-time digital signal processing applications based on the C programming language. It incorporates a C compiler, an assembler, and a linker. It has graphical capabilities and supports real-time debugging.

The C compiler compiles a C source program with extension .c to produce an assembly source file with extension .asm. The assembler assembles an .asm source file to produce a machine language object file with extension .obj. The linker combines object files and object libraries as input to produce an executable file with extension .out. This executable file represents a linked common object file format (COFF), popular in Unix-based systems and adopted by several makers of digital signal processors [44]. This executable file can be loaded and run directly on the digital signal processor. Chapter 3 introduces the linear assembly source file with extension .sa, which is a "cross" between C and assembly code. A linear optimizer optimizes this source file to create an assembly file with extension .asm (similar to the task of the C compiler).

A Code Composer Studio project comprises all of the files (or links to all of the files) required in order to generate an executable file. A variety of options enabling files of different types to be added to or removed from a project are provided. In addition, a Code Composer Studio project contains information about exactly how files are to be used in order to generate an executable file. Compiler/linker options can be specified. A number of debugging features are available, including setting breakpoints and watching variables, viewing memory, registers, and mixed C and assembly code, graphing results, and monitoring execution time. One can step through a program in different ways (step into, or over, or out).

Real-time analysis can be performed using CCS's real-time data exchange (RTDX) facility. This allows for data exchange between the host PC and the target DSK as well as analysis in real-time without halting the target. The use of RTDX is illustrated in Chapter 9.

1.3.1 CCS Version 3.1 Installation and Support

Instructions for installation of CCS Version 3.1 are supplied with the DSKs. The default location for CCS files is *c:\CCStudio_v3.1* and the following instructions assume that that you have used this default. An icon with the label *6713 DSK CCStudio v3.1* (or *6416 DSK CCStudio v3.1*) should appear on the desktop.

CCS Version 3.1 provides extensive help facilities and a number of examples and tutorials are included with the DSK package. Further information (e.g., data sheets and application notes) are available on the Texas Instruments website http://www.ti.com.

1.3.2 Installation of Files Supplied with This Book

The great majority of the examples described in this book will run on either the C6713 or the C6416 DSK. However, there are differences, particularly concerning the library files used by the different processors, and for that reason a complete set of files is provided on the CD for each DSK. Depending on whether you are using a C6713 or a C6416 DSK, copy all of the subfolders, and their contents, supplied on the CD accompanying this book in folders C6416 or C6713 into the folder *c:\CCStudio_v3.1\MyProjects* so that, for example, the source file *sine8_LED.c* will be located at *c:\CCStudio_v3.1\MyProjects\sine8_LED\sine8_LED.c*.

Change the properties of all the files copied so that they are not read-only (all the folders can be highlighted to change the properties of their contents at once).

1.3.3 File Types

You will be working with a number of files with different extensions. They include:

1. file.pjt: to create and build a project named file.
2. file.c: C source program.
3. file.asm: assembly source program created by the user, by the C compiler, or by the linear optimizer.
4. file.sa: linear assembly source program. The linear optimizer uses *file.sa* as input to produce an assembly program *file.asm*.
5. file.h: header support file.
6. file.lib: library file, such as the run-time support library file rts6700.lib.
7. file.cmd: linker command file that maps sections to memory.
8. file.obj: object file created by the assembler.
9. file.out: executable file created by the linker to be loaded and run on the C6713 or C6416 processor.
10. file.cdb: configuration file when using DSP/BIOS.

1.4 QUICK TESTS OF THE DSK (ON POWER ON AND USING CCS)

1. On power on, a power on self-test (POST) program, stored by default in the onboard flash memory, uses routines from the board support library (BSL) to test the DSK. The source file for this program, post.c, is stored in folder

`c:\CCStudio_v3.1\examples\dsk6713\bsl\post`. It tests the internal, external, and flash memory, the two multichannel buffered serial ports (McBSP), DMA, the onboard codec, and the LEDs. If all tests are successful, all four LEDs blink three times and stop (with all LEDs on). During the testing of the codec, a 1-kHz tone is generated for 1 second.

2. Launch CCS from the icon on the desktop. A USB enumeration process will take place and the Code Composer Studio window will open.

3. Click on *Debug→Connect* and you should see the message "The target is now connected" appear (for a few seconds) in the bottom left-hand corner of the CCS window.

4. Click on *GEL→Check DSK→QuickTest*. The Quick Test can be used for confirmation of correct operation and installation. A message of the following form should then be displayed in a new window within CCS:

Switches: 15   *Board Revision*: 2   *CPLD Revision*: 2

The value displayed following the label Switches reflects the state of the four DIP switches on the edge of the DSK circuit board. A value of 15 corresponds to all four switches in the up position. Change the switches to $(1110)_2$, that is, the first three switches (0,1,2) up and the fourth switch (3) down. Click again on *GEL→Check DSK→QuickTest* and verify that the value displayed is now 7 ("Switches: 7"). You can set the value represented by the four user switches from 0 to 15. Programs running on the DSK can test the state of the DIP switches and react accordingly. The values displayed following the labels Board Revision and CPLD Revision depend on the type and revision of the DSK circuit board.

Alternative Quick Test of DSK Using Code Supplied with This Book

1. Open/launch CCS from the icon on the desktop if not done already.

2. Select *Debug→Connect* and check that the symbol in the bottom left-hand corner of the CCS window indicates connection to the DSK.

3. Select *File→Load Program* and load the file `c:\CCStudio_v3.1\MyProjects\` `sine8_LED\Debug\`*sine8_LED.out*. This loads the executable file *sine8_LED.out* into the digital signal processor. (This assumes that you have already copied all the folders on the accompanying CD into the folder: `c:\CCStudio_v3.1\MyProjects`.)

4. Select *Debug→Run*.

Check that the DSP is running. The word RUNNING should be displayed in the bottom left-hand corner of the CCS window.

Press DIP switch #0 down. LED #0 should light and a 1-kHz tone should be generated by the codec. Connect the LINE OUT (or the HEADPHONE) socket on the DSK board to a speaker, an oscilloscope, or headphones and verify the generation of the 1-kHz tone. The four connectors on the DSK board for input and output (MIC, LINE IN, LINE OUT, and HEADPHONE) each use a 3.5-mm jack audio cable. halt execution of program `sine8_LED.out` by selecting *Debug→Halt*.

1.5 PROGRAMMING EXAMPLES TO TEST THE DSK TOOLS

Three programming examples are introduced to illustrate some of the features of CCS and the DSK board. The aim of these examples is to enable the reader to become familiar with both the software and hardware tools that will be used throughout this book. It is strongly suggested that you complete these three examples before proceeding to subsequent chapters. The examples will be described assuming that a C6713 DSK is being used.

Example 1.1: Sine Wave Generation Using Eight Points with DIP Switch Control (*sine8_LED*)

This example generates a sinusoidal analog output waveform using a table-lookup method. More importantly, it illustrates some of the features of CCS for editing source files, building a project, accessing the code generation tools, and running a program on the C6713 processor. The C source file *sine8_LED.c* listed in Figure 1.2 is included in the folder *sine8_LED*.

Program Description

The operation of program *sine8_LED.c* is as follows. An array, `sine_table`, of eight 16-bit signed integers is declared and initialized to contain eight samples of exactly one cycle of a sinusoid. The value of `sine_table[i]` is equal to

$$1000\sin(2\pi i/8) \quad \text{for } i = 1, 2, 3, \ldots, 7$$

Within function *main()*, calls to functions *comm_poll()*, *DSK6713_LED_init()*, and *DSK6713_DIP_init()* initialize the DSK, the AIC23 codec onboard the DSK, and the two multichannel buffered serial ports (McBSPs) on the C6713 processor. Function *comm_poll()* is defined in the file *c6713dskinit.c*, and functions *DSK6713_LED_init()* and *DSK6713_DIP_init()* are supplied in the board support library (BSL) file `dsk6713bsl.lib`.

The program statement `while(1)` within the function *main()* creates an infinite loop. Within that loop, the state of DIP switch #0 is tested and if it is pressed down, LED #0 is switched on and a sample from the lookup table is output. If DIP switch #0 is not pressed down then LED #0 is switched off. As long as DIP switch #0 is pressed down, sample values read from the array `sine_table` will be output and a sinusoidal analog output waveform will be generated via the left-hand channel of the AIC23 codec and the LINE OUT and HEADPHONE sockets. Each time a sample value is read from the array `sine_table`, multiplied by the value of the variable `gain`, and written to the codec, the index, `loopindex`, into the array

```
//sine8_LED.c  sine generation with DIP switch control

#include "dsk6713_aic23.h"              //codec support
Uint32 fs = DSK6713_AIC23_FREQ_8KHZ;   //set sampling rate
#define DSK6713_AIC23_INPUT_MIC 0x0015
#define DSK6713_AIC23_INPUT_LINE 0x0011
Uint16 inputsource=DSK6713_AIC23_INPUT_MIC; //select input
#define LOOPLENGTH 8
short loopindex = 0;                    //table index
short gain = 10;                        //gain factor
short sine_table[LOOPLENGTH]=
  {0,707,1000,707,0,-707,-1000,-707};  //sine values

void main()
{
  comm_poll();                         //init DSK,codec,McBSP
  DSK6713_LED_init();                  //init LED from BSL
  DSK6713_DIP_init();                  //init DIP from BSL
  while(1)                             //infinite loop
  {
    if(DSK6713_DIP_get(0)==0)          //if DIP #0 pressed
    {
      DSK6713_LED_on();                //turn LED #0 ON
      output_left_sample(sine_table[loopindex++]*gain); //output
      if (loopindex >= LOOPLENGTH) loopindex = 0; //reset index
    }
    else DSK6713_LED_off(0);           //else turn LED #0 OFF
  }                                    //end of while(1)
}                                      //end of main
```

FIGURE 1.2. Sine wave generation program using eight points with DIP switch control (sine8_LED.c).

is incremented and when its value exceeds the allowable range for the array (LOOPLENGTH-1), it is reset to zero.

Each time the function output_left_sample(), defined in source file *C6713dskinit.c*, is called to output a sample value, it waits until the codec, initialized by the function comm_poll() to output samples at a rate of 8 kHz, is ready for the next sample. In this way, once DIP switch #0 has been pressed down it will be tested at a rate of 8 kHz. The sampling rate at which the codec operates is set by the program statement

```
Uint32 fs = DSK6713_AIC23_FREQ_8KHZ;
```

One cycle of the sinusoidal analog output waveform corresponds to eight output samples and hence the frequency of the sinusoidal analog output waveform is equal to the codec sampling rate (8 kHz) divided by eight, that is, 1 kHz.

Creating a Project

This section illustrates how to create a project, adding the necessary files to generate an executable file **sine8_LED.out**. As supplied on the CD, folder sine8_LED contains a suitable project file named sine8_LED.pjt. However, for the purposes of gaining familiarity with CCS, this section will illustrate how to create that project file from scratch.

1. **Delete the existing project file sine8_LED.pjt** in folder c:\CCStudio_v3.1\ myprojects\sine8_LED. Do this from outside CCS. Remember, a copy of the file sine8_LED.pjt still exists on the CD.

2. **Launch CCS** by double-clicking on its desktop icon.

3. **Create a new project file sine8_LED.pjt** by selecting *Project→New* and typing *sine8_LED* as the project name, as shown in Figure 1.3. **Set *Target* to *TMS320C67XX*** before clicking on *Finish*. The new project file will be saved in the folder c:\CCStudio_v3.1\myprojects*sine8_LED*. The .pjt file stores project information on build options, source filenames, and dependencies. The names of the files used by a project are displayed in the *Project View* window, which, by default, appears at the left-hand side of the Code Composer window.

4. **Add the source file sine8_LED.c to the project.** sine8_LED.c is the top level C source file containing the definition of function main(). This source file is stored in the folder sine8_LED and must be added to the project if it is to be used to generate the executable file sine8_LED.out. Select *Project→Add Files to Project* and look for *Files of Type C Source Files (*.c, *.ccc). Open,*

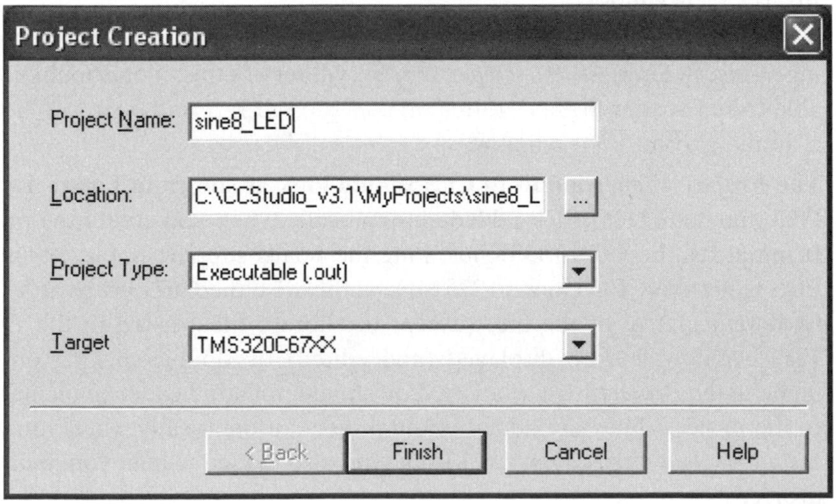

FIGURE 1.3. CCS *Project Creation* window for project sine8_LED.

or double-click on, sine8_LED.c. It should appear in the *Project View* window in the *Source* folder.

5. **Add the source file c6713dskinit.c to the project.** c6713dskinit.c contains the function definitions for a number of low level routines including comm._poll() and output_left_sample(). This source file is stored in the folder c:\CCStudio_v3.1\myprojects\Support. Select *Project→Add Files to Project* and look for *Files of Type C Source Files (*.c, *.ccc). Open*, or double-click on, c6713dskinit.c. It should appear in the *Project View* window in the *Source* folder.

6. **Add the source file vectors__poll.asm to the project.** vectors_poll.asm contains the interrupt service table for the C6713. This source file is stored in the folder c:\CCStudio_v3.1\myprojects\Support. Select *Project→Add Files to Project* and look for *Files of Type ASM Source Files (*.a*). Open*, or double-click on, vectors_poll.asm. It should appear in the *Project View* window in the *Source* folder.

7. **Add library support files *rts6700.lib, dsk6713bsl.lib*, and *csl6713.lib* to the project.** Three more times, select *Project→Add Files to Project* and look for *Files of Type Object and Library Files (*.o*, *.l*)* The three library files are stored in folders c:\CCStudio_v3.1\c6000\cgtools\lib, c:\CCStudio_v3.1\c6000\dsk6713\lib, and c:\CCStudio_v3.1\c6000\csl\lib, respectively. These are the run-time support (for C67x architecture), board support (for C6713 DSK), and chip support (for C6713 processor) library files.

8. **Add the linker command file c6713dsk.cmd to the project.** This file is stored in the folder c:\CCStudio_v3.1\myprojects\Support. Select *Project→Add Files to Project* and look for *Files of Type Linker Command File (*.cmd;*.lcf). Open*, or double-click on, c6713dsk.cmd. It should then appear in the *Project View* window.

9. No header files will be shown in the *Project View* window at this stage. Selecting *Project→Scan All File Dependencies* will rectify this. You should now be able to see header files c6713dskinit.h, dsk6713.h, and dsk6713_aic23.h, in the *Project View* window.

10. **The *Project View* window in CCS should look as shown in Figure 1.4.** The GEL file dsk6713.gel is added automatically when you create the project. It initializes the C6713 DSK invoking the board support library to use the PLL to set the CPU clock to 225 MHz (otherwise the C6713 runs at 50 MHz by default). Any of the files (except the library files) listed in the *Project View* window can be displayed (and edited) by double-clicking on their name in the *Project View* window. You should not add header or include files to the project. They are added to the project automatically when you select *Scan All File Dependencies*. (They are also added when you build the project.)

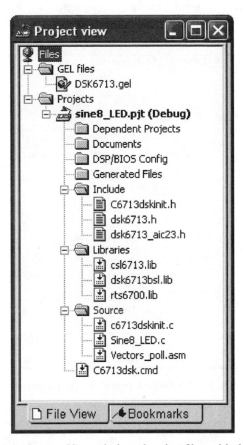

FIGURE 1.4. *Project View* window showing files added at step 10.

Verify from the *Project View* window that the project (.pjt) file, the linker command (.cmd) file, the three library (.lib) files, the two C source (.c) files, and the assembly (.asm) file have been added to the project.

Code Generation and Build Options

The code generation tools underlying CCS, that is, C compiler, assembler, and linker, have a number of options associated with each of them. These options must be set appropriately before attempting to build a project. Once set, these options will be stored in the project file.

Setting Compiler Options

Select *Project→Build Options* and click on the *Compiler* tab. Set the following options, as shown in Figures 1.5, 1.6, and 1.7. In the *Basic* category set *Target Version* to *C671x (-mv6710)*. In the *Advanced* category set *Memory Models* to *Far (–mem_*

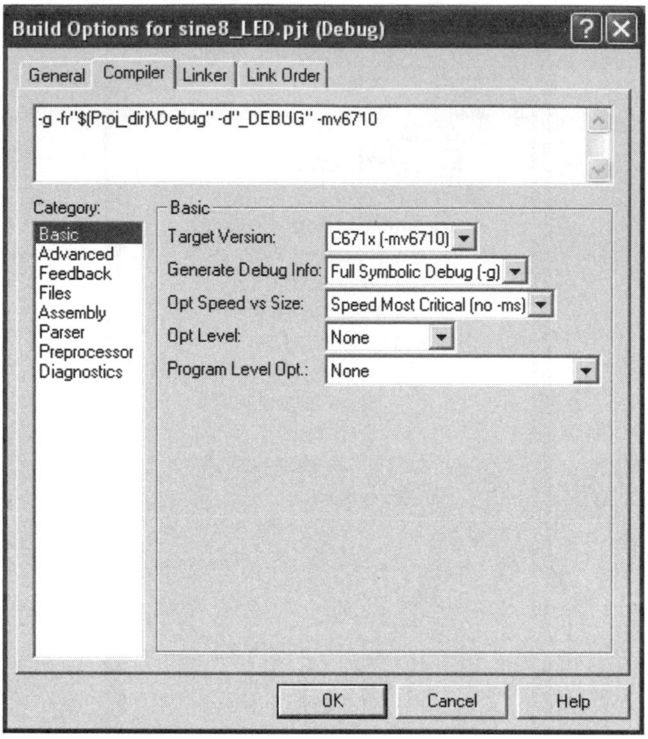

FIGURE 1.5. CCS Build Options: Basic compiler settings.

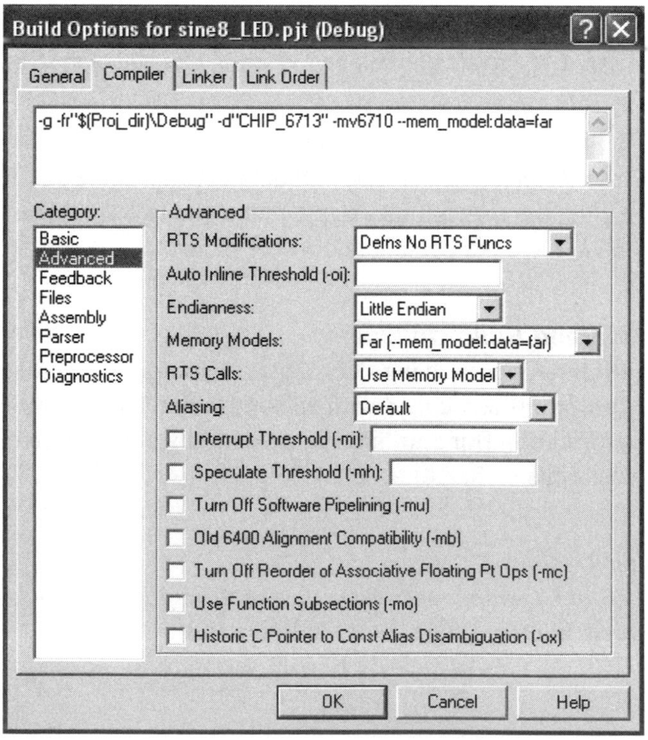

FIGURE 1.6. CCS Build Options: Advanced compiler settings.

FIGURE 1.7. CCS Build Options: Preprocessor compiler settings.

model:data=far). In the *Preprocessor* category set *Pre-Define Symbol* to *CHIP_6713* and *Include Search Path* to *c:\CCStudio_v3.1\C6000\dsk6713\include.* Compiler options are described in more detail in Ref. 28. Click on *OK.*

Setting Linker Options

Click on the *Linker* tab in the *Build Options* window, as shown in Figure 1.8. The *Output Filename* should default to *.\Debug\sine8_LED.out* based on the name of the project file and the *Autoinit Model* should default to *Run-Time Autoinitialization.* Set the following options (all in the *Basic* category). Set *Library Search Path* to *c:\CCStudio_v3.1\C6000\dsk6713\lib* and set *Include Libraries* to *rts6700.lib; dsk6713bsl.lib;csl6713.lib.* The map file can provide useful information for debugging (memory locations of functions, etc.). The –c option is used to initialize variables at run time, and the –o option is to name the linked executable output file sine8_LED.out. Click on *OK.*

Building, Downloading, and Running the Project

The project sine8_LED can now be built, and the executable file sine8_LED.out can be downloaded to the DSK and run.

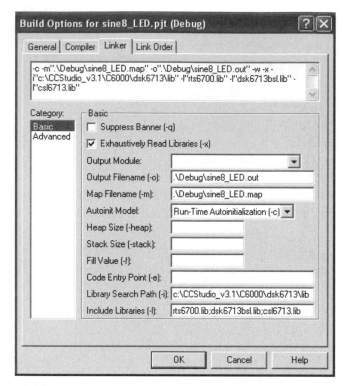

FIGURE 1.8. CCS Build Options: Basic Linker settings.

1. Build this project as **sine8_LED**. Select *Project→Rebuild All*. Or press the toolbar button with the three downward arrows. This compiles and assembles all the C files using `cl6x` and assembles the assembly file `vectors_poll.asm` using `asm6x`. The resulting object files are then linked with the library files using `lnk6x`. This creates an executable file `sine8_LED.out` that can be loaded into the C6713 processor and run. Note that the commands for compiling, assembling, and linking are performed with the Build option. A log file `cc_build_Debug.log` is created that shows the files that are compiled and assembled, along with the compiler options selected. It also lists the support functions that are used. The building process causes all the dependent files to be included (in case one forgets to scan for all the file dependencies). You should see a number of diagnostic messages, culminating in the message "Build Complete, 0 Errors, 0 Warnings, 0 Remarks" appear in an output window in the bottom left-hand side of the CCS window. It is possible that a warning about the Stack Size will have appeared. This can be ignored or can be suppressed by unchecking the *Warn About Output Sections* option in the *Advanced* category of *Linker Build Options*. Alternatively, it can be eliminated by setting the *Stack*

Size option in the *Advanced* category of *Linker Build Options* to a suitable value (e.g., 0x1000).

Connect to the DSK. Select *Debug→Connect* and check that the symbol in the bottom left-hand corner of the CCS window indicates connection to the DSK.

2. Select *File→Load Program* in order to load sine8_LED.out. It should be stored in the folder c:\CCStudio_v3.1\MyProjects\sine8_LED\Debug. Select *Debug→Run*. In order to verify that a sinusoidal output waveform with a frequency of 1 kHz is present at both the LINE OUT and HEADPHONE sockets on the DSK, when DIP switch #0 is pressed down, use an oscilloscope connected to the LINE OUT socket and a pair of headphones connected to the HEADPHONE socket.

Editing Source Files Within CCS

Carry out the following actions in order to practice editing source files.

1. Halt execution of the program (if it is running) by selecting *Debug→Halt*.
2. Double-click on the file sine8_LED.c in the *Project View* window. This should open a new window in CCS within which the source file is displayed and may be edited.
3. Delete the semicolon in the program statement

```
short gain = 10;
```

4. Select *Debug→Build* to perform an incremental build or use the toolbar button with the two (not three) downward arrows. The incremental build is chosen so that only the C source file sine8_LED.c is compiled. Using the Rebuild option (the toolbar button with three downward arrows), files compiled and/or assembled previously would again go through this unnecessary process.
5. Two error messages, highlighted in red, stating

```
"Sine8_LED.c", Line 11: error: expected a ";"
"Sine8_LED.c", Line 23: error: identifier "sine_table" is
undefined
```

should appear in the *Build* window of CCS (lower left). You may need to scroll-up the *Build* window for a better display of these error messages. Double-click on the first highlighted error message line. This should bring the cursor to the section of code where the error occurs. Make the appropriate correction (i.e. replace the semicolon) *Build* again, *Load*, and *Run* the program and verify your previous results.

Monitoring the Watch Window

Ensure that the processor is still running (and that DIP switch #0 is pressed down). Note the message "RUNNING" displayed at the bottom left of CCS. The *Watch* window allows you to change the value of a parameter or to monitor a variable:

1. Select *View→Quick Watch*. Type *gain*, then click on *Add to Watch*. The gain value of 10 set in the program in Figure 1.2 should appear in the Watch window.

2. Change `gain` from 10 to 30 in the *Watch* window. Press enter. Verify that the amplitude of the generated tone has increased (with the processor still running and DIP switch #0 pressed down). The amplitude of the sine wave should have increased from approximately 0.9 V p-p to approximately 2.5 V p-p.

Using a GEL Slider to Control the Gain

The General Extension Language (GEL) is an interpreted language similar to (a subset of) C. It allows you to change the value of a variable (e.g., gain) while the processor is running.

1. Select *File→Load GEL* and load the file `gain.gel` (in folder `sine8_LED`). Double-click on the filename `gain.gel` in the *Project View* window to view it within CCS. The file is listed in Figure 1.9. The format of a slider GEL function is

```
slider param_definition( minVal, maxVal, increment,
pageIncrement, paramName )
{
  statements
}
```

where `param_definition` identifies the slider and is displayed as the name of the slider window, `minVal` is the value assigned to the GEL variable `param-Name` when the slider is at its lowest level, `maxVal` is the value assigned to the

```
/*gain.gel GEL slider to vary amplitude of sine wave*/
/*generated by program sine8_LED.c*/

menuitem "Sine Gain"

slider Gain(0,30,4,1,gain_parameter) /*incr by 4, up to 30*/
{
  gain = gain_parameter;                 /*vary gain of sine*/
}
```

FIGURE 1.9. Listing of GEL file `gain.gel`.

GEL variable `paramName` when the slider is at its highest level, increment specifies the incremental change to the value of the GEL variable `paramName` made using the up- or down-arrow keys, and `pageIncrement` specifies the incremental change to the value of the GEL variable `paramName` made by clicking in the slider window.

In the case of `gain.gel`, the statement

```
gain = gain_parameter;
```

assigns the value of the GEL variable `gain_parameter` to the variable `gain` in program `sine8_LED`. The line

```
menuitem "Sine Gain"
```

sets the text that will appear as an option in the CCS *GEL* menu when `gain.gel` is loaded.

2. Select *GEL→Sine Gain→Gain*. This should bring out the slider window shown in Figure 1.10, with the minimum value of 0 set for the gain.
3. Press the up-arrow key three times to increase the gain value from 0 to 12. Verify that the peak-to-peak value of the sine wave generated is approximately 1.05 V. Press the up-arrow key again to continue increasing the slider, incrementing by 4 each time. The amplitude of the sine wave should be about 2.5 V p-p with the value of `gain` set to 30. Clicking in the *Gain* slider window above or below the current position of the slider will increment or decrement its value by 1. The slider can also be dragged up and down. Changes to the value of `gain` made using the slider are reflected in the *Watch* window.

Figure 1.11 shows several windows within CCS for the project `sine8_LED`.

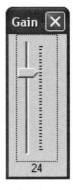

FIGURE 1.10. GEL slider used to vary gain in program `sine8_LED.c`.

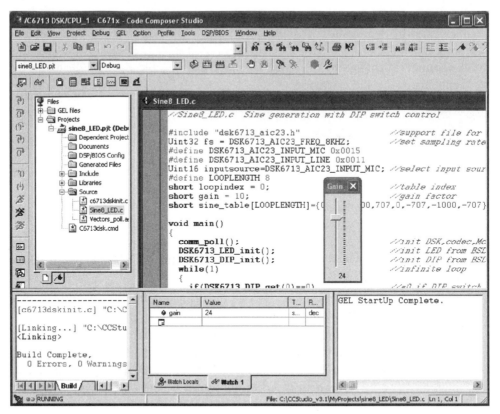

FIGURE 1.11. CCS windows for project sin8_LED, including *Watch* window and GEL slider.

Changing the Frequency of the Generated Sinusoid

There are several different ways in which the frequency of the sinusoid generated by program sine8_LED.c can be altered.

1. Change the AIC23 codec sampling frequency from 8 kHz to 16 kHz by changing the line that reads

   ```
   Uint32 fs = DSK6713_AIC23_FREQ_8KHZ;
   ```

   to read

   ```
   Uint32 fs = DSK6713_AIC23_FREQ_16KHZ;
   ```

   Rebuild (use incremental build) the project, load and run the new executable file, and verify that the frequency of the generated sinusoid is 2 kHz. The sampling frequencies supported by the AIC23 codec are 8, 16, 24, 32, 44.1, 48, and 96 kHz.

2. Change the number of samples stored in the lookup table to four. By changing the lines that read

```
#define LOOPLENGTH 8
short sine_table[LOOPLENGTH]={0,707,1000,707,0,-707,0,-1000,
-707};
```

to read

```
#define LOOPLENGTH 4
short sine_table[LOOPLENGTH]={0,1000,0,-1000};
```

Verify that the frequency of the sinusoid generated is 2 kHz (assuming an 8-kHz sampling frequency).

Remember that the sinusoid is no longer generated if the DIP switch #0 is not pressed down. A different DIP switch can be used to control whether or not a sinusoid is generated by changing the value of the parameter passed to the functions `DSK6713_DIP_get()`, `DSK6713_LED_on()`, and `DSK6713_LED_off()`. Suitable values are 0, 1, 2, and 3.

Two sliders can readily be used, one to change the gain and the other to change the frequency. A different signal frequency can be generated, by changing the incremental changes applied to the value of `loopindex` within the C program (e.g., stepping through every two points in the table). When you exit CCS after you build a project, all changes made to the project can be saved. You can later return to the project with the status as you left it before. For example, when returning to the project, after launching CCS, select *Project→Open* to open an existing project such as `sine8_LED.pjt` (with all the necessary files for the project already added).

Example 1.2: Generation of Sinusoid and Plotting with CCS (`sine8_buf`)

This example generates a sinusoidal analog output signal using eight precalculated and prestored sample values. However, it differs fundamentally from `sine8_LED` in that its operation is based on the use of interrupts. In addition, it uses a buffer to store the BUFFERLENGTH most recent output samples. It is used to illustrate the capabilities of CCS for plotting data in both time and frequency domains.

All the files necessary to build and run an executable file `sine8_BUF.out` are stored in folder `sine8_buf`. Program file *sine8_buf.c* is listed in Figure 1.12. Because a project file `sine8_buf.pjt` is supplied, there is no need to create a new

```
//sine8_buf.c sine generation with output stored in buffer

#include "DSK6713_AIC23.h"                    //codec support
Uint32 fs=DSK6713_AIC23_FREQ_8KHZ;           //set sampling rate
#define DSK6713_AIC23_INPUT_MIC 0x0015
#define DSK6713_AIC23_INPUT_LINE 0x0011
Uint16 inputsource=DSK6713_AIC23_INPUT_MIC; // select input
#define LOOPLENGTH 8
#define BUFFERLENGTH 256
int loopindex = 0;                              //table index
int bufindex = 0;                               //buffer index
short sine_table[LOOPLENGTH]={0,707,1000,707,0,-707,-1000,-707};
int out_buffer[BUFFERLENGTH];                  //output buffer
short gain = 10;
interrupt void c_int11()                  //interrupt service routine
   short out_sample;

   out_sample = sine_table[loopindex++]*gain;
   output_left_sample(out_sample);         //output sample value
   out_buffer[bufindex++] = out_sample;    //store in buffer
   if (loopindex >= LOOPLENGTH) loopindex = 0; //check end table
   if (bufindex >= BUFFERLENGTH) bufindex = 0; //check end buffer
   return;
}                                         //return from interrupt

void main()
{
   comm_intr();                            //initialise DSK
   while(1);                               //infinite loop
}
```

FIGURE 1.12. Listing of program sine8_buf.c.

project file, add files to it, or alter compiler and linker build options. In order to build, download and run program sine8_buf.c.

1. Close any open projects in CCS.

2. Open project *sine8_buf.pjt* by selecting *Project→Open* and double-clicking on file sine8_buf.pjt in folder sine8_buf. Because this program uses interrupt-driven input/output rather than polling, the file vectors_intr.asm is used in place of vectors_poll.asm. The interrupt service table specified in vectors_intr.asm associates the interrupt service routine c_int11() with hardware interrupt INT11, which is asserted by the AIC23 codec on the DSK at each sampling instant.

Within function main(), function comm_intr() is used in place of comm_poll(). This function is defined in file c6713dskinit.c and is described in more detail in Chapter 2. Essentially, it initializes the DSK hardware, including the AIC23 codec, such that the codec sampling rate is set according to the value of the variable `fs` and the codec interrupts the processor at every sampling instant. The statement `while(1)` in function `main()` creates an infinite loop, during which the processor waits for interrupts. On interrupt, execution proceeds to the interrupt service routine (ISR) `c_int11()`, which reads a new sample value from the array `sine_table` and writes it both to the array `out_buffer` and to the DAC using function `output_left_sample()`. Interrupts are discussed in more detail in Chapter 3.

Build this project as **sine8_buf**. Load and run the executable file *sine8_buf.out* and verify that a 1-kHz sinusoid is generated at the LINE OUT and HEADPHONE sockets (as in Example 1.1).

Graphical Displays in CCS

The array `out_buffer` is used to store the BUFFERLENGTH most recently output sample values. Once program execution has been halted, the data stored in `out_buffer` can be displayed graphically in CCS.

1. Select *View→Graph→Time/Frequency* and set the *Graph Property Dialog* properties as shown in Figure 1.13a. Figure 1.13b shows the resultant *Graphical Display* window.
2. Figure 1.14a shows the *Graph Property Dialog* window that corresponds to the frequency domain representation of the contents of `out_buffer` shown in Figure 1.14b. The spike at 1 kHz represents the frequency of the sinusoid generated by program `sine8_buf.c`.

Viewing and Saving Data from Memory into File

To view the contents of `out_buffer`, select *View→Memory*. Specify `out_buffer` as the *Address* and select *32-bit Signed Integer* as the *Format*, as shown in Figure 1.15a. The resultant *Memory* window is shown in Figure 1.15b.

To save the contents of `out_buffer` to a file, select *File→Data→Save*. Save the file as `sine8_buf.dat`, selecting data type *Integer*, in the folder `sine8_buf`. In the *Storing Memory into File* window, specify `out_buffer` as the *Address* and a *Length* of 256. The resulting file is a text file and you can plot this data using other applications (e.g., MATLAB). Although the values stored in array `sine_table` and passed to function `output_left_sample()` are 16-bit signed integers, array `out_buffer` is declared as type `int` (32-bit signed integer) in program `sine8_buf.c` to allow for the fact that there is no 16-bit Signed Integer data type option in the *Save Data* facility in CCS.

Example 1.3: Dot Product of Two Arrays (`dotp4`)

This example illustrates the use of breakpoints and single stepping within CCS. In addition, it illustrates the use of Code Composer's *Profile Clock* in order to estimate the time taken to execute a section of code.

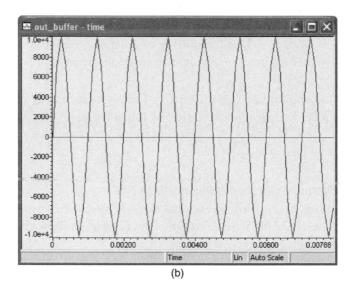

(a)

(b)

FIGURE 1.13. (a) *Graph Property* window and (b) Time domain plot of data stored in
out_buffer.

Multiply/accumulate is a very important operation in digital signal processing. It
is a fundamental part of digital filtering, correlation, and fast Fourier transform
algorithms. Since the multiplication operation is executed so commonly and is
essential for most digital signal processing algorithms, it is important that it executes
in a single instruction cycle. The C6713 and C6416 processors can perform two
multiply/accumulate operations within a single instruction cycle.

The C source file dotp4.c, listed in Figure 1.16, calculates the dot products of
two arrays of integer values. The first array is initialized using the four values 1, 2,
3, and 4, and the second array using the four values 0, 2, 4, and 6. The dot product
is $(1 \times 0) + (2 \times 2) + (3 \times 4) + (4 \times 6) = 40$.

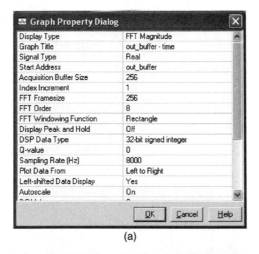

(a)

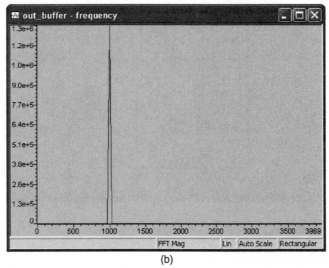

(b)

FIGURE 1.14. (a) *Graph Property* window and (b) Frequency domain plot of data stored in `out_buffer`.

The program can readily be modified to handle larger arrays. No real-time input or output is used in this example, and so the real-time support files `c6713dskinit.c` and `vectors_intr.asm` are not needed.

Build this project as **dotp4** ensuring that the following files are included in the project:

1. `dotp4.c`: C source file.
2. `6713dsk.cmd`: generic linker command file.
3. `rts6700.lib`: library file.

The *Project View* window should appear as shown in Figure 1.17.

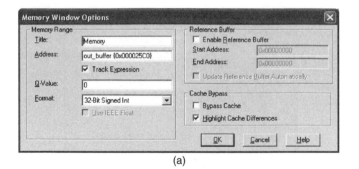

(a)

(b)

FIGURE 1.15. (a) *Memory* window settings and (b) *Memory* window view of data stored in out_buffer.

Implementing a Variable Watch

1. Select *Project→Build Options* and verify that the *Basic Compiler* settings are as shown in Figure 1.18. In this example it is important to **ensure that the optimization is disabled** (*Opt Level None*).

2. Build the project by clicking on the toolbar button with the three downward arrows (or select *Project→Build*). Load the executable file dotp4.out.

3. Select *View→Quick Watch*. Type sum to watch the variable sum, and click on *Add to Watch*. The message "identifier not found: sum" should be displayed in the *Watch* window. The variable sum is declared locally in function dotp() and until that function is called it does not exist.

4. Set a breakpoint at the line of code

```
sum += a[i] * b[i];
```

by clicking on that line in the source file dotp4.c and then either right-clicking and selecting *Toggle Software Breakpoint*, or clicking on the *Toggle Breakpoint* toolbar button. A red dot should appear to the left of that line of code.

```
//dotp4.c dot product of two vectors

int dotp(short *a, short *b, int ncount); //function prototype
#include <stdio.h>              //for printf
#define count 4                 //# of data in each array
short x[count] = {1,2,3,4};     //declaration of 1st array
short y[count] = {0,2,4,6};     //declaration of 2nd array

main()
{
  int result = 0;                      //result sum of products

  result = dotp(x, y, count);    //call dotp function
  printf("result = %d (decimal) \n", result); //print result
}

int dotp(short *a, short *b, int ncount) //dot product function
{
  int i;
  int sum = 0;
  for (i = 0; i < ncount; i++)
    sum += a[i] * b[i];          //sum of products
  return(sum);                   //return sum as result
}
```

FIGURE 1.16. Listing of program dotp4.c.

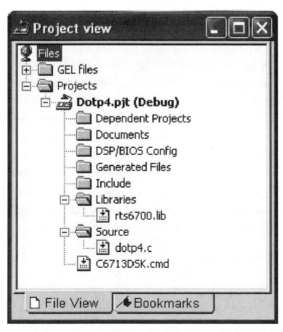

FIGURE 1.17. *Project View* window for project dotp4.

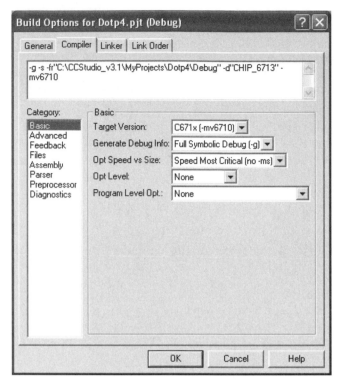

FIGURE 1.18. *Build Options* for project `dotp4`.

5. Select *Debug→Run* (or use the "running man" toolbar button). The program will execute up to, but not including, the line of code at which the breakpoint has been set. A yellow arrow will appear to the left of that line of code. At this point, a value of 0 for the variable `sum` should appear in the *Watch* window. `sum` is a variable that is local to function `dotp()`. Now that the function is being executed, the variable exists and its value can be displayed.

6. Continue program execution by selecting *Debug→Step Into*, or by using function key F11. Continue to single-step and watch the variable `sum` in the *Watch* window change in value through 0, 4, 16, and 40 (See Figure 1.19.).

7. Once the value of the variable `sum` has reached 40, select *Debug→Run* in order to complete execution of the program, and verify that the value returned by function `dotp()` is displayed as

```
result = 40 (decimal)
```

in the *Stdout* window. At this point, the message "identifier not found: sum" should be displayed in the *Watch* window again, reflecting the fact that execution of function `dotp()` has ended and that the local variable `sum` no longer exists.

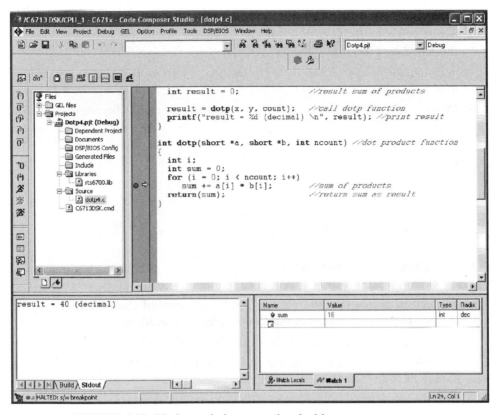

FIGURE 1.19. Various windows associated with program dotp4.c.

The printf() function is useful for debugging but its use should be avoided in real-time programs since it takes over 6000 instruction cycles to execute.

Animating

1. Select *File→Reload Program* to reload the executable file dotp4.out (alternatively, select *Debug→Restart*). After the executable file is loaded, or following restart, the program counter is set to the address labeled c_int00. This can be verified by looking at the *Disassembly* window.

2. The breakpoint set previously should still be set at the same line of code as before. Select *Debug→Animate* and watch the value of the variable *sum* displayed in the *Watch* window change. The speed of animation can be controlled by selecting *Option→Customize→Animate Speed* (by default, the maximum speed setting of 0 seconds is set).

Estimating Execution Time for Function dotp() Using the Profile Clock

The time taken to execute function dotp() can be estimated using Code Composer's *Profile Clock*.

1. Open project `dotp4.pjt`.

2. Select *Project→Build Options*. In the *Compiler* tab in the *Basic* category set the *Opt Level* to *none*.

3. Select *Project→Build* and then *File→Load Program* in order to create and load file `dotp4.out`.

4. Open source file `dotp4.c` and clear all breakpoints. Set breakpoints at the lines

```
result = dotp(x, y, count);
```

and

```
printf("result = %d (decimal) \n", result);
```

5. Select *Profile→Clock→Enable*.

6. Select *Profile→Clock View*. A small clock icon and the number of processor instruction cycles that the *Profile Clock* has counted should appear in the bottom right-hand corner of the Code Composer window.

7. Run the program. It should halt at the first breakpoint.

8. Reset the *Profile Clock* by double-clicking on its icon in the bottom right-hand corner of the Code Composer window.

9. Run the program. It should stop at the second breakpoint.

The number of instruction cycles counted by the *Profile Clock* between the two breakpoints, that is, during execution of function `dotp()`, should be displayed next to the icon. On a 225-MHz C6713 processor, each instruction cycle takes 4.44 ns. Repeat the experiment having set the compiler optimization level to *Function (–o2)* and you should see a reduction in the number of instruction cycles used by function `dotp()` by a factor of approximately 2. Using breakpoints and the *Profile Clock* can give an indication of the execution times of sections of program but it does not always work with higher levels of compiler optimization, for example, *File (-o3)*. More detailed profiling of program execution can be achieved using a simulator.

1.6 SUPPORT FILES

The support files *c6713dskinit.c*, *vectors_intr.asm* or *vectors_poll.asm*, and *c6713dsk.cmd* are used by nearly all of the examples in this book.

1.6.1 Initialization/Communication File (*c6713dskinit.c*)

Source file *c6713dskinit.c*, supplied on the CD accompanying this book and listed in Figure 1.20, contains the definitions of a number of functions used to initialize

```
//c6713dskinit.c
//includes functions from TI in the C6713 CSL and C6713DSK BSL

#include "C6713dskinit.h"
#define using_bios
extern Uint32 fs;              //sampling frequency
extern Uint16 inputsource;     //input source (MIC or LINE)

void c6713_dsk_init()          //initialize DSK
{
  DSK6713_init();              //BSL routine to init DSK

  hAIC23_handle=DSK6713_AIC23_openCodec(0, &config);
  DSK6713_AIC23_setFreq(hAIC23_handle, fs); //set sampling rate
  // choose MIC or LINE IN on AIC23
  DSK6713_AIC23_rset(hAIC23_handle, 0x0004, inputsource);
  MCBSP_config(DSK6713_AIC23_DATAHANDLE,&AIC23CfgData);
  MCBSP_start(DSK6713_AIC23_DATAHANDLE, MCBSP_XMIT_START |
              MCBSP_RCV_START | MCBSP_SRGR_START |
              MCBSP_SRGR_FRAMESYNC, 220); //restart data channel
}

void comm_poll()               //for communication using polling
{
  poll=1;                      //1 if using polling
  c6713_dsk_init();            //init DSP and codec
}

void comm_intr()               //for communication using interrupt
{
  poll=0;                      //0 since not polling
  IRQ_globalDisable();         //globally disable interrupts
  c6713_dsk_init();            //init DSP and codec
  CODECEventId=MCBSP_getXmtEventId(DSK6713_AIC23_codecdatahandle);

#ifndef using_bios             //if not using DSP/BIOS
  IRQ_setVecs(vectors);        //use interrupt vector table
#endif                         //set up in vectors_intr.asm

  IRQ_map(CODECEventId, 11); //map McBSP1 Xmit to INT11
  IRQ_reset(CODECEventId);   //reset codec INT 11
  IRQ_globalEnable();        //globally enable interrupts
  IRQ_nmiEnable();           //enable NMI interrupt
  IRQ_enable(CODECEventId);  //enable CODEC eventXmit INT11

  output_sample(0);          //start McBSP by outputting a sample
}

void output_sample(int out_data) //output to both channels
{
  short CHANNEL_data;

  AIC_data.uint=0;             //clear data structure
```

FIGURE 1.20. Listing of support file c6713dskinit.c.

```
    AIC_data.uint=out_data;      //write 32-bit data

//The existing interface defaults to right channel.
//To default instead to the left channel and use
//output_sample(short), left and right channels are swapped.
//In main source program use LEFT 0 and RIGHT 1
//(opposite of what is used here)

  CHANNEL_data=AIC_data.channel[RIGHT]; //swap channels
  AIC_data.channel[RIGHT]=AIC_data.channel[LEFT];
  AIC_data.channel[LEFT]=CHANNEL_data;
  // if polling, wait for ready to transmit
  if (poll) while(!MCBSP_xrdy(DSK6713_AIC23_DATAHANDLE));
  // write data to AIC23 via MCBSP
  MCBSP_write(DSK6713_AIC23_DATAHANDLE,AIC_data.uint);
}

void output_left_sample(short out_data) //output to left channel
{
  AIC_data.uint=0;                   //clear data structure
  AIC_data.channel[LEFT]=out_data; //write 16-bit data
  // if polling, wait for ready to transmit
  if (poll) while(!MCBSP_xrdy(DSK6713_AIC23_DATAHANDLE));
  // write data to AIC23 via MCBSP
  MCBSP_write(DSK6713_AIC23_DATAHANDLE,AIC_data.uint);
}

void output_right_sample(short out_data)//output to right channel
{
  AIC_data.uint=0;                   //clear data structure
  AIC_data.channel[RIGHT]=out_data; //write 16-bit data
  // if polling, wait for ready to transmit
  if (poll) while(!MCBSP_xrdy(DSK6713_AIC23_DATAHANDLE));
  // write data to AIC23 via MCBSP
  MCBSP_write(DSK6713_AIC23_DATAHANDLE,AIC_data.uint);
}

Uint32 input_sample()          //input from both channels
{
  short CHANNEL_data;

  // if polling, wait for ready to receive
  if (poll) while(!MCBSP_rrdy(DSK6713_AIC23_DATAHANDLE));
  //read data from AIC23 via MCBSP
  AIC_data.uint=MCBSP_read(DSK6713_AIC23_DATAHANDLE);

  //Swap left and right channels (see comments in output_sample())
  CHANNEL_data=AIC_data.channel[RIGHT]; //swap channels
  AIC_data.channel[RIGHT]=AIC_data.channel[LEFT];
  AIC_data.channel[LEFT]=CHANNEL_data;
  return(AIC_data.uint);
}
```

FIGURE 1.20. (*Continued*)

```
short input_left_sample()      //input from left channel
{
    // if polling, wait for ready to receive
    if (poll) while(!MCBSP_rrdy(DSK6713_AIC23_DATAHANDLE));
    //read data from AIC23 via MCBSP
    AIC_data.uint=MCBSP_read(DSK6713_AIC23_DATAHANDLE);
    return(AIC_data.channel[LEFT]);  //return left channel data
}
short input_right_sample()     //input from right channel
{
    // if polling, wait for ready to receive
    if (poll) while(!MCBSP_rrdy(DSK6713_AIC23_DATAHANDLE));
    //read data from AIC23 via MCBSP
    AIC_data.uint=MCBSP_read(DSK6713_AIC23_DATAHANDLE);
    return(AIC_data.channel[RIGHT]); //return right channel data
}
```

FIGURE 1.20. (*Continued*)

the DSK. Calls are made from these functions to lower level functions provided with CCS in the board support library (BSL) and chip support library (CSL) files dsk6713bsl.lib and csl6713.lib.

Functions *comm_intr()* and *comm_poll()* initialize communications between the C6713 processor and the AIC23 codec for either interrupt-driven or polling-based input and output. In the case of interrupt-driven input and output, interrupt #11 (INT11), generated by the codec via the serial port (McBSP), is configured and enabled (selected). The nonmaskable interrupt bit must be enabled as well as the global interrupt enable (GIE) bit.

Functions *input_sample()*, *input_left_sample()*, *input_right_sample()*, *output_sample()*, *output_left_sample()*, and *input_right_sample()* are used to read and write data to and from the codec. In the case of polling-based input and output, these functions wait until the next sampling instant (determined by the codec) before reading or writing, using the lower level functions MCBSP_ read() or MCBSP_write(). They do this by polling (testing) the receive ready (RRDY) or transmit ready (XRDY) bits of the McBSP control register (SPCR). In the case of interrupt-driven input and output, the processor is interrupted by the codec at each sampling instant and when either *input_sample()* or *output_ sample()* is called from within the interrupt service routine, reading or writing proceeds without RRDY or XRDY being tested. Interrupts are discussed further in Chapter 3.

1.6.2 Header File (*c6713dskinit.h*)

The corresponding header support file *c6713dskinit.h* contains function proto-types as well as initial settings for the control registers of the AIC23 codec. Nearly all of the example programs in this book use the same AIC23 control register settings. However, two codec parameters—namely, sampling frequency and selection of ADC input (LINE IN or MIC IN)—are changed more often, from one program example to another, and for that reason the following mechanism has been adopted. During initialization of the DSK (in function `dsk_init()`, defined in file `c6713dskinit.c`), the AIC23 codec control registers are initialized using the `DSK6713_AIC23_Config` type data structure `config` defined in header file `c6713dskinit.h`. Immediately following this initialization, two functions `DSK6713_AIC23_setFreq()` and `DSK6713_AIC23_rset()` are called and these set the sampling frequency and select the input source according to the values of the variables `fs` and `inputsource`. These values are set in the first few lines of every top level source file; for example,

```
Uint32 fs = DSK6713_AIC23_FREQ_8KHZ; //set sampling rate
#define DSK6713_AIC23_INPUT_MIC 0x0015
#define DSK6713_AIC23_INPUT_LINE 0x0011
Uint16 inputsource=DSK6713_AIC23_INPUT_MIC; //select input source
```

In this way, the sampling frequency and input source can be changed without having to edit either `c6713sdkinit.h` or `c6713dskinit.c`. (See Figure 1.21.)

1.6.3 Vector Files (`vectors_intr.asm`, `vectors_poll.asm`)

To make use of interrupt INT11, a branch instruction (jump) to the interrupt service routine (ISR) `c_int11()` defined in a C program, for example, `sine8_buf.c`, must be placed at the appropriate point in the interrupt service table (IST). Assembly language file `vectors_intr.asm`, which sets up the IST, is listed in Figure 1.22. Note the underscore preceding the name of the routine or function being called. By convention, this indicates a C function.

For a polling-based program, file `vectors_poll.asm` is used, in place of `vectors_intr.asm`. The main difference between these files is that there is no branch to `c_int11()` in the IST set up by `vectors_poll.asm`. Common to both files is a branch to `c_int00()`, the start of a C program, associated with reset. (See Figure 1.23.)

1.6.4 Linker Command File (`c6713dsk.cmd`)

Linker command file `c6713dsk.cmd` is listed in Figure 1.24. It specifies the memory configuration of the internal and external memory available on the DSK and the mapping of sections of code and data to absolute addresses in that memory. For example, the `.text` section, produced by the C compiler, is mapped into IRAM, that is, the internal memory of the C6713 digital signal processor, starting at address `0x00000220`. The section `.vectors` created by `vectors_intr.asm` or by `vectors_poll.asm` is mapped into IVECS, that is, internal memory starting at address

```
/*c6713dskinit.h include file for c6713dskinit.c */

#include "dsk6713.h"\
#include "dsk6713_aic23.h"

#define LEFT  1
#define RIGHT 0

union {
        Uint32 uint;
        short channel[2];
        } AIC_data;

extern far void vectors();         //external function

static Uint32 CODECEventId, poll;

// needed to modify the BSL data channel McBSP configuration
MCBSP_Config AIC23CfgData = {
        MCBSP_FMKS(SPCR, FREE, NO)              |
        MCBSP_FMKS(SPCR, SOFT, NO)              |
        MCBSP_FMKS(SPCR, FRST, YES)             |
        MCBSP_FMKS(SPCR, GRST, YES)             |
        MCBSP_FMKS(SPCR, XINTM, XRDY)           |
        MCBSP_FMKS(SPCR, XSYNCERR, NO)          |
        MCBSP_FMKS(SPCR, XRST, YES)             |
        MCBSP_FMKS(SPCR, DLB, OFF)              |
        MCBSP_FMKS(SPCR, RJUST, RZF)            |
        MCBSP_FMKS(SPCR, CLKSTP, DISABLE)       |
        MCBSP_FMKS(SPCR, DXENA, OFF)            |
        MCBSP_FMKS(SPCR, RINTM, RRDY)           |
        MCBSP_FMKS(SPCR, RSYNCERR, NO)          |
        MCBSP_FMKS(SPCR, RRST, YES),

        MCBSP_FMKS(RCR, RPHASE, SINGLE)         |
        MCBSP_FMKS(RCR, RFRLEN2, DEFAULT)       |
        MCBSP_FMKS(RCR, RWDLEN2, DEFAULT)       |
        MCBSP_FMKS(RCR, RCOMPAND, MSB)          |
        MCBSP_FMKS(RCR, RFIG, NO)               |
        MCBSP_FMKS(RCR, RDATDLY, 0BIT)          |
        MCBSP_FMKS(RCR, RFRLEN1, OF(0))         |
        MCBSP_FMKS(RCR, RWDLEN1, 32BIT)         |
        MCBSP_FMKS(RCR, RWDREVRS, DISABLE),

        MCBSP_FMKS(XCR, XPHASE, SINGLE)         |
        MCBSP_FMKS(XCR, XFRLEN2, DEFAULT)       |
        MCBSP_FMKS(XCR, XWDLEN2, DEFAULT)       |
```

FIGURE 1.21. Listing of support header file c6713dskinit.h.

```
*Vectors_poll.asm  Vector file for polling
    .global _vectors
    .global _c_int00
    .global _vector1
    .global _vector2
    .global _vector3
    .global _vector4
    .global _vector5
    .global _vector6
    .global _vector7
    .global _vector8
    .global _vector9
    .global _vector10
    .global _vector11
    .global _vector12
    .global _vector13
    .global _vector14
    .global _vector15

    .ref _c_int00                           ;entry address

VEC_ENTRY .macro addr
    STW   B0,*--B15
    MVKL  addr,B0
    MVKH  addr,B0
    B     B0
    LDW   *B15++,B0
    NOP   2
    NOP
    NOP
    .endm

_vec_dummy:
  B     B3
  NOP   5

 .sect ".vectors"
 .align 1024

_vectors:
_vector0:     VEC_ENTRY _c_int00      ;RESET
_vector1:     VEC_ENTRY _vec_dummy    ;NMI
_vector2:     VEC_ENTRY _vec_dummy    ;RSVD
_vector3:     VEC_ENTRY _vec_dummy
_vector4:     VEC_ENTRY _vec_dummy
_vector5:     VEC_ENTRY _vec_dummy
_vector6:     VEC_ENTRY _vec_dummy
_vector7:     VEC_ENTRY _vec_dummy
_vector8:     VEC_ENTRY _vec_dummy
_vector9:     VEC_ENTRY _vec_dummy
_vector10:    VEC_ENTRY _vec_dummy
_vector11:    VEC_ENTRY _vec_dummy
_vector12:    VEC_ENTRY _vec_dummy
_vector13:    VEC_ENTRY _vec_dummy
_vector14:    VEC_ENTRY _vec_dummy
_vector15:    VEC_ENTRY _vec_dummy
```

FIGURE 1.23. Listing of vector file vectors_poll.asm.

```
/*C6713dsk.cmd  Linker command file*/

MEMORY
{
   IVECS:    org=0h,           len=0x220
   IRAM:     org=0x00000220,  len=0x0002FDE0 /*internal memory*/
   SDRAM:    org=0x80000000,  len=0x01000000 /*external memory*/
   FLASH:    org=0x90000000,  len=0x00020000 /*flash memory*/
}

SECTIONS
{
   .EXT_RAM :> SDRAM
   .vectors :> IVECS /*in vector file*/
   .text    :> IRAM  /*Created by C Compiler*/
   .bss     :> IRAM
   .cinit   :> IRAM
   .stack   :> IRAM
   .sysmem  :> IRAM
   .const   :> IRAM
   .switch  :> IRAM
   .far     :> IRAM
   .cio     :> IRAM
   .csldata :> IRAM
}
```

FIGURE 1.24. Listing of linker command file `c6713dsk.cmd`.

4. Write a program that reads input samples from the left-hand channel of the AIC23 codec ADC at a sampling frequency of 16 kHz using function `input_left_sample()` and, just after it has been read, writes each sample value to the right-hand channel of the AIC23 codec DAC using function `output_right_sample()`. Verify the effective connection of the left-hand channel of the LINE IN socket to the right-hand channel of the LINE OUT socket using a signal generator and an oscilloscope. Gradually increase the frequency of the input signal until the amplitude of the output signal is reduced drastically. This frequency corresponds to the bandwidth of the DSP system (illustrated in more detail in Chapter 2).

REFERENCES

1. R. Chassaing, *DSP Applications Using C and the TMS320C6x DSK*, Wiley, Hoboken, NJ, 2002.

2. R. Chassaing, *Digital Signal Processing Laboratory Experiments Using C and the TMS320C31 DSK*, Wiley, Hoboken, NJ, 1999.

3. R. Chassaing, *Digital Signal Processing with C and the TMS320C30*, Wiley, Hoboken NJ, 1992.

4. R. Chassaing and D. W. Horning, *Digital Signal Processing with the TMS320C25*, Wiley, Hoboken NJ, 1990.

5. N. Kehtarnavaz and M. Keramat, *DSP System Design Using the TMS320C6000*, Prentice Hall, Upper Saddle River, NJ, 2001.

6. N. Kehtarnavaz and B. Simsek, *C6x-Based Digital Signal Processing*, Prentice Hall, Upper Saddle River, NJ, 2000.

7. N. Dahnoun, *DSP Implementation Using the TMS320C6x Processors*, Prentice Hall, Upper Saddle River, NJ, 2000.

8. Steven A. Tretter, *Communication System Design Using DSP Algorithms With Laboratory Experiments for the TMS320C6701 and TMS320C6711*, Kluwer Academic, New York, 2003.

9. J. H. McClellan, R. W. Schafer, and M. A. Yoder, *DSP First: A Multimedia Approach*, Prentice Hall, Upper Saddle River, NJ, 1998.

10. C. Marven and G. Ewers, *A Simple Approach to Digital Signal Processing*, Wiley, Hoboken NJ, 1996.

11. J. Chen and H. V. Sorensen, *A Digital Signal Processing Laboratory Using the TMS320C30*, Prentice Hall, Upper Saddle River, NJ, 1997.

12. S. A. Tretter, *Communication System Design Using DSP Algorithms*, Plenum Press, New York, 1995.

13. A. Bateman and W. Yates, *Digital Signal Processing Design*, Computer Science Press, New York, 1991.

14. Y. Dote, *Servo Motor and Motion Control Using Digital Signal Processors*, Prentice Hall, Upper Saddle River, NJ, 1990.

15. J. Eyre, The newest breed trade off speed, energy consumption, and cost to vie for an ever bigger piece of the action, *IEEE Spectrum*, June 2001.

16. J. M. Rabaey, Ed., VLSI design and implementation fuels the signal-processing revolution, *IEEE Signal Processing*, Jan. 1998.

17. P. Lapsley, J. Bier, A. Shoham, and E. Lee, *DSP Processor Fundamentals: Architectures and Features*, Berkeley Design Technology, Berkeley, CA, 1996.

18. R. M. Piedra and A. Fritsh, Digital signal processing comes of age, *IEEE Spectrum*, May 1996.

19. R. Chassaing, The need for a laboratory component in DSP education: a personal glimpse, *Digital Signal Processing*, Jan. 1993.

20. R. Chassaing, W. Anakwa, and A. Richardson, Real-time digital signal processing in education, *Proceedings of the 1993 International Conference on Acoustics, Speech and Signal Processing (ICASSP)*, Apr. 1993.

21. S. H. Leibson, DSP development software, *EDN Magazine*, Nov. 8, 1990.

22. D. W. Horning, An undergraduate digital signal processing laboratory, *Proceedings of the 1987 ASEE Annual Conference*, June 1987.

23. *TMS320C6000 Programmer's Guide*, SPRU198G, Texas Instruments, Dallas, TX, 2002.

24. *TMS320C6211 Fixed-Point Digital Signal Processor–TMS320C6711 Floating-Point Digital Signal Processor*, SPRS073C, Texas Instruments, Dallas, TX, 2000.

25. *TMS320C6000 CPU and Instruction Set Reference Guide*, SPRU189F, Texas Instruments, Dallas, TX, 2000.

26. *TMS320C6000 Assembly Language Tools User's Guide*, SPRU186K, Texas Instruments, Dallas, TX, 2002.

27. *TMS320C6000 Peripherals Reference Guide*, SPRU190D, Texas Instruments, Dallas, TX, 2001.

28. *TMS320C6000 Optimizing C Compiler User's Guide*, SPRU187K, Texas Instruments, Dallas, TX, 2002.

29. *TMS320C6000 Technical Brief*, SPRU197D, Texas Instruments, Dallas, TX, 1999.

30. *TMS320C64x Technical Overview*, SPRU395, Texas Instruments, Dallas, TX, 2000.

31. *TMS320C6x Peripheral Support Library Programmer's Reference*, SPRU273B, Texas Instruments, Dallas, TX, 1998.

32. *Code Composer Studio User's Guide*, SPRU328B, Texas Instruments, Dallas, TX, 2000.

33. *Code Composer Studio Getting Started Guide*, SPRU509, Texas Instruments, Dallas, TX, 2001.

34. *TMS320C6000 Code Composer Studio Tutorial*, SPRU301C, Texas Instruments, Dallas, TX, 2000.

35. *TLC320AD535C/I Data Manual Dual Channel Voice/Data Codec*, SLAS202A, Texas Instruments, Dallas, TX, 1999.

36. *TMS320C6713 Floating-Point Digital Signal Processor*, SPRS186L, 2005.

37. *TMS320C6414T, TMS320C6415T, TMS320C6416T Fixed-Point Digital Signal Processors (Rev. J)*, SPRS226J, 2006.

38. TLV320AIC23 Stereo Audio Codec, 8- to 96-kHz, with Integrated Headphone Amplifier Data Manual, SLWS106G, 2003.

39. TMS320C6000 DSP Phase-Locked Loop (PLL) Controller Peripheral Reference Guide, SPRU233.

40. *Migrating from TMS320C6211/C6711 to TMS320C6713*, SPRA851.

41. *How to Begin Development Today with the TMS320C6713 Floating-Point DSP*, SPRA809.

42. TMS320C6000 DSP/BIOS User's Guide, SPRU423, 2002.

43. TMS320C6000 Optimizing C Compiler Tutorial, SPRU425A, 2002.

44. G. R. Gircys, *Understanding and Using COFF*, O'Reilly & Associates, Newton, MA, 1988.

2

Input and Output with the DSK

· Input and output with the onboard AIC23 stereo codec
· Programming examples using C code

2.1 INTRODUCTION

A basic DSP system, suitable for processing audio frequency signals, comprises a digital signal processor and analog interfaces as shown in Figure 2.1. The C6713 and C6416 DSKs provide just such a system, using either the TMS320C6713 (C6713) floating-point processor or the TMS320C6416 (C6416) fixed-point processor and the TLV320AIC23 (AIC23) codec [1]. The term codec refers to the *coding* of analog waveforms as digital signals and the *decoding* of digital signals as analog waveforms. The AIC23 codec performs both the analog-to-digital conversion (ADC) and digital-to-analog conversion (DAC) functions shown in Figure 2.1.

Alternatively, I/O daughter cards, plugged into the External Peripheral Interface 80-pin connector J3 on the DSK board can be used for analog input and output. However, the programming examples in this book use only the onboard AIC23 codec.

The programming examples in this chapter will run on either the C6713 or the C6416 DSK but for the most part only the C6713 DSK will be referred to.

Sampling, Reconstruction, and Aliasing

Within digital signal processors, signals are represented as sequences of discrete samples and whenever signals are sampled, the possibility of aliasing arises. Later

Digital Signal Processing and Applications with the TMS320C6713 and TMS320C6416 DSK,
Second Edition   By Rulph Chassaing and Donald Reay
Copyright © 2008 John Wiley & Sons, Inc.

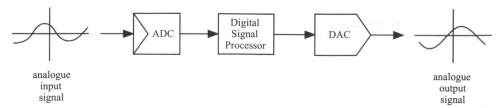

FIGURE 2.1. Basic digital signal processing system.

in this chapter, the phenomenon of aliasing is explored in some detail. Suffice it to say at this stage that aliasing is undesirable and that it may be avoided by the use of an antialiasing filter placed at the input to the system shown in Figure 2.1 and by suitable design of the DAC. In a baseband system, an effective antialiasing filter is one that allows frequency components below half of the sampling frequency to pass but which attenuates greatly, or stops, frequency components equal to or greater than half of the sampling frequency. A suitable DAC for a baseband system essentially comprises a lowpass filter having characteristics similar to the aforementioned antialiasing filter. The AIC23 codec contains digital antialiasing and reconstruction filters.

2.2 TLV320AIC23 (AIC23) ONBOARD STEREO CODEC FOR INPUT AND OUTPUT

Both the C6713 and C6416 DSKs make use of the TLV320AIC23 (AIC23) codec for analog input and output. The analog-to-digital converter (ADC), or coder, part of the codec converts an analog input signal into a sequence of sample values (16-bit signed integer) to be processed by the digital signal processor. The digital-to-analog converter (DAC), or decoder, part of the codec reconstructs an analog output signal from a sequence of sample values (16-bit signed integer) that have been processed by the digital signal processor.

The AIC23 is a stereo audio codec based on sigma–delta technology [1–5]. Its functional block diagram is shown in Figure 2.2.

A 12-MHz crystal supplies the clock to the AIC23 codec (also to the DSP and the USB interface). Using this 12-MHz master clock, with oversampling rates of $250Fs$ and $272Fs$, exact audio sample rates of 48 kHz (12 MHz/250) and the CD rate of 44.1 kHz (12 MHz/272) can be obtained. The sampling rate of the AIC23 can be configured to be 8, 16, 24, 32, 44.1, 48, or 96 kHz.

Communication with the AIC23 codec for input and output uses two multichannel buffered serial ports (McBSPs) on the C6713 or C6416. McBSP0 is used as a unidirectional channel to send a 16-bit control word to the AIC23. McBSP1 is used as a bidirectional channel to send and receive audio data (McBSP1 and McBSP2 are used on the C6416 DSK). The codec can be configured for data-transfer word-lengths of 16, 20, 24, or 32 bits.

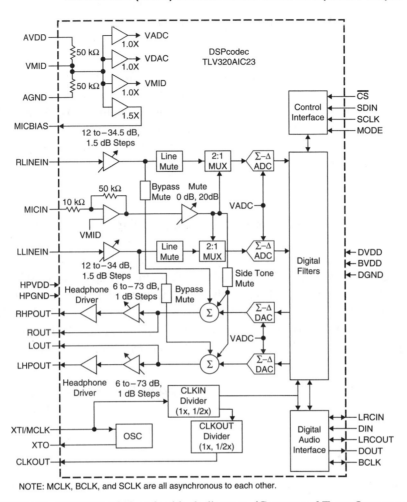

FIGURE 2.2. TLV320AIC23 codec block diagram. (Courtesy of Texas Instruments.)

The LINE IN and HEADPHONE OUT signal paths within the codec contain configurable gain elements with ranges of 12 to −34 dB in steps of 1.5 dB, and 6 to −73 dB in steps of 1 dB, respectively. A diagram of the AIC23 codec interfaced to the C6713 DSK is shown in *6713_dsk_schem.pdf*, included with the CCS package. With few exceptions, the programming examples in this book configure the codec for a sampling rate of 8 kHz, 32-bit data transfer, and 0-dB gain in the LINE IN and HEADPHONE OUT signal paths.

The maximum allowable input signal level at the LINE IN inputs to the codec is 1 V rms. However, the C6713 and C6416 DSKs contain a potential divider circuit with a gain of 0.5 between their LINE IN sockets and the codec itself with the effect that the maximum allowable input signal level at the LINE IN sockets on the DSKs is 2 V rms. Above this level, input signals will be distorted. Input and output sockets on the DSKs are ac coupled to the codec.

2.3 PROGRAMMING EXAMPLES USING C CODE

The following examples illustrate analog input and output using the DSK. They are included in order to introduce both the DSK hardware and the CCS development environment. The example programs demonstrate some important concepts associated with analog-to-digital conversion and digital-to-analog conversion, including sampling, aliasing, and reconstruction. In addition, they illustrate the use of interrupts in order to implement real-time applications using the DSK. Many of the concepts and techniques described in this chapter are used again in subsequent chapters.

Example 2.1: Basic Input and Output Using Polling (`loop_poll`)

The C language source file for a program, `loop_poll.c`, that simply copies input samples read from the AIC23 codec ADC back to the AIC23 codec DAC as output samples is listed in Figure 2.3. Effectively, the MIC input socket is connected straight through to the HEADPHONE OUT socket on the DSK via the AIC23 codec and the digital signal processor. `loop_poll.c` uses the same polling technique for real-time input and output as program `sine8_LED.c`, presented in Chapter 1.

Input and Output Functions Defined in Support File `c6713dskinit.c`
The functions `input_left_sample()`, `output_left_sample()`, and `comm_poll()` are defined in the support file `c6713dskinit.c`. This way the C source file `loop_`

```
//loop_poll.c loop program using polling

#include "DSK6713_AIC23.h"            //codec support
Uint32 fs=DSK6713_AIC23_FREQ_8KHZ;    //set sampling rate
#define DSK6713_AIC23_INPUT_MIC 0x0015
#define DSK6713_AIC23_INPUT_LINE 0x0011
Uint16 inputsource=DSK6713_AIC23_INPUT_MIC; // select input

void main()
{
  short sample_data;

  comm_poll();                        //init DSK, codec, McBSP
  while(1)                            //infinite loop
  {
    sample_data = input_left_sample(); //input sample
    output_left_sample(sample_data);   //output sample
  }
}
```

FIGURE 2.3. Loop program using polling (`loop_poll.c`).

poll.c is kept as small as possible and potentially distracting low level detail is hidden. The implementation details of these, and other, functions defined in c6713dskinit.c need not be studied in detail in order to carry out the examples presented in this book but are described here for completeness.

Further calls are made by input_left_sample() and output_left_sample() to lower level functions contained in the board support library DSK6713bsl.lib.

Function comm_poll() initializes the DSK and, in particular, the AIC23 codec such that its sample rate is set according to the value of the variable fs (assigned in loop_poll.c), its input source according to the value of the variable input-source (assigned in loop_poll.c), and polling mode is selected. Other AIC23 configuration settings are determined by the parameters specified in file c6713dskinit.h. These parameters include the gain settings in the LINE IN and HEADPHONE out signal paths, the digital audio interface format, and so on. Similar values for all of these parameters are used by almost all of the program examples in this book. Only rarely will they be changed and so it is convenient to hide them out of the way in file c6713dskinit.h.

The two settings, sampling rate and input source, are changed sufficiently frequently, from one program example to another, that their values are set in each example program by initializing the values of the variables fs and inputsource. In function dsk6713_init() in file c6713dskinit.c, these values are used by functions DSK6713_AIC23_setFreq() and DSK6713_AIC23_rset(), respectively.

In polling mode, function input_left_sample() polls, or tests, the receive ready bit (RRDY) of the McBSP serial port control register (SPCR) until this indicates that newly converted data is available to be read using function MCBSP_read(). Function output_left_sample() polls, or tests, the transmit ready bit (XRDY) of the McBSP serial port control register (SPCR) until this indicates that the codec is ready to receive a new output sample. A new output sample is sent to the codec using function McBSP_write().

Although polling is simpler than the interrupt technique used in sine8_buf.c (and in nearly all the other programs in subsequent chapters of this book), it is less efficient since the processor spends nearly all of its time repeatedly testing whether the codec is ready either to transmit or to receive data.

Running the Program

Project file loop_poll.pjt is stored in folder loop_poll. *Open* project loop_poll.pjt and load the executable file loop_poll.out. Run the program and use a microphone and headphones to verify that the program operates as intended.

For a closer examination of the characteristics of the program you can use a signal generator and oscilloscope. Prior to connecting a signal generator to the LINE IN socket, you will need to *Rebuild* the program having changed the line that reads

```
Uint16 inputsource=DSK6713_AIC23_INPUT_MIC;
```

to read

```
Uint16 inputsource=DSK6713_AIC23_INPUT_LINE;
```

in order to select the LINE IN rather than the MIC socket on the DSK.

Testing the Allowable Input Signal Amplitude

Input a sinusoidal waveform to the LINE IN connector on the DSK, with an amplitude of approximately 2.0 V p-p and a frequency of approximately 1 kHz. Connect the output of the DSK, LINE OUT, to an oscilloscope, and verify the presence of a tone of the same frequency, but attenuated to approximately 1.0 V p-p. This attenuation is due to the potential divider network, comprising two resistors, on the DSK circuit board between the LINE IN socket and the codec input.

The full scale range of the ADC and of the DAC in the codec is 1 V rms (2.83 V p-p). Increase the amplitude of the input sinusoidal waveform (at the LINE IN socket) beyond 2 V rms (5.66 V p-p) and verify that the output signal becomes distorted.

Changing the LINE IN Gain of the AIC23 Codec

The AIC23 codec allows for the gain on left- and right-hand line-in input channels to be adjusted independently in steps of 1.5 dB by writing different values to the left and right line input channel volume control registers. The values assigned to these registers by function comm_poll() are defined in the header file c6713dskinit.h. In order to change the values written, that file must be modified.

1. Copy the files c6713dskinit.h and C6713dskinit.c from the Support folder into the folder loop_poll so that you don't modify the original header file.

2. Remove these two files from the loop_poll project by right-clicking on c6713dskinit.c in the *Project View* window and then selecting *Project→ Remove from Project*.

3. Add the copy of the file c6713dskinit.c in folder loop_poll to the project by selecting *Project → Add Files to Project*.

4. Check that you have added the *copy* of file c6713dskinit.c to the project by right-clicking on it in the *Project View* window and selecting *Properties*.

5. Select *Project→Scan all Dependencies* in order to replace the file c6713dskinit.h with the copy in folder loop_poll.

6. Edit the copy of file *c6713dskinit.h* included in the project (and stored in folder loop_poll), changing the line that reads

```
0x0017 /* Set-Up Reg 0 Left line volume control */
```

to read

```
0x001B /* Set-Up Reg 0 Left line volume control */
```

This modifies the value written to the AIC23 left line input channel gain register from 0x0017 to 0x001B and this increases the gain from 0 dB to 6 dB.

7. *Build* the project, making sure that the copy of the file `c6713dskinit.c` used in the project is the copy in folder `loop_poll`. The header file `c6713dskinit.h` that will be included will come from that same folder.

8. Load and run the executable file `loop_poll.out` and verify that the output signal is not attenuated, but has the same amplitude as the input signal, that is, 2 V p-p. The changes you have just made are to a copy of `c6713dskinit.h` in folder `loop_poll` and are limited in the scope of their effect to that project.

Example 2.2: Basic Input and Output Using Interrupts (`loop_intr`)

Program `loop_intr.c` is functionally equivalent to program `loop_poll.c` but makes use of interrupts. This simple program is important because many of the other example programs in this book are based on the same interrupt-driven model. Instead of simply copying the sequence of samples representing an input signal to the codec output, a digital filtering operation can be performed each time a new input sample is received. It is worth taking time to ensure that you understand how program `loop_intr.c` works. In function `main()`, the initialization function `comm_intr()` is called. `comm_intr()` is very similar to `comm_poll()` but in addition to initializing the DSK, codec, and McBSP, and *not* selecting polling mode, it sets up interrupts such that the AIC23 codec will sample the analog input signal and interrupt the C6713 processor, at the sampling frequency defined by the line

```
Uint32 fs=DSK6713_AIC23_FREQ_8KHZ; //set sampling rate
```

It also initiates communication with the codec via the McBSP.

In this example, a sampling rate of 8 kHz is used and interrupts will occur every 0.125 ms. (Sampling rates of 16, 24, 32, 44.1, 48, and 96 kHz are also possible.)

Following initialization, function `main()` enters an endless while loop, doing nothing but waiting for interrupts. The functions that will act as interrupt service routines for the various different interrupts are specified in the interrupt service table contained in file `vectors_intr.asm`. This assembly language file differs from the file `vectors_poll.asm` in that function `c_int11()` is specified as the interrupt service routine for interrupt INT11.

On interrupt, the interrupt service routine (ISR) `c_int11()` is called and it is within that routine that the most important program statements are executed.

```
//loop_intr.c loop program using interrupts

#include "DSK6713_AIC23.h"              //codec support
Uint32 fs=DSK6713_AIC23_FREQ_8KHZ;     //set sampling rate
#define DSK6713_AIC23_INPUT_MIC 0x0015
#define DSK6713_AIC23_INPUT_LINE 0x0011
Uint16 inputsource=DSK6713_AIC23_INPUT_MIC; //select input

interrupt void c_int11()                    //interrupt service routine
{
  short sample_data;

  sample_data = input_left_sample(); //input data
  output_left_sample(sample_data);   //output data
  return;
}

void main()
{
  comm_intr();                              //init DSK, codec, McBSP
  while(1);                                 //infinite loop
}
```

FIGURE 2.4. Loop program using interrupts (loop_intr.c).

Function output_left_sample() is used to output a value read from the codec using function input_left_sample().

Format of Data Transferred to and from AIC23 Codec

The AIC23 ADC converts left- and right-hand channel analog input signals into 16-bit signed integers and the DAC converts 16-bit signed integers to left- and right-hand channel analog output signals. Left- and right-hand channel samples are combined to form 32-bit values that are communicated via the multichannel buffered serial port (McBSP) to and from the C6713. Access to the ADC and DAC from a C program is via the functions Uint32 input_sample(), short input_left_sample(), short input_right_sample(), void output_sample(int out_data), and void output_left_sample(short out_data), and void output_right_sample(short out_data).

The 32-bit unsigned integers (Uint32) returned by input_sample() and passed to output_sample() contain both left and right channel samples. The statement

```
union {
 Uint32 uint;
 Short channel [2];
} AIC_data;
```

in file `dsk6713init.h` declares a variable that may be handled either as one 32-bit unsigned integer (`AIC_data.uint`) containing left and right channel sample values, or as two 16-bit signed integers (`AIC_data.channel[0]` and `AIC_data.channel[1]`).

Most of the program examples in this book use only one channel for input and output and for clarity most use the functions `input_left_sample()` and `output_left_sample()`. These functions are defined in the file `c6713dskinit.c`, where the unpacking and packing of the signed 16-bit integer left-hand channel sample values out of and into the 32-bit words received and transmitted from and to the codec are carried out.

Running the Program

Create and build this project as **loop_intr** ensuring that the support files `c6713dskint.c` and `vectors_intr.asm` have been added to the project. Verify the same results as obtained using program `loop_poll`.

Example 2.3: Modifying Program `loop_intr.c` to Create a Delay (`delay`)

Some simple, yet striking, effects can by achieved simply by delaying the samples as they pass from input to output. Program `delay.c`, listed in Figure 2.5, demonstrates this. A delay line is implemented using the array `buffer` to store samples as they are read from the codec. Once the array is full, the program overwrites the oldest stored input sample with the current, or newest, input sample. Just prior to overwriting the oldest stored input sample in `buffer`, that sample is retrieved, added to the current input sample, and output to the codec. Figure 2.6 shows a block diagram representation of the operation of program `delay.c` in which the block labeled *T* represents a delay of *T* seconds.

Build and run the project as **delay**, using microphone and headphones to verify its operation.

Example 2.4: Modifying Program `loop_intr.c` to Create an Echo (`echo`)

By feeding back a fraction of the output of the delay line to its input, a fading echo effect can be realized. Program `echo.c`, listed in Figure 2.7, does this. Figure 2.8 shows a block diagram representation of the operation of program `echo.c`.

The value of the constant `BUF_SIZE` determines the number of samples stored in the array `buffer` and hence the duration of the delay. The value of the constant `GAIN` determines the fraction of the output that is fed back into the delay line and hence the rate at which the echo effect fades away. Setting the value of `GAIN` equal to or greater than unity would cause instability of the loop.

Build and run this project as **echo**. Experiment with different values of `GAIN` (between 0.0 and 1.0) and `BUF_SIZE` (between 100 and 8000). Source file `echo.c` must be edited and the project rebuilt in order to make these changes.

```
//delay.c Basic time delay

#include "DSK6713_AIC23.h"                //codec support
Uint32 fs=DSK6713_AIC23_FREQ_8KHZ;        //set sampling rate
#define DSK6713_AIC23_INPUT_MIC 0x0015
#define DSK6713_AIC23_INPUT_LINE 0x0011
Uint16 inputsource=DSK6713_AIC23_INPUT_MIC; //select input


#define BUF_SIZE 8000
short input,output,delayed;
short buffer[BUF_SIZE];
int i;

interrupt void c_int11()         //interrupt service routine
{
  input = input_left_sample();   //read new input sample
  delayed = buffer[i];           //read output of delay line
  output = input + delayed;      //output sum of new and delayed
  buffer[i] = input;             //replace delayed sample with
  if(++i >= BUF_SIZE) i=0;       //new input sample then increment
  output_left_sample(output);    //buffer index
  return;                        //return from ISR
}

void main()
{
  for(i=0 ; i<BUF_SIZE ; i++)
    buffer[i] = 0;
  comm_intr();                   //init DSK, codec, McBSP
  while(1);                      //infinite loop
}
```

FIGURE 2.5. Delay program using interrupts (delay.c).

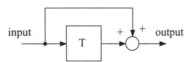

FIGURE 2.6. Block diagram representation of program delay.c.

Example 2.5: Echo with GEL Slider Control of Delay and Feedback (echo_control)

This example extends Example 2.4 to allow real-time adjustment of the gain and delay parameters of the echo effect. Two GEL sliders, defined in file echo_control. gel, are used. Program echo_control.c, listed in Figure 2.9, differs from program echo.c in the following respects.

```
//echo.c echo with fixed delay and feedback

#include "DSK6713_AIC23.h"              //codec support
Uint32 fs=DSK6713_AIC23_FREQ_8KHZ;      //set sampling rate
#define DSK6713_AIC23_INPUT_MIC 0x0015
#define DSK6713_AIC23_INPUT_LINE 0x0011
Uint16 inputsource=DSK6713_AIC23_INPUT_MIC; //select input

#define GAIN 0.6                    //fraction of output fed back
#define BUF_SIZE 2000               //this sets length of delay
short buffer[BUF_SIZE];             //storage for previous samples
short input,output,delayed;
int i;                             //index into buffer

interrupt void c_int11()            //interrupt service routine
{
  input = input_left_sample();     //read new input sample from ADC
  delayed = buffer[i];             //read delayed value from buffer
  output = input + delayed;        //output sum of input and delayed
  output_left_sample(output);
  buffer[i] = input + delayed*GAIN; //store new input and
                                   //fraction of delayed value
  if(++i >= BUF_SIZE) i=0;         //test for end of buffer
  return;                          //return from ISR
}

void main()
{
  comm_intr();                     //init DSK, codec, McBSP
  for(i=0 ; i<BUF_SIZE ; i++)      //clear buffer
    buffer[i] = 0;
  while(1);                        //infinite loop
}
```

FIGURE 2.7. Fading echo program (echo.c).

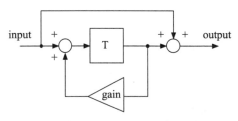

FIGURE 2.8. Block diagram representation of program echo.c.

```
//echo_control.c echo with variable delay and feedback

#include "DSK6713_AIC23.h"                //codec support
Uint32 fs=DSK6713_AIC23_FREQ_8KHZ;        //set sampling rate
#define DSK6713_AIC23_INPUT_MIC 0x0015
#define DSK6713_AIC23_INPUT_LINE 0x0011
Uint16 inputsource=DSK6713_AIC23_INPUT_MIC; //select input

#define MAX_BUF_SIZE 8000       //set maximum length of delay
float gain = 0.5;
short buflength = 1000;
short buffer[MAX_BUF_SIZE];     //storage for previous samples
short input,output,delayed;
int i = 0;                      //index into buffer

interrupt void c_int11()        //interrupt service routine
{
  input = input_left_sample(); //read new input sample from ADC
  delayed = buffer[i];          //read delayed value from buffer
  output = input + delayed;     //output sum of input and delayed
  output_left_sample(output);
  buffer[i] = input + delayed*gain; //store new input and
                                //fraction of delayed value
  if(++i >= MAX_BUF_SIZE)       //test for end of buffer
    i = MAX_BUF_SIZE - buflength;
  return;                       //return from ISR
}

void main()
{
  for(i=0 ; i<MAX_BUF_SIZE ; i++) //clear buffer
    buffer[i] = 0;
  comm_intr();                  //init DSK, codec, McBSP
  while(1);                     //infinite loop
}
```

FIGURE 2.9. Echo program with variable delay and feedback gain (echo_control.c).

1. Array buffer is declared to be the maximum size required, MAX_BUF_SIZE.

2. To achieve a variable delay, integer variable buflength is used to control the length of the circular buffer implemented using array buffer. When the value of the index i, used to access elements of the array buffer, is incremented beyond the maximum value allowable (MAX_BUF_SIZE), it is reset not to zero as in program echo.c but to (MAX_BUF_SIZE - buflength).

Build the project as **echo_control**. Load and run file echo_control.out and then select *File→Load GEL* and load echo_control.gel. Select *GEL→echo control* to bring up the gain and delay sliders.

```
//echo_control.gel

menuitem "echo control"

slider gain(0,18,1,1,gain_parameter)
{
  gain = gain_parameter*0.05;
}

slider delay(1,20,1,1,delay_parameter)
{
  buflength = delay_parameter*100;
}
```

FIGURE 2.10. GEL file (echo_control.gel) for slider control of delay and feedback gain in program echo_control.c.

```
//loop_buf.c loop program with storage

#include "DSK6713_AIC23.h"            //codec support
Uint32 fs=DSK6713_AIC23_FREQ_8KHZ;   //set sampling rate
#define DSK6713_AIC23_INPUT_MIC 0x0015
#define DSK6713_AIC23_INPUT_LINE 0x0011
Uint16 inputsource=DSK6713_AIC23_INPUT_LINE; //select input
#define BUFSIZE 512

int buffer[BUFSIZE];
int buf_ptr = 0;

interrupt void c_int11()         //interrupt service routine
{
  int sample_data;

  sample_data = input_left_sample();    //read input sample
  buffer[buf_ptr] = sample_data;        //store in buffer
  if(++buf_ptr >= BUFSIZE) buf_ptr = 0; //update buffer index
  output_left_sample(sample_data);      // write output sample
  return;
}

void main()
{
  comm_intr();                          //init DSK, codec, McBSP
  while(1);                             //infinite loop
}
```

FIGURE 2.11. Loop program with input data stored in memory (loop_buf.c).

Example 2.6: Loop Program with Input Data Stored in a Buffer (loop_buf)

Program *loop_buf.c*, listed in Figure 2.11, is an interrupt-based program and is stored in folder loop_buf. It is similar to program loop_intr.c except that it maintains a circular buffer in array `buffer` containing the `BUF_SIZE` most recent input sample values. Consequently, it is possible to display this data in CCS after halting the program.

Build this project as **`loop_buf`**. Use a signal generator connected to the LINE IN socket to input a sinusoidal signal with a frequency between 100 and 3500 Hz. Halt the program after a short time and select *View→Graph→Time/Frequency* in order to display the contents of array `buffer`. Figures 2.12 and 2.13 show

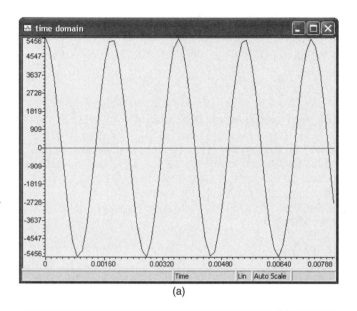

(a)

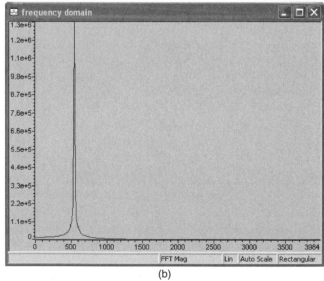

(b)

FIGURE 2.12. Input samples corresponding to 550-Hz sine wave, obtained using program `loop_buf.c`, plotted using Code Composer: (a) time domain and (b) frequency domain.

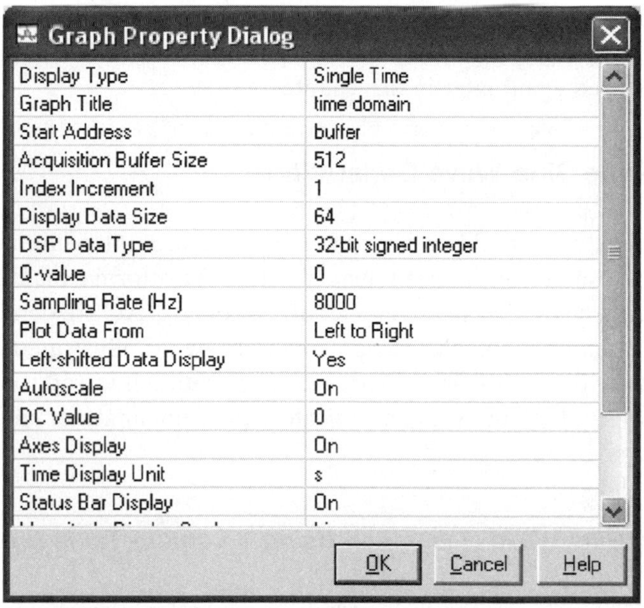

(a)

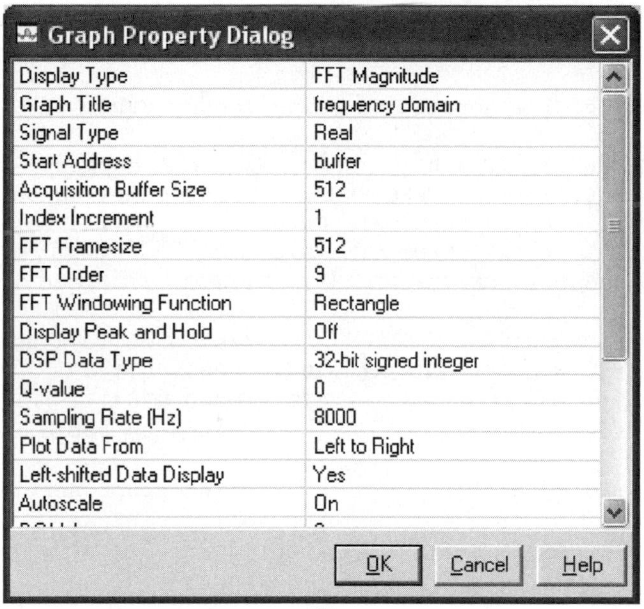

(b)

FIGURE 2.13. *Graph Property Dialog* windows showing properties for use with program `loop_buf.c`: (a) time domain and (b) frequency domain.

examples of time- and frequency-domain representations of that data and the *Graph Properties* used in each case. An input frequency of 550 Hz was used. Program loop_buf.c is used again later in this chapter.

2.3.1 Real-Time Sine Wave Generation

The following examples build on program sine8_buf.c, introduced in Chapter 1. By generating a variety of different analog output waveforms, including sinusoids of different frequencies, the characteristics of the codec DAC are demonstrated and the concepts of sampling, reconstruction, and aliasing are illustrated.

In addition, use of the *Goldwave* shareware application is introduced. This virtual instrument is a useful alternative to a dedicated spectrum analyzer and is used again in later chapters.

Example 2.7: Sine Wave Generation Using a Lookup Table (sine8_intr)

Program sine8_intr.c, listed in Figure 2.14, generates a sinusoidal signal using interrupts and a table lookup method. Its operation is as follows. An eight point

```
//sine8_intr.c Sine generation using lookup table

#include "DSK6713_AIC23.h"              //codec support
Uint32 fs=DSK6713_AIC23_FREQ_8KHZ;     //set sampling rate
#define DSK6713_AIC23_INPUT_MIC 0x0015
#define DSK6713_AIC23_INPUT_LINE 0x0011
Uint16 inputsource=DSK6713_AIC23_INPUT_MIC; //select input

#define LOOPLENGTH 8                    //size of look up table
short sine_table[LOOPLENGTH]={0,7071,10000,7071,0,
                              -7071,-10000,-7071};
short loopindex = 0;                    //look up table index

interrupt void c_int11()                //interrupt service routine
{
  output_left_sample(sine_table[loopindex]); //output value
  if (++loopindex >= LOOPLENGTH) loopindex = 0;
  return;                               //return from interrupt\
}

void main()
{
  comm_intr();                          //initialise DSK
  while(1);                             //infinite loop
}
```

FIGURE 2.14. Sine wave generation program using lookup table (sine8_intr.c).

lookup table is initialized in the array `sine_table` such that the value of `sine_table[i]` is equal to

$$10,000 \sin(2\pi i/8) \quad \text{for } i = 0, 1, 2, \ldots, 7$$

In this example, a sampling rate of 8 kHz is used and interrupts will occur every 0.125 ms. On interrupt, the interrupt service routine (ISR) `c_int11()` is called and within that routine the most important program statements are executed. Function `output_left_sample()` is used to output a value read from the array `sine_table` to the DAC and the index variable `loopindex` is incremented to point to the next value in the array. If the incremented value of `loopindex` is greater than or equal to the number of sample values in the table (`LOOPLENGTH`), it is reset to zero. The 1-kHz frequency of the sinusoidal output signal corresponds to the eight samples per cycle output at a rate of 8 kHz.

The DAC converts the output sample values into a sinusoidal analog output signal. Build and run the project as **sine8_intr** and verify a 1 kHz output waveform.

Example 2.8: Sine Wave Generation Using `sin()` Function Call (`sine_intr`)

Different sine wave frequencies can be generated using the table lookup method used by program `sine8_intr.c`. For example, a 3-kHz tone can be generated by changing the line that reads

```
short sine_table[LOOPSIZE] =
  {0, 7071, 10000, 7071, 0, -7071, -10000, -7071};
```

to read

```
short sine_table[LOOPSIZE] =
  {0, 7071, -10000, 7071, 0, -7071, 10000, -7071};
```

However, changing the contents and/or size of the lookup table is not a flexible way of generating sinusoids of arbitrary frequencies. Program `sine_intr.c`, listed in Figure 2.15, takes a different approach. At each sampling instant, that is, within function `c_int11()`, a new output sample value is calculated using a call to the math library function `sin()`. The floating-point parameter, `theta`, passed to that function is incremented at each sampling instant by the value `theta_increment = 2*PI*frequency/SAMPLING_FREQ` and when value of `theta` exceeds 2π the value 2π is subtracted from it.

While program `sine_intr.c` has the advantage of flexibility, it also has the disadvantage, relative to program `sine8_intr.c`, that it requires far greater computational effort, which is important in real-time applications.

```c
//sine_intr.c Sine generation using sin() function

#include <math.h>
#include "DSK6713_AIC23.h"                //codec support
Uint32 fs=DSK6713_AIC23_FREQ_8KHZ;        //set sampling rate
#define DSK6713_AIC23_INPUT_MIC 0x0015
#define DSK6713_AIC23_INPUT_LINE 0x0011
Uint16 inputsource=DSK6713_AIC23_INPUT_MIC; //select input

#define SAMPLING_FREQ 8000
#define PI 3.14159265358979

float frequency = 1000.0;
float amplitude = 10000.0;
float theta_increment;
float theta = 0.0;

interrupt void c_int11()
{
  theta_increment = 2*PI*frequency/SAMPLING_FREQ;
  theta += theta_increment;
  if (theta > 2*PI) theta -= 2*PI;
  output_left_sample((short)(amplitude*sin(theta)));
  return;
}

void main()
{
  comm_intr();
  while(1);
}
```

FIGURE 2.15. Sine wave generation program using call to math function `sin()` (sine_intr.c).

Build and run this project as **sine_intr** and experiment by changing the value assigned to the variable `frequency` (within the range 100–3800).

Example 2.9: Sine Wave Generation with Stereo Output (*sine_stereo*)

Source file `sine_stereo.c`, stored in the folder `sine_stereo`, is listed in Figure 2.16. It illustrates the use of both left- and right-hand channels of the AIC23 codec. Build and run this project as **sine_stereo**. Verify that a 1-kHz sinusoid is output through the right-hand channel and a 3-kHz sinusoid is output through the left-hand channel.

```
//sine_stereo Sine generation to both LEFT and RIGHT channels

#include "dsk6713_aic23.h"          //codec support
Uint32 fs=DSK6713_AIC23_FREQ_8KHZ;  //set sampling rate
#define DSK6713_AIC23_INPUT_MIC 0x0015
#define DSK6713_AIC23_INPUT_LINE 0x0011
Uint16 inputsource=DSK6713_AIC23_INPUT_MIC; //select input

#define LEFT  0
#define RIGHT 1
union {Uint32 uint; short channel[2];} AIC23_data;

#define LOOPLENGTH 8                 //size of look up table
short sine_table_left[LOOPLENGTH]={0,7071,10000,7071,0,
                                   -7071,-10000,-7071};
short sine_table_right[LOOPLENGTH]={0,-7071,10000,-7071,
                                    0,7071,-10000,7071};
short loopindex = 0;                 //look up table index

interrupt void c_int11()             //interrupt service routine
{
  AIC23_data.channel[RIGHT]=sine_table_right[loopindex];
  AIC23_data.channel[LEFT]=sine_table_left[loopindex];

  output_sample(AIC23_data.uint); //output to both channels
  if (++loopindex >= LOOPLENGTH)
    loopindex = 0;                   //check for end of table
  return;
}

void main()
{
  comm_intr();                       //init DSK,codec,McBSP
  while(1) ;                         //infinite loop
}
```

FIGURE 2.16. Program to generate two sinusoids of different frequencies using left- and right-hand channels (sine_stereo.c).

Example 2.10: Sine Wave Generation with Two Sliders for Amplitude and Frequency Control (sine2sliders)

The polling-based program sine2sliders.c, listed in Figure 2.17, generates a sine wave. Two sliders are used to vary both the amplitude (gain) and the frequency of the sinusoid generated. Using a lookup table containing 32 samples of exactly one cycle of a sine wave, the frequency of the output waveform is varied by selecting different numbers of points per cycle. The gain slider scales the volume/amplitude of the waveform signal. The appropriate GEL file sine2sliders.gel is listed in Figure 2.18.

```
//sine2sliders.c Sine generation with different # of points

#include "DSK6713_AIC23.h"                    //codec support
Uint32 fs=DSK6713_AIC23_FREQ_8KHZ;           //set sampling rate
#define DSK6713_AIC23_INPUT_MIC 0x0015
#define DSK6713_AIC23_INPUT_LINE 0x0011
Uint16 inputsource=DSK6713_AIC23_INPUT_LINE; //select input

short loop = 0;
short sine_table[32]={0,195,383,556,707,831,924,981,1000,
                      981,924,831,707,556,383,195,0,-195,
                      -383,-556,-707,-831,-924,-981,-1000,
                      -981,-924,-831,-707,-556,-383,-195};
short gain = 1;                          //slider gain
short frequency = 2;                     //slider frequency

void main()
{
  comm_poll();                           //init DSK, codec, McBSP
  while(1)                               //infinite loop
  {
    output_left_sample(sine_table[loop]*gain); //output value
    loop += frequency;                   //incr frequency index
    loop = loop % 32;                    //modulo to reset index
  }
}
```

FIGURE 2.17. Sinusoid generation with GEL slider controls for gain and frequency (sine2sliders.c).

```
/*sine2sliders.gel Two sliders to vary gain and frequency*/

menuitem "Sine Parameters"

slider Gain(1,8,1,1,gain_parameter)
{
  gain = gain_parameter;
}

slider Frequency(2,8,2,2,frequency_parameter)
{
  frequency = frequency_parameter;
}
```

FIGURE 2.18. GEL slider controls for gain and frequency (sine2sliders.gel).

The 32 sine data values in the table or buffer correspond to

$$1000\sin(2\pi i/32) \quad \text{for } i = 0, 1, 2, \ldots, 31$$

The frequency slider sets the value of the variable `frequency` to 2, 4, 6, or 8. This value is added to the index into the lookup table, `loop`, at each sampling instant. The modulo operator is used to test when the end of the lookup table is reached. When the loop index reaches 32, it is reset to zero. For example, with the frequency slider at position 2, the loop or frequency index steps through every other value in the table. This corresponds to 16 data values within one cycle.

Build this project as **sine2sliders**, using the appropriate support files for a polling-driven program. The main C source file *sine2sliders.c* is contained in the folder `sine2sliders`. Verify that initially the frequency generated is 500 Hz (frequency = 2). Increase the slider position (the use of a slider was introduced in Example 1.1) to 4, 6, and 8, and verify that the signal frequencies generated are 1000, 1500, and 2000 Hz, respectively. Note that when the slider is at position 4, the loop or frequency index steps through the table selecting the eight values (per cycle)—sin[0], sin[4], sin[8], . . . , sin[28]—that correspond to the data values 0, 707, 1000, 707, 0, −707, −1000, and −707. The resulting frequency generated is then 1 kHz (as in Example 1.1).

Example 2.11: Sweep Sinusoid Using Table with 8000 Points (sweep8000)

Figure 2.19 shows a listing of the program `sweep8000.c`, which generates a sweeping sinusoidal signal using a table lookup with 8000 points. The header file `sine8000_table.h` contains the 8000 data points that represent exactly one cycle of a sine wave. The file `sine8000_table.h` (stored in folder `sweep8000`) was generated using the MATLAB command

```
x = 1000*sin(2*pi*[0:7999]/8000)
```

Figure 2.20 shows a partial listing of the file *sine8000_table.h*.

At each sampling instant, program `sweep8000.c` reads an output sample value from the array `sine8000`, using the value of `float_index`, cast as an integer, as an index, and increments the value of `float_index` by the value `float_incr`. With N points in the lookup table representing one cycle of a sinusoid, the frequency of the output waveform is equal to `SAMPLING_FREQ*float_incr/N`.

A fixed value of `float_incr` would result in a fixed output frequency. In program `sweep8000.c`, the value of `float_incr` itself is incremented at each sampling instant by the value `DELTA_INCR` and hence the frequency of the output waveform increases gradually from `START_FREQ` to `STOP_FREQ`. The output waveform generated by the program can be altered by changing the values of the constants `START_FREQ`, `STOP_FREQ`, and `SWEEPTIME`, from which the value of `DELTA_INCR` is calculated.

```
//sweep8000.c sweep sinusoid using table with 8000 points

#include "DSK6713_AIC23.h"                    //codec support
Uint32 fs=DSK6713_AIC23_FREQ_8KHZ;            //set sampling rate
#define DSK6713_AIC23_INPUT_MIC 0x0015
#define DSK6713_AIC23_INPUT_LINE 0x0011
Uint16 inputsource=DSK6713_AIC23_INPUT_LINE; // select input
#include "sine8000_table.h"        //one cycle with 8000 points
#define SAMPLING_FREQ 8000.0
#define N 8000
#define START_FREQ 100.0
#define STOP_FREQ 3500.0
#define START_INCR START_FREQ*N/SAMPLING_FREQ
#define STOP_INCR STOP_FREQ*N/SAMPLING_FREQ
#define SWEEPTIME 5.0
#define DELTA_INCR (STOP_INCR - START_INCR)/(N*SWEEPTIME)

short amplitude = 10;                  //amplitude
float float_index = 0.0;
float float_incr = START_INCR;
short i;

void main()
{
  comm_poll();                          //init DSK, codec, McBSP
  while(1)                              //infinite loop
  {
    float_incr += DELTA_INCR;
      if (float_incr > STOP_INCR) float_incr = START_INCR;
    float_index += float_incr;
    if (float_index > N) float_index -= N;
    i = (short)(float_index);
    output_left_sample(amplitude*sine8000[i]); //output
  }
}
```

FIGURE 2.19. Program to generate sweeping sinusoid using table lookup with 8000 points (sweep8000.c).

Build and run this project as **sweep8000**. Verify the output as a sweeping sinusoid taking SWEEPTIME seconds to increase in frequency from START_FREQ to STOP_FREQ. Note that the source program *sweep8000.c* is polling-driven (use the appropriate interrupt vector file vectors_poll.asm).

Example 2.12: Generation of DTMF Tones Using a Lookup Table (sineDTMF_intr)

Program sineDTMF_intr.c, listed in Figure 2.21, uses a lookup table containing 512 samples of a single cycle of a sinusoid together with two independent pointers to generate a dual tone multifrequency (DTMF) waveform. DTMF waveforms are

```
//sine8000_table.h Sine table with 8000 points generated with MATLAB

short sine8000[8000]=
{0, 1, 2, 2, 3, 4, 5, 5,
6, 7, 8, 9, 9, 10, 11, 12,
13, 13, 14, 15, 16, 16, 17, 18,
19, 20, 20, 21, 22, 23, 24, 24,
25, 26, 27, 27, 28, 29, 30, 31,
31, 32, 33, 34, 35, 35, 36, 37,
38, 38, 39, 40, 41, 42, 42, 43,
44, 45, 46, 46, 47, 48, 49, 49,
50, 51, 52, 53, 53, 54, 55, 56,
57, 57, 58, 59, 60, 60, 61, 62,
63, 64, 64, 65, 66, 67, 67, 68,
69, 70, 71, 71, 72, 73, 74, 75,
75, 76, 77, 78, 78, 79, 80, 81,
    .
    .
    .
-13, -12, -11, -10, -9, -9, -8, -7,
-6, -5, -5, -4, -3, -2, -2, -1};
```

FIGURE 2.20. Partial listing of sine with 8000 data points (sine8000_table.h).

used in telephone networks to indicate key presses. A DTMF waveform is the sum of two sinusoids of different frequencies. A total of 16 different combinations of frequencies each comprising one of four low frequency components (697, 770, 852, or 941 Hz) and one of four high frequency components (1209, 1336, 1477, or 1633) are used. Program sineDTMF_intr.c uses two independent pointers into a single lookup table, each updated at the same rate (16 kHz) but each stepping through the values in the table at a different rate.

A pointer that stepped through every single one of the TABLESIZE samples stored in the lookup table at a sampling rate of 16 kHz would generate a sinusoidal tone with a frequency of $f = (16000 \ / \ TABLESIZE)$. A pointer that stepped through the samples stored in the lookup table, incremented by a value STEP, would generate a sinusoidal tone with a frequency of $f = (16000 \ * \ STEP \ / \ TABLESIZE)$.

From this it is possible to calculate the required step size for any desired frequency f. For example, in order to generate a sinusoid with frequency 770 Hz, the required step size is STEP = TABLESIZE * 770/16000 = 512 * 770/16000 = 24.64.

In other words, at each sampling instant, the pointer into the lookup table should be incremented by 24.64. The pointer value, or index, into the lookup table must in practice be an integer value ((short)loopindexlow) but the floating-point value of the pointer, or index, loopindexlow, can be maintained and incremented by 24.64, wrapping around 0.0 when its value exceeds 512.0 using the statements

```
loopindexlow += 24.64;
if(loopindexlow>(float)TABLESIZE)loopindexlow-=(float)TABLESIZE;
```

```
//sinedtmf_intr.c DTMF tone generation using lookup table

#include "DSK6713_AIC23.h"                //codec support
Uint32 fs=DSK6713_AIC23_FREQ_16KHZ;       //set sampling rate
#define DSK6713_AIC23_INPUT_MIC 0x0015
#define DSK6713_AIC23_INPUT_LINE 0x0011
Uint16 inputsource=DSK6713_AIC23_INPUT_MIC; //select input
#include <math.h>
#define PI 3.14159265358979

#define TABLESIZE 512                 // size of look up table
#define SAMPLING_FREQ 16000
#define STEP_770 (float)(770 * TABLESIZE)/SAMPLING_FREQ
#define STEP_1336 (float)(1336 * TABLESIZE)/SAMPLING_FREQ
#define STEP_697 (float)(697 * TABLESIZE)/SAMPLING_FREQ
#define STEP_852 (float)(852 * TABLESIZE)/SAMPLING_FREQ
#define STEP_941 (float)(941 * TABLESIZE)/SAMPLING_FREQ
#define STEP_1209 (float)(1209 * TABLESIZE)/SAMPLING_FREQ
#define STEP_1477 (float)(1477 * TABLESIZE)/SAMPLING_FREQ
#define STEP_1633 (float)(1633 * TABLESIZE)/SAMPLING_FREQ

short sine_table[TABLESIZE];
float loopindexlow = 0.0;             // look up table index
float loopindexhigh = 0.0;
short i;

interrupt void c_int11()              //interrupt service routine
{
  output_left_sample(sine_table[(short)loopindexlow]
  + sine_table[(short)loopindexhigh]);
  loopindexlow += STEP_697;
  if (loopindexlow > (float)TABLESIZE)
    loopindexlow -= (float)TABLESIZE;
  loopindexhigh += STEP_1477;
  if (loopindexhigh > (float)TABLESIZE)
    loopindexhigh -= (float)TABLESIZE;

  return;                             //return from interrupt
}

void main()
{
  comm_intr();                        // initialise DSK
  for (i=0 ; i< TABLESIZE ; i++)
    sine_table[i] = (short)(10000.0*sin(2*PI*i/TABLESIZE));
  while(1);
}
```

FIGURE 2.21. Program to generate DTMF tone using lookup table (sineDTMF_intr.c).

In program sineDTMF_intr.c, the floating-point values by which the table lookup indices are incremented are predefined using, for example, line

```
#define STEP_770 (float)(770 * TABLESIZE) / SAMPLING_FREQ
```

In order to change the DTMF tone generated, and simulate a different key press, edit the file sineDTMF.c and change the lines that read

```
loopindexlow  += STEP_697;
loopindexhi   += STEP_1477;
```

to, for example,

```
loopindexlow  += STEP_770;
loopindexhi   += STEP_1209;
```

An example of the output generated by program sineDTMF_intr.c is shown in Figure 2.22.

Example 2.13: Sine Wave Generation with Table Values Generated Within Program (*sinegen_table*)

This example creates one period of sine data values for a table. Then these values are output to generate a sine wave. Figure 2.23 shows a listing of the program *sinegen_table.c*, which implements this project. The frequency generated is

$$f = F_S/(\text{number of points}) = 8000/10 = 800\,\text{Hz}$$

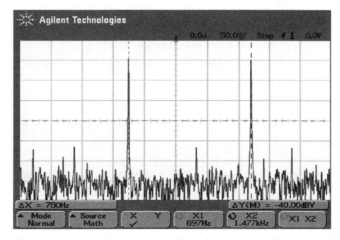

FIGURE 2.22. Frequency-domain representation of output signal generated using program sineDTMF_intr.c.

```
//sinegen_table.c generates a sinusoid using a look-up table

#include "DSK6713_AIC23.h"              //codec-DSK support file
Uint32 fs=DSK6713_AIC23_FREQ_8KHZ;     //set sampling rate
#define DSK6713_AIC23_INPUT_MIC 0x0015
#define DSK6713_AIC23_INPUT_LINE 0x0011
Uint16 inputsource=DSK6713_AIC23_INPUT_MIC; //select mic in
#include <math.h>
#define table_size (short)10           //set table size
short sine_table[table_size];          //sine table array
int i;

interrupt void c_int11()               //interrupt service routine
{
  output_left_sample(sine_table[i]); //output each sine value
  if (i < table_size - 1) ++i;       //incr index until end of table
  else i = 0;                        //reset index if end of table
  return;                            //return from interrupt
}

void main()
{
  float pi=3.14159;

  for(i = 0; i < table_size; i++)
    sine_table[i]=10000*sin(2.0*pi*i/table_size);
  i = 0;
  comm_intr();                         //init DSK, codec, McBSP
  while(1);                            //infinite loop
}
```

FIGURE 2.23. Sine wave generation program using table generated within program sinegen_table.c.

Build and run this project as **sinegen_table**. Verify a sine wave generated with a frequency of 800 Hz. Change the number of points to generate a 400-Hz sine wave (only table_size needs to be changed).

Example 2.14: Sine Wave Generation with Table Created by MATLAB (*sin1500MATL*)

This example illustrates the generation of a sinusoid using a lookup table created with MATLAB. Figure 2.24 shows a listing of the MATLAB program sin1500.m, which generates a file with 128 data points with 24 cycles. The sine wave frequency generated is

$$f = F_S(\text{number of cycles})/(\text{number of points}) = 1500\,\text{Hz}$$

```
%sin1500.m Generates 128 points representing sin(1500) Hz
%Creates file sin1500.h
for i=1:128
   sine(i) = round(1000*sin(2*pi*(i-1)*1500/8000)); %sin(1500)
end

fid = fopen('sin1500.h','w');           %open/create file
fprintf(fid,'short sin1500[128]={');    %print array name,"={"
fprintf(fid,'%d, ',sine(1:127));        %print 127 points
fprintf(fid,'%d' ,sine(128));           %print 128th point
fprintf(fid,'};\n');                    %print closing bracket
fclose(fid);                            %close file
```

FIGURE 2.24. MATLAB program to generate a lookup table for sine wave data (sin1500.m).

```
short sin1500[128]=
{0, 924, 707, -383, -1000, -383, 707, 924,
 0, -924, -707, 383, 1000, 383, -707, -924,
 0, 924, 707, -383, -1000, -383, 707, 924,
 0, -924, -707, 383, 1000, 383, -707, -924,
 0, 924, 707, -383, -1000, -383, 707, 924,
 0, -924, -707, 383, 1000, 383, -707, -924,
 0, 924, 707, -383, -1000, -383, 707, 924,
 0, -924, -707, 383, 1000, 383, -707, -924,
 0, 924, 707, -383, -1000, -383, 707, 924,
 0, -924, -707, 383, 1000, 383, -707, -924,
 0, 924, 707, -383, -1000, -383, 707, 924,
 0, -924, -707, 383, 1000, 383, -707, -924,
 0, 924, 707, -383, -1000, -383, 707, 924,
 0, -924, -707, 383, 1000, 383, -707, -924,
 0, 924, 707, -383, -1000, -383, 707, 924,
 0, -924, -707, 383, 1000, 383, -707, -924};
```

FIGURE 2.25. Sine table lookup header file generated by MATLAB (sin1500.h).

Run sin1500.m within MATLAB and verify the creation of header file sin1500.h with 128 points, as shown in Figure 2.25. Different numbers of points representing sinusoidal signals of different frequencies can readily be obtained with minor changes to the MATLAB program sin1500.m.

Figure 2.26 shows a listing of the C source file sin1500MATL.c, which implements this project in real time. This program includes the header file generated by MATLAB. See also Example 2.13, which generates the table within the main C source program rather than using MATLAB.

Build and run this project as **sin1500MATL**. Verify that the output is a 1500-Hz sine wave signal. Within CCS, be careful when you view the header file sin1500.h so as not to truncate it.

```
//Sin1500MATL.c Generates sine from table created with MATLAB

#include "DSK6713_AIC23.h"                //codec-DSK support file
Uint32 fs=DSK6713_AIC23_FREQ_8KHZ;   //set sampling rate
#define DSK6713_AIC23_INPUT_MIC 0x0015
#define DSK6713_AIC23_INPUT_LINE 0x0011
Uint16 inputsource=DSK6713_AIC23_INPUT_LINE; //select input
#include "sin1500.h"                     //created with MATLAB
int i=0;

interrupt void c_int11()
{
  output_sample(sin1500[i]*10);         //output each sine value
  if (i < 127) ++i;                     //incr until end of table
    else i = 0;
  return;                               //return from interrupt
}

void main()
{
  comm_intr();                          //init DSK, codec, McBSP
  while(1);                             //infinite loop
}
```

FIGURE 2.26. Sine generation program using header file with sine data values generated by MATLAB (sin1500MATL.c).

Example 2.15: Sine Wave Generation with DIP Switch Control (sine_led_ctrl)

The program sine_led_ctrl.c, shown in Figure 2.27, implements sine wave generation using a DIP switch to control for how long the sine wave is generated. When DIP switch #0 is pressed and held down, LED #0 toggles and a 1-kHz sine wave is generated, for as long as DIP switch #0 is held down or for a duration determined by the value of the variable on_time. A GEL slider (sine_led_ctrl.gel) can be used to vary the value of on_time between 1 and 10 seconds. Unlike Example 1.1, after DIP switch #0 is pressed down, a sine wave is generated but *only* for on_time seconds.

Build and run this project as **sine_led_ctrl**. Press DIP switch #0 and verify both that LED #0 toggles and that a 1-kHz sine wave is generated for 1 second (with on_time set at 1). Load the GEL file sine_led_ctrl.gel and select *GEL→ Delay Control* to obtain the slider. Increase the slider value to 8. The sine wave should be generated and LED #0 should toggle for approximately 8 seconds after DIP switch #0 is pressed.

//**sine_led_ctrl.c** Sine generation with DIP Switch control

```c
#include "dsk6713_aic23.h"              //codec support
Uint32 fs=DSK6713_AIC23_FREQ_8KHZ;  //set sampling rate
#define DSK6713_AIC23_INPUT_MIC 0x0015
#define DSK6713_AIC23_INPUT_LINE 0x0011
Uint16 inputsource=DSK6713_AIC23_INPUT_MIC; //select input
short sine_table[8]={0,707,1000,707,0,-707,-1000,-707};
short loop=0, gain=10;
short j=0, k = 0;                       //delay counter
short flag = 0;                         //for LED on
short const delay = 800;                //for delay loop
short on_time = 1;                      //led is on for on_time secs

void main()
{
 comm_poll();                           //init BSL
 DSK6713_LED_init();                    //init LEDs
 DSK6713_DIP_init();                    //init DIP SWs
 while(1)                               //infinite loop
 {
  if(DSK6713_DIP_get(0)==0 &&(k<=(on_time*5))) //if SW0 pressed
  {
   if(flag==0) DSK6713_LED_toggle(0);     //LED0 toggles
   else DSK6713_LED_off(0);               //turn LED0 off
   output_sample(sine_table[loop]*gain); //output with gain
   if(loop < 7) loop++;                   //increment loop index
   else loop = 0;                         //reset if end of table
   if (j < delay) ++j;                    //delay counter
   else
   {
    j = 0;                                //reset delay counter
    if (flag == 1)
    {
     flag = 0;                            //if flag=1 toggle LED
     k++;
    }
    else flag = 1;                        //toggle flag
   }
  }
  else
  {
   DSK6713_LED_off(0);                    //turn off LED0
   if(DSK6713_DIP_get(0)==1) k=0;   //if LED0 off reset counter
  }
 }
}
```

FIGURE 2.27. Sine generation with DIP switch control program (sine_led_ctrl.c).

Example 2.16: Signal Reconstruction, Aliasing, and the Properties of the AIC23 Codec

Generating analog output signals using programs such as sine_intr.c (Figure 2.15) is a useful means of investigating the characteristics of the AIC23 codec. Change the value of the variable frequency in program sine_intr.c to an arbitrary value between 100.0 and 3500.0 and you should find that a sine wave of that frequency is generated. Change the value of the variable frequency to 7000.0, however, and you will find that a 1-kHz sine wave is generated. Verify that the same is true if the value of frequency is changed to 9000.0 or 15000.0.

These effects are due to the phenomenon of aliasing. Sequences of samples calculated using function sin() at frequencies $8000n \pm 1000\,\mathrm{Hz}$, where $n = 0, \pm1, \pm2, \pm3, \ldots$ are identical and all are reconstructed by the codec as a 1-kHz sine wave.

A graphical representation of this is shown in Figure 2.28.

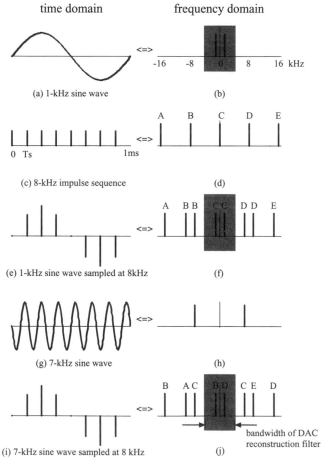

FIGURE 2.28. Graphical representation of equivalence of signals generated by sampling 1-kHz and 7-kHz sine waves at 8-kHz.

In the time domain, the sampling process may be represented by multiplication of the analog input waveform sin(2*pi*1000*t) (Figure 2.28a) by a sequence of impulses at intervals of T_s = 0.125 ms (Figure 2.28c), resulting in a sequence of weighted impulses (Figure 2.28e).

In the frequency domain, the analog input waveform is represented by two discrete values at ±1 kHz (Figure 2.28b) and the sequence of time-domain impulses by a sequence of impulses in the frequency domain at intervals of $1/T_s$ = 8 kHz (Figure 2.28d).

Multiplication in the time domain is equivalent to convolution in the frequency domain. Convolving the signals of Figures 2.28b and 2.28d, the frequency-domain representation of the sampled sinusoid (Figure 2.28e) is an infinitely repeated sequence of copies of the two impulses at ±1 kHz centered at 0 Hz, ±8 kHz, ±16 kHz, ... (Figure 2.28f).

Next, consider the case of a 7-kHz sine wave sampled at 8 kHz. Time- and frequency-domain representations of the analog input signal sin(2*pi*7000*t) are shown in Figures 2.28g and 2.28h.

Convolving the signal shown in Figure 2.28h with that shown in Figure 2.28d results in the signal shown in Figure 2.28j. This comprises an infinitely repeated sequence of copies of the two impulses at ±7 kHz centered at 0 Hz, ±8 kHz, ±16 kHz, ... Despite their different derivations, Figures 2.28f and 2.28j are identical. This corresponds to the equivalence of the time-domain sample sequences shown in Figures 2.28e and 2.28i.

The lowpass characteristic of the DAC can be represented by the attenuation, or blocking, of frequency components outside the range ±4 kHz. This results, in this example, in the lowpass filtered or reconstructed versions of the signals in Figures 2.28f and 2.28j being identical to that shown in Figure 2.28b.

Since the reconstruction (digital-to-analog conversion) process is one of lowpass filtering, it follows that the bandwidth of signals output by the codec is limited. This can be demonstrated in a number of different ways.

For example, run program `sine_intr.c` with the value of the variable `frequency` set to 3500.0. Verify that the output waveform generated has a frequency of 3500 Hz. Next, change the value of the variable `frequency` to 4500.0. The frequency of the output waveform should again be equal to 3500 Hz. Try any value for the variable `frequency`. You should find that it is impossible to generate an output waveform with a frequency greater than 4000 Hz (assuming a sampling frequency of 8 kHz).

Example 2.17: Square Wave Generation Using Lookup Table (squarewave)

Program `squarewave.c`, listed in Figure 2.29 and stored in folder `squarewave`, differs from program `sine8_intr.c` only insofar as it uses a lookup table containing 64 samples of one cycle of a square wave of frequency 125 Hz rather than 8 samples of one cycle of a sine wave. Build and run the program and, using an oscilloscope,

```c
//squarewave.c 125 Hz square wave generated using lookup table
#include "dsk6713_aic23.h"            //codec support
Uint32 fs=DSK6713_AIC23_FREQ_8KHZ;  // set sampling rate
#define DSK6713_AIC23_INPUT_MIC 0x0015
#define DSK6713_AIC23_INPUT_LINE 0x0011
Uint16 inputsource=DSK6713_AIC23_INPUT_LINE; // select input
#define LOOPLENGTH 64
short square_table[LOOPLENGTH] =
{10000, 10000, 10000, 10000, 10000, 10000, 10000, 10000,
10000, 10000, 10000, 10000, 10000, 10000, 10000, 10000,
10000, 10000, 10000, 10000, 10000, 10000, 10000, 10000,
10000, 10000, 10000, 10000, 10000, 10000, 10000, 10000,
-10000,-10000,-10000,-10000,-10000,-10000,-10000,-10000,
-10000,-10000,-10000,-10000,-10000,-10000,-10000,-10000,
-10000,-10000,-10000,-10000,-10000,-10000,-10000,-10000,
-10000,-10000,-10000,-10000,-10000,-10000,-10000,-10000};
short loopindex = 0;

interrupt void c_int11()
{
  output_sample(square_table[loopindex++]);
  if (loopindex >= LOOPLENGTH)
    loopindex = 0;
  return;
}

void main()
{
  comm_intr();
  while(1);
}
```

FIGURE 2.29. Program to generate square wave (`squarewave.c`).

you should see an output waveform as shown in Figure 2.30a. This waveform is equivalent to a square wave (represented by the samples in the lookup table) passed through a lowpass filter (the DAC). The symmetrical "ringing" at each edge of the square wave is indicative of the presence of a digital FIR filter, which is exactly how the DAC implements the lowpass reconstruction filter. Figure 2.31 shows the magnitude frequency response of that filter as specified in the AIC23 datasheet. The drooping of the level of the waveform between transients seen in the oscilloscope trace of Figure 2.30a is due to the ac coupling of the codec to the LINE OUT socket.

The lowpass characteristic of the reconstruction filter can further be highlighted by looking at the frequency content of the output waveform. Only harmonics below 3.8 kHz are present in the analog output waveform as shown in Figure 2.30(b).

The following examples demonstrate a number of alternative approaches to observing the lowpass characteristic of the DAC reconstruction filter.

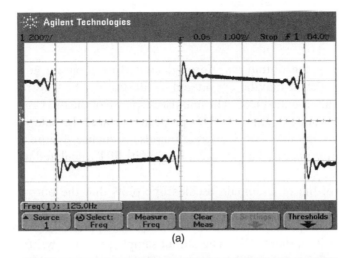

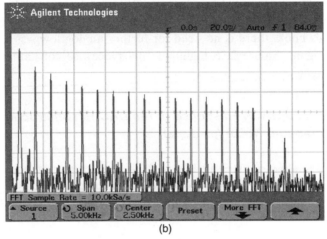

FIGURE 2.30. Waveform generated using program `squarewave.c`: (a) time domain and (b) frequency domain.

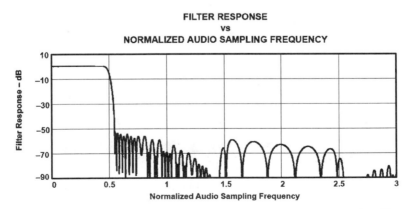

FIGURE 2.31. Magnitude frequency response of AIC23 codec reconstruction filter. (Courtesy of Texas Instruments.)

Example 2.18: Step and Impulse Responses of the DAC Reconstruction Filter (`dimpulse`)

The frequency of the output waveform generated by program `squarewave.c` is sufficiently low that it may be considered as illustrating the step response of the reconstruction filter in the DAC. The impulse response of the filter can be illustrated by running program `dimpulse.c`. This program replaces the samples of a square wave in the lookup table with a discrete impulse sequence. Figure 2.32 shows the output waveform generated by `dimpulse.c` and its FFT calculated using an *Agilent 54621A* oscilloscope.

Note in all of the time-domain oscilloscope plots that the output waveform is piecewise constant with a step length of $12.5\,\mu$s. This, together with the symmetrical form of impulse response, is indicative of an oversampling digital lowpass reconstruction filter within the AIC23. The output sample sequence written to the DAC at a rate of 8 kHz is upsampled to a rate of 80 kHz and passed through a lowpass digital filter and then a zero order hold. For many applications, including the example programs in this book, there is no need for further analog lowpass filtering of the codec output signal. Figure 2.33 highlights the piecewise constant nature of the codec output signal.

Example 2.19: Frequency Response of the DAC Reconstruction Filter Using Pseudorandom Noise (`prandom`)

The program `prandom.c`, listed in Figure 2.34, generates a pseudorandom noise sequence. It uses a software-based implementation of a maximal-length sequence

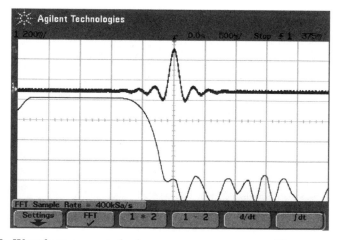

FIGURE 2.32. Waveform generated using program `dimpulse.c`. Upper trace shows output signal in time domain ($5\,\mu$s/div); lower trace shows output signal in frequency domain (1-kHz/div).

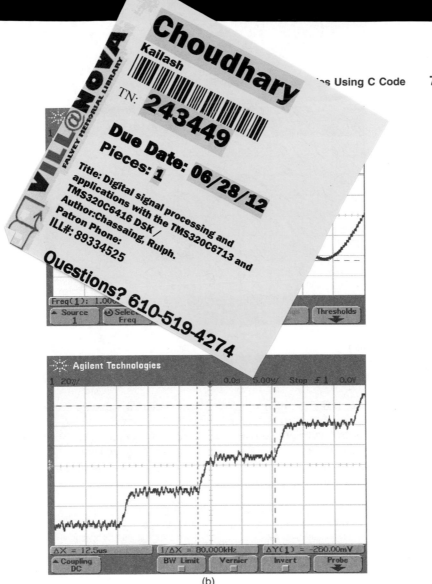

(b)

FIGURE 2.33. Output waveform generated using program `sine_intr.c`: (a) 1-kHz sine wave and (b) detailed view emphasizing piecewise constant nature.

technique for generating a pseudorandom binary sequence. An initial 16-bit seed is assigned to a register. Bits b0, b1, b11, and b13 are XORed and the result is placed into a feedback variable. The register with the initial seed value is then shifted one bit to the left. The feedback variable is then assigned to bit b0 of the register. A scaled minimum or maximum is assigned to `prnseq`, depending on whether the register's bit b0 is zero or 1. This scaled value corresponds to the noise-level amplitude. The header file `noise_gen.h` defines the shift register bits.

Build and run this project as **prandom**. Figure 2.35 shows the output waveform displayed using an oscilloscope and using *Goldwave*. The output spectrum is relatively flat until the cutoff frequency of approximately 3800 Hz, which represents the bandwidth of the reconstruction filter on the AIC23 codec.

```
// prandom.c program to test response of AIC23 codec
// using pseudo-random noise
#include "DSK6713_AIC23.h"          // codec support
Uint32 fs=DSK6713_AIC23_FREQ_8KHZ; //set sampling rate
#define DSK6713_AIC23_INPUT_MIC 0x0015
#define DSK6713_AIC23_INPUT_LINE 0x0011
Uint16 inputsource=DSK6713_AIC23_INPUT_MIC; //select mic in

#include "noise_gen.h"              //support file for noise
#define NOISELEVEL 8000             //scale for  1 sequence
int fb;                            //feedback variable
shift_reg sreg;                    //shift register
float yn;                          //output sample

int prand(void)                    //pseudo-random noise
{
  int prnseq;
  if(sreg.bt.b0)
    prnseq = -NOISELEVEL;          //scaled -ve noise level
  else
    prnseq = NOISELEVEL;           //scaled  noise level
  fb =(sreg.bt.b0)^(sreg.bt.b1);   //XOR bits 0,1
  fb^=(sreg.bt.b11)^(sreg.bt.b13); //with bits 11,13 -> fb
  sreg.regval<<=1;                 //shift register to left
  sreg.bt.b0=fb;                   //close feedback path
  return prnseq;
}

void resetreg(void)                //reset shift register
{
  sreg.regval=0xFFFF;              //initial seed value
  fb = 1;                          //initial feedback value
}

interrupt void c_int11()           //interrupt service routine
{
  yn = (float)(prand());           //get new sample value
  output_left_sample((short)(yn)); //output to codec
  return;                          //return from interrupt
}

void main()
{
  resetreg();                      //reset shift register
  comm_intr();                     //initialise McBSP and codec
  while (1);                       //infinite loop
}
```

FIGURE 2.34. Pseudorandom binary sequence generation (prandom.c).

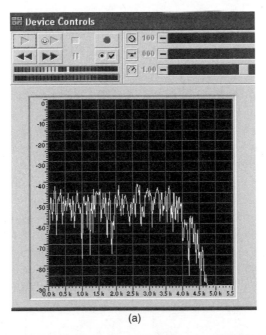

(a)

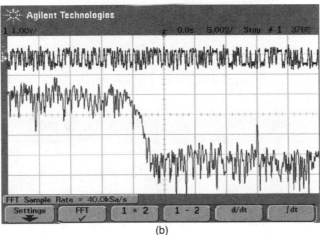

(b)

FIGURE 2.35. Waveform generated using program `prandom.c`: (a) using *Goldwave* and (b) using oscilloscope. Upper trace shows output signal in time domain (5ms/div); lower trace shows output signal in frequency domain (1kHz/div).

Aliasing

So far we have seen that the AIC23 codec cannot generate signal components having frequencies greater than 4kHz. This is true however we produce output sample sequences. It follows that it is inadvisable to allow analog input signal components having frequencies greater than 4kHz to be sampled at the input to a DSP system. This can be prevented by passing analog input signals through a lowpass antialiasing filter prior to sampling. An oversampling digital antialiasing filter with characteristics similar to those of the reconstruction filter in the digital-to-analog converter is built in to the analog-to-digital converter in the AIC23 codec.

Example 2.20: Step Response of the AIC23 Codec Antialiasing Filter (`loop_buf`)

In order to investigate the step response of the *antialiasing* filter on the AIC23, connect a signal generator to the DSK LINE IN socket. Adjust the signal generator to give a square wave output of frequency 270 Hz and amplitude 0.2 V. Load and run program `loopbuf.c` on the DSK, halting the DSP after a few seconds. You can view the most recent 64 input sample values by selecting *View→Graph* and setting the *Graph Properties* as shown in Figure 2.36. You should see something similar to the display shown in Figure 2.37b. Figure 2.37a shows the square wave input signal that produced the display of Figure 2.37b. The ringing on either side of edges of the square wave and the drooping of the level between transients are due to the antialiasing filter and the ac coupling of the codec to the LINE IN socket on the DSK.

Example 2.21: Demonstration of Aliasing (`aliasing`)

The digital antialiasing filters in the AIC23 codec cannot be bypassed or disabled. However, as mentioned previously, aliasing is a potential problem whenever sampling takes place. Program `aliasing.c` uses a sampling rate of 16 kHz for the codec

Graph Property Dialog	
Display Type	Single Time
Graph Title	loop_buf
Start Address	buffer
Acquisition Buffer Size	64
Index Increment	1
Display Data Size	64
DSP Data Type	32-bit signed integer
Q-value	0
Sampling Rate (Hz)	8000
Plot Data From	Left to Right
Left-shifted Data Display	Yes
Autoscale	Off
DC Value	0
Maximum Y-value	5000
Axes Display	On
Time Display Unit	s

FIGURE 2.36. *Graph Property* settings for use with program `loop_buf.c`.

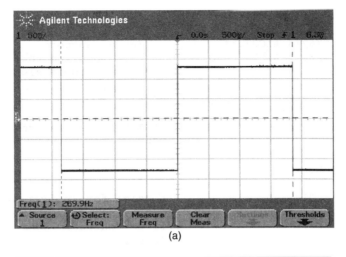

(a)

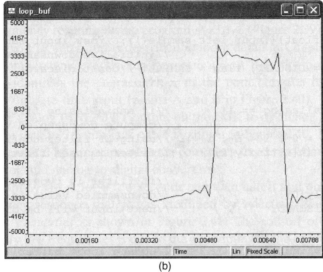

(b)

FIGURE 2.37. Step response of AIC23 codec antialiasing filter: (a) square wave input signal and (b) input samples captured using program `loop_buf.c`.

but then resamples the sequence of samples produced by the ADC at the lower rate of 8 kHz (downsampling). The sequence of samples generated at a rate of 16 kHz by the ADC may contain frequency components at frequencies greater than 4 kHz and therefore if that sample sequence is downsampled to a rate of 8 kHz simply by discarding every second sample, aliasing may occur. To avoid aliasing, the 16-kHz sample sequence output by the ADC must be passed through a digital antialiasing filter before downsampling. Program `aliasing.c` uses an FIR filter (see Chapter 4) for this task (Figure 2.38). For the purposes of this example, it is sufficient to consider that the program demonstrates the effect of sampling at a frequency of 8 kHz with and without using an antialiasing filter.

```
//sysid.c
#include "DSK6713_AIC23.h"              //codec support
Uint32 fs=DSK6713_AIC23_FREQ_8KHZ; //set sampling rate
#define DSK6713_AIC23_INPUT_MIC 0x0015
#define DSK6713_AIC23_INPUT_LINE 0x0011
Uint16 inputsource=DSK6713_AIC23_INPUT_LINE; //select line in

#include "noise_gen.h"                  //support for noise gen
#define beta 1E-12                      //learning rate
#define WLENGTH 256
#define NOISELEVEL 8000

float w[WLENGTH+1];                      //coeffs for adaptive FIR
float dly_adapt[WLENGTH+1];             //samples of adaptive FIR
int fb;                                  //feedback variable
shift_reg sreg;                          //shift register

int prand(void)                          //gen pseudo-random sequence
{
  int prnseq;
  if(sreg.bt.b0)
   prnseq = -NOISELEVEL;                 //scaled negative noise level
  else
   prnseq = NOISELEVEL;                  //scaled positive noise level
  fb =(sreg.bt.b0)^(sreg.bt.b1);    //XOR bits 0,1
  fb^=(sreg.bt.b11)^(sreg.bt.b13); //with bits 11,13 -> fb
  sreg.regval<<=1;
  sreg.bt.b0=fb;                         //close feedback path
  return prnseq;                         //return noise value
}

interrupt void c_int11()                 //interrupt service routine
{
 int i;
 float adaptfir_out = 0.0;               //init adaptive filter output
 float fir_out = 0.0;
 float E;                                //output error

 fir_out = (float)(input_left_sample()); //unknown system output
 dly_adapt[0]=prand();                   //pseudo-random noise used as
 output_left_sample((short)(dly_adapt[0])); //input to filter
                                         //and to unknown system
 for (i = 0; i < WLENGTH; i++)
  adaptfir_out +=(w[i]*dly_adapt[i]); //compute filter output

 E = fir_out - adaptfir_out;             //compute output error signal
```

FIGURE 2.39. C source program used to identify characteristics of AIC23 codec antialiasing and reconstruction filters (sysid.c).

```
for (i = WLENGTH-1; i >= 0; i--)
  {
   w[i] = w[i]+(beta*E*dly_adapt[i]); //update adaptive weights
   dly_adapt[i+1] = dly_adapt[i];    //update adaptive samples
  }
 return;
}

void main()
{
 int i = 0;
 for (i = 0; i <= WLENGTH; i++)
  {
   w[i] = 0.0;                     //init coeffs for adaptive FIR
   dly_adapt[i] = 0.0;             //init buffer for adaptive FIR
  }

 sreg.regval=0xFFFF;               //initial seed value
 fb = 1;                           //initial feedback value
 comm_intr();                      //init DSK, codec, McBSP
 while (1);                        //infinite loop
}
```

FIGURE 2.39. (*Continued*)

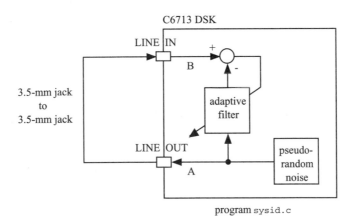

program sysid.c

FIGURE 2.40. Connection diagram for identification of antialiasing and reconstruction filter characteristics using adaptive filter (sysid.c).

Connect two DSKs together as shown in Figure 2.45. Make sure that program loop_intr.c is running on one DSK before running program sysid16.c for a short time on the other. After running and halting program sysid16.c, use the *View→Graph* facility with parameters set as shown in Figure 2.46 in order to view the magnitude frequency response of the DSK running program loop_intr.c. Compare the frequency response shown in Figure 2.47 with that shown in the AIC23 datasheet (Figure 2.31). (See also Figures 2.48 and 2.49).

☑ Graph Property Dialog	
Display Type	FFT Magnitude
Graph Title	sysid
Signal Type	Real
Start Address	w
Acquisition Buffer Size	128
Index Increment	1
FFT Framesize	128
FFT Order	8
FFT Windowing Function	Rectangle
Display Peak and Hold	Off
DSP Data Type	32-bit floating point
Sampling Rate (Hz)	8000
Plot Data From	Left to Right
Left-shifted Data Display	Yes
Autoscale	Off
DC Value	-20
Maximum Y-value	0
Axes Display	On
Frequency Display Unit	Hz
Status Bar Display	On
Magnitude Display Scale	Logarithmic

OK   Cancel   Help

FIGURE 2.41. *Graph Property* settings (frequency domain) for use with program `sysid.c`.

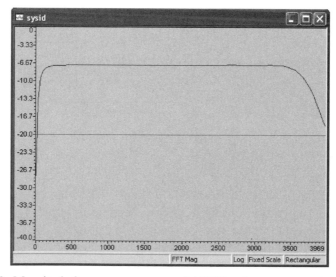

FIGURE 2.42. Magnitude frequency response of AIC23 codec reconstruction and antialiasing filters identified using program `sysid.c`.

```
mp.c Generates a ramp

lude "dsk6713_aic23.h"          //codec support
32 fs=DSK6713_AIC23_FREQ_8KHZ; //set sampling freq
ine DSK6713_AIC23_INPUT_MIC 0x0015
ine DSK6713_AIC23_INPUT_LINE 0x0011
16 inputsource=DSK6713_AIC23_INPUT_MIC; //select mic in
t output;

rrupt void c_int11()            //interrupt service routine

put_left_sample(output);        //output each sample period
out += 2000;                    //increment output value
(output >= 30000)               //if peak is reached
tput = -30000;                  //reinitialize
urn;                            //return from interrupt


 main()

ut = 0;                         //init output to zero
_intr();                        //init DSK, codec, McBSP
e(1);                           //infinite loop
```

FIGURE 2.50. Ramp generation program (`ramp.c`).

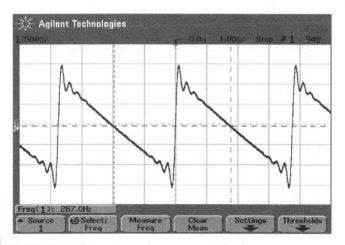

Output waveform generated using program `ramp.c`.

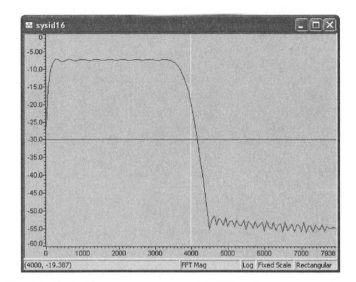

FIGURE 2.47. Magnitude frequency response of AIC23 codec reconstruction and antialiasing filters identified using program `sysid16.c`.

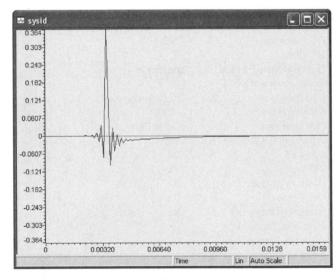

FIGURE 2.48. *Graph Property* settings (time domain) for use with program `sysid16.c`.

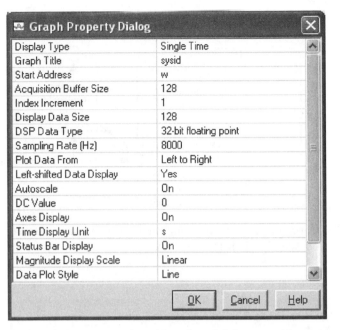

FIGURE 2.43. *Graph Property* settings (time domain) for use with program `sysid.c`.

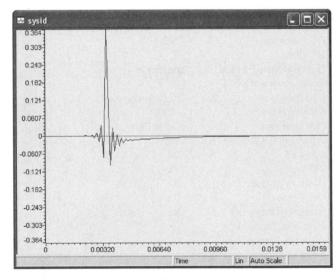

FIGURE 2.44. Impulse response of AIC23 codec reconstruction and antialiasing filters identified using program `sysid.c`.

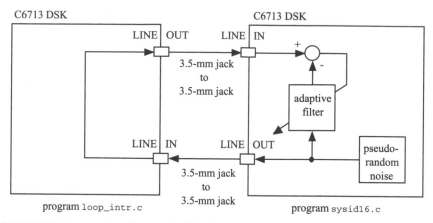

FIGURE 2.45. Connection diagram for identification of antialiasing and reconstruction filter characteristics using adaptive filter (sysid16.c).

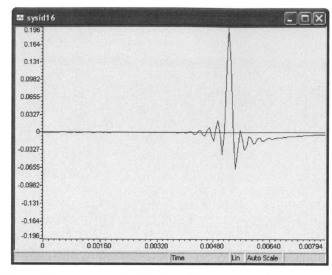

FIGURE 2.46. *Graph Property* settings (frequency domain) for use with program sysid16.c.

FIGURE 2.49. Impulse response of AIC23 codec reconstruction and antialiasing filters identified using program sysid16.c.

Example 2.24: Ramp Generation (ramp)

Figure 2.50 shows a listing of the program ramp.c, which generates a ramp, or sawtooth, output waveform. The value of the output sample is incremented by 2000 every sampling instant until it reaches the value 30,000, at which point it is reset to the value −30,000. The range of output sample values is constrained to be less than the full signed 16-bit integer range in order to prevent the sharp ringing effect seen in the output waveform overloading the codec output circuits.

Build and run this project as **ramp**. Figure 2.51 shows the output waveform captured using an oscilloscope.

Example 2.25: Amplitude Modulation (am)

This example illustrates an amplitude modulation (AM) scheme. Figure 2.52 shows a listing of the program am.c, which generates an AM signal. The array baseband holds 20 samples of one cycle of a cosine waveform with a frequency of fs/20 = 400 Hz. The array carrier holds 20 samples of five cycles of a sinusoidal carrier signal with a frequency of 5 fs/20 = 2000 Hz. Output sample values are calculated by multiplying the baseband signal by the carrier signal. In this way, the baseband signal modulates the carrier signal. The variable amp is used to set the modulation index. Program am.c uses the polling method for input and output.

Build this project as **am**. Verify that the output consists of the 2-kHz carrier signal and two sideband signals as shown in Figure 2.53. The sideband signals are at the frequency of the carrier signal ± the frequency of the sideband signal, or at 1600 and 2400 Hz.

```
//am.c AM using table for carrier and baseband signals

#include "DSK6713_AIC23.h"              // codec suppo
Uint32 fs=DSK6713_AIC23_FREQ_8KHZ;      //set sampling
#define DSK6713_AIC23_INPUT_MIC 0x0015
#define DSK6713_AIC23_INPUT_LINE 0x0011
Uint16 inputsource=DSK6713_AIC23_INPUT_LINE; //select mic

short amp = 1;                          //modulation

void main()
{

  short baseband[20]={1000,951,809,587,309,0,
              -309,-587,-809,-951,-1000,
              -951,-809,-587,-309,0,309,
              587,809,951};             //400-Hz base
  short carrier[20] ={1000,0,-1000,0,1000,0,-1000,
              0,1000,0,-1000,0,1000,0,-1000,
              0,1000,0,-1000,0};        //2-kHz carri
  short output[20];
  short k;

  comm_poll();                          //init DSK,
  while(1)                              //infinite l
  {
    for (k=0; k<20; k++)
    {
      output[k]=carrier[k]+((amp*baseband[k]*carrier[k]/10
      output_left_sample(20*output[k]); //scale outp
    }
  }
}
```

FIGURE 2.52. Amplitude modulation program (am.c).

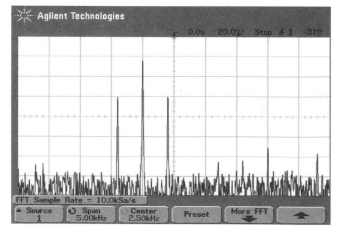

FIGURE 2.53. Frequency-domain representation of output waveform gen program am.c.

Load the GEL file `am.gel`, and verify that the modulation index can be controlled using the GEL slider. Projects on modulation are included in Chapter 10.

Example 2.26: Use of External Memory to Record Voice (`record`)

This example illustrates the use of the *pragma* directive in a C source program to store data in external memory. The C6713 processor contains a total of 264 kB of internal memory but the DSK board includes 16 MB of SDRAM external memory. Figure 2.54 shows the C source program `record.c` that implements this project example. It defines a buffer size of 2,400,000 allowing approximately 300 seconds of speech to be recorded and stored in external memory, sampling at 8 kHz.

The *pragma* directive, used in the line.

```
#pragma DATA_SECTION(buffer,"EXT_RAM")
```

specifies that the array `buffer` is allocated to a memory section named `.EXT_RAM`. Within the linker command file, `c6713dsk.cmd`, that section is mapped into the external SDRAM on the DSK (starting at address `0x080000000`). Without the use of the *pragma* directive, array `buffer` would have been allocated to the memory section `.bss` along with the other variables declared in program `record.c`.

Build this project as **record**. Load and run the program. Connect a microphone to the MIC socket and headphones to the LINE OUT socket.

1. When DIP switch #3 is pressed, and while it remains down, input samples from the microphone are stored in `buffer`, starting at `buffer[0]`. LED #3 should light, indicating that recording is in progress. Lift DIP switch #3 up to stop recording.
2. When DIP switch #0 is pressed, and while it remains down, the samples stored in `buffer` are replayed. LED #0 should light, indicating that playing is in progress. Lift DIP switch #0 up to stop replaying.

The same recording can be replayed as many times as desired, but when DIP switch #3 is pressed down again, recording will overwrite the existing contents of `buffer`.

Example 2.27: Use of Flash Memory to Run an Application on Power Up (`flash_sine`)

By default, the C6713 DSP on the DSK uses external memory interface (EMIF) boot mode. On power up or reset, the first 1000 bytes of data stored in nonvolatile flash memory, starting at address `0x90000000`, are copied to internal RAM starting at address `0x00000000` and execution starts from address `0x00000000` (reset vector). As supplied, the program executed on the DSK at this point is a short power on

```
//record.c record/play input using external memory

#include "dsk6713_aic23.h"                //codec support
Uint32 fs=DSK6713_AIC23_FREQ_8KHZ;        //set sampling rate
#define DSK6713_AIC23_INPUT_MIC 0x0015
#define DSK6713_AIC23_INPUT_LINE 0x0011
Uint16 inputsource=DSK6713_AIC23_INPUT_MIC; //select input
#define N 2400000                          //buffer size 300 secs
long i;
short buffer[N];
#pragma DATA_SECTION(buffer,".EXT_RAM") //buffer in ext memory

void main()
{
 short recording = 0;
 short playing = 0;
 for (i=0 ; i<N ; i++) buffer[i] = 0;
 DSK6713_DIP_init();
 DSK6713_LED_init();
 comm_poll();                             //init DSK, codec
 while(1)                                 //infinite loop
 {
  if(DSK6713_DIP_get(3) == 0)             //if SW#3 is pressed
  {
   i=0;
   recording = 1;                         //start recording
   while (recording == 1)
   {
    DSK6713_LED_on(3);                    //turn on LED#3
    buffer[i++] = input_left_sample();    //input data
    if (i>2000)
      if (DSK6713_DIP_get(3)==1)          //if SW#3 lifted
      {
       recording = 0;                     //stop recording
       DSK6713_LED_off(3);                //turn LED#3 off
      }
   }
  }
  if(DSK6713_DIP_get(0)==0)               //if SW#0 is pressed
  {
   i=0;
   playing = 1;                           //start playing
   while (playing == 1)
   {
    DSK6713_LED_on(0);                    //turn on LED#0
    output_left_sample(buffer[i++]);      //output data
    if (i>2000)
      if (DSK6713_DIP_get(0) == 1)        //if SW#1 is lifted
      {
       playing = 0;                       //stop playing
       DSK6713_LED_off(0);                //turn LED#0 off
      }
   }
  }
 }
}
```

FIGURE 2.54. C source program to illustrate use of external memory to store samples (record.c).

self-test (POST) procedure that, among other functions, flashes the onboard LEDs, tests memory, and outputs a burst of 1-kHz sine wave via the codec. By reprogramming the flash memory we can get the processor to run a different program on power up. In the following example, we will program the flash memory so that the following sequence of events will take place on power up or reset.

1. In EMIF boot mode, the contents of the first 1000 bytes of flash memory (0x90000000 through 0x900003FF) will be copied to internal RAM (0x00000000 through 0x000003FF). Addresses 0x00000000 through 0x000001FF comprise the interrupt service table. Address 0x00000200 is the start of a small boot program, or function, boot_start, pointed to by the reset vector.

2. Program execution will start from the reset vector (address 0x00000000) and branch immediately to the program boot_start. That program will then load an application program from flash memory (starting after the first 1000 bytes, at address 0x90000400) into internal memory and then execute it.

In order to get an application to run at power up, in addition to the application code we must provide the small boot program and an interrupt service table in which the reset vector points to the small boot program. This can be done by replacing the files vectors_intr.asm and c6713dsk.cmd in a Code Composer project.

Utilities hex6x and *FlashBurn* are required in order to:

1. Convert an executable application program (.out file) from COFF to a hex file format suitable for storage in flash memory.

2. Reprogram the flash memory.

The application used for this example is flash_sine.c, which generates a 1-kHz sine wave (Figure 2.55).

1. **Verify that program flash_sine.c works as intended.** Build the executable file flash_sine.out using the standard support files c6713dskinit.c and vectors_intr.asm and linker command file c6713dsk.cmd. These files are included in the project flash_sine.pjt in the folder flash_sine. Load and run the executable file flash_sine.out, and verify that a 1-kHz sine wave is generated.

2. **Remove the files vectors_intr.asm and c6713dsk.cmd from the project** and replace them with the files vecs_int_flash.asm and c6713dsk_flash.cmd. vecs_int_flash.asm is a modified version of vectors_intr.asm. In addition to the vector table it contains the small boot program boot_start, which copies the code from flash to internal memory upon boot up. The address of the code in flash memory (0x90000400) and the code size (0x00003000) are hard-coded into the file vecs_int_flash.asm. c6713dsk_flash.cmd is a new linker command file. It sets up a section called bootload starting at address 0x200 with a length of 0x200 into which program boot_start will be loaded.

```
//flash_sine.c Sine generation to illustrate use of flash

#include "dsk6713_aic23.h"              //codec-DSK support file
Uint32 fs=DSK6713_AIC23_FREQ_8KHZ;    //set sampling rate
#define DSK6713_AIC23_INPUT_MIC 0x0015
#define DSK6713_AIC23_INPUT_LINE 0x0011
Uint16 inputsource=DSK6713_AIC23_INPUT_LINE; //select line in
short loop = 0,  gain = 10;
short sine_table[8] = {0,707,1000,707,0,-707,-1000,-707};

interrupt void c_int11()                    //interrupt service routine
{
  output_sample(sine_table[loop]*gain);
  if (++loop > 7)                           //if end of buffer
    loop = 0;                               //reset index
  return;
}

void main()
{
  comm_intr();                              //init DSK, codec, McBSP
  while(1);                                 //infinite loop
}
```

FIGURE 2.55. C source program to illustrate use of flash memory to store application program (flash_sine.c).

These two files are stored in folder flash_sine. Rebuild the project and verify again that the 1-kHz tone is generated using the new executable (.out) file (also named sine_flash.out).

This test verifies that the INT11 vector specified in file vecs_int_flash.asm is correct. Neither the reset vector nor the boot_start routine are tested since when a .out file is run from Code Composer, the program counter is loaded with the start address of the application.

Creating a .hex File

In order to be loaded into flash memory, the executable file flash_sine.out must be converted from a COFF to a hex file format. The COFF-to-hex converter file hex6x.exe is included with Code Composer in the directory c:\CCStudio_v3.1\c6000\cgtools\bin. Copy hex6x.exe into the folder flash_sine. Access DOS, and from the folder flash_sine, type

hex6x flash_sine_hex.cmd

to create flash_sine.hex. Within the file flash_sine_hex.cmd, the executable file flash_sine.out is specified as input and flash_sine.hex as output. A flash length of 0x40000 is specified, which should be at least the length of the actual code (this can be found in the `.map` file). If this length is not great enough, you will be prompted to increase it.

Configuring the FlashBurn (.cdd) Utility

The *FlashBurn* utility is stored in folder `c:\CCStudio_v3.1\bin\utilities\flashburn`. If it has not been installed already, install it using the DSK tools CD. If a *FlashBurn* option is not present in CCS, then start *FlashBurn* directly by double-clicking on its icon in *Windows Explorer*. Within *FlashBurn*, select *File→New*, to configure the *FlashBurn* utility and create `flash_sine.cdd`, as shown in Figure 2.56, with the following fields:

1. *Conversion cmd File*:

   `c:\CCStudio_v3.1\myprojects\flash_sine\flash_sine_hex.cmd`

2. *File to Burn*:

   `c:\CCStudio_v3.1\myprojects\flash_sine\flash_sine.hex`

3. *FBTC Program File*:

   `c:\CCStudio_v3.1\bin\utilities\flashburn\c6000\dsk6713\`
   `FBTC6713.out`

Save this file as *flash_sine.cdd* in directory `c:\CCStudio_v3.1\myprojects\flash_sine`.

Select *Program→Download FBTC*. This will connect *FlashBurn* to the DSK. We now need to erase and reprogram the flash memory.

Erasing and Programming the Flash Memory

Within the *FlashBurn* utility shown in Figure 2.56, select *Program→Erase Flash*. This erases any program stored in the flash memory. Still within the *Flashburn* utility, select *Program→Program Flash*. This loads `flash_sine.hex` into the flash memory. To verify that the sine generation program is stored into the flash memory, close the (.cdd) *Flashburn* utility, exit CCS, and unplug the power to the DSK. Turn the power to the DSK back on. The *post* program no longer runs. Instead, verify that a 1-kHz sine wave is now generated continuously.

Recovering the post Program

Launch CCS and select *Debug→Connect*. Launch *FlashBurn* and open the configuration file `post.cdd`, stored in folder *c:\CCStudio_v3.1\examples\dsk6713\ bsl\post*, by selecting *File→Open*. The *FlashBurn* window should appear as shown in Figure 2.57. *Select Program→Erase Flash* to erase the program `flash_sine.hex`

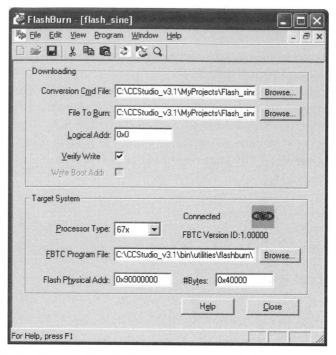

FIGURE 2.56. *Flashburn* utility during programming of flash memory with program `flash_sine.c`.

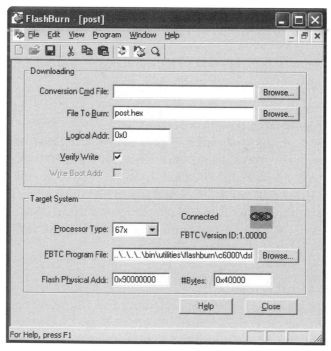

FIGURE 2.57. *Flashburn* utility during programming of flash memory with program `post.c`.

currently stored in the flash memory. Then select *Program→Program Flash* to download *post.hex* into the flash memory.

2.4 ASSIGNMENTS

1. Implement a suppressed carrier amplitude modulation scheme using an external input to modulate a 2-kHz carrier signal generated using a lookup table. Use a sampling frequency of 8 kHz. Test your results using a sinusoidal input signal with an amplitude less than 0.35 V and a frequency less than 2 kHz (a higher frequency input signal will cause aliasing). Use an oscilloscope with an FFT function or a spectrum analyzer to display the signal generated.

2. Write a program to generate a sine wave of frequency 666 Hz and turn on LED #0 while DIP switch #0 is pressed down, generate a sine wave of frequency 1.33 kHz and turn on LED #1 while DIP switch #1 is pressed down, generate a sine wave of frequency 2 kHz and turn on LED #2 while DIP switch #2 is pressed down, and generate a sine wave of frequency 2.667 kHz and turn on LED #3 while DIP switch #3 is pressed down. Use a sampling frequency of 8 kHz and a 12-point lookup table to generate the sine waves.

REFERENCES

1. TLV320AIC23 Stereo Audio Codec, 8-to 96-kHz, with Integrated Headphone Amplifier Data Manual, SLWS106G, 2003.

2. S. Norsworthy, R. Schreier, and G. Temes, *Delta–Sigma Data Converters: Theory, Design and Simulation*, IEEE Press, Piscataway, NJ, 1997.

3. P. M. Aziz, H. V. Sorensen, and J. Van Der Spiegel, An overview of sigma delta converters, *IEEE Signal Processing*, Jan. 1996.

4. J. C. Candy and G. C. Temes, Eds., *Oversampling Delta–Sigma Data Converters: Theory, Design and Simulation*, IEEE Press, Piscataway, NJ, 1992.

5. C. W. Solomon, Switched-capacitor filters, *IEEE Spectrum*, June 1988.

3

Architecture and Instruction Set of the C6x Processor

- Architecture and instruction set of the TMS320C6x processor
- Addressing modes
- Assembler directives
- Linear assembler
- Programming examples using C, assembly, and linear assembly code

3.1 INTRODUCTION

Texas Instruments introduced the first-generation TMS32010 DSP in 1982, the TMS320C25 in 1986 [1], and the TMS320C50 in 1991. Several versions of each of these processors—C1x, C2x, and C5x—are available with different features, such as faster execution speed. These 16-bit processors are all fixed-point processors and are code compatible.

In a von Neumann architecture, program instructions and data are stored in a single memory space. A processor with a von Neumann architecture can make a read or a write to memory during each instruction cycle. Typical DSP applications require several accesses to memory within one instruction cycle. The fixed-point processors C1x, C2x, and C5x are based on a modified Harvard architecture with separate memory spaces for data and instructions that allow concurrent accesses.

Quantization error or round-off noise from an ADC is a concern with a fixed-point processor. An ADC uses only a best-estimate digital value to represent an

Digital Signal Processing and Applications with the TMS320C6713 and TMS320C6416 DSK, Second Edition By Rulph Chassaing and Donald Reay
Copyright © 2008 John Wiley & Sons, Inc.

input. For example, consider an ADC with a word length of 8 bits and an input range of ± 1.5 V. The steps represented by the ADC are: input range/$2^8 = 3/256 = 11.72$ mV. This produces errors that can be up to $\pm(11.72 \text{ mV})/2 = \pm 5.86$ mV. Only a best estimate can be used by the ADC to represent input values that are not multiples of 11.72 mV. With an 8-bit ADC, 2^8 or 256 different levels can represent the input signal. An ADC with a larger word length, such as a 16-bit ADC (or larger, currently very common), can reduce the quantization error, yielding a higher resolution. The more bits an ADC has, the better it can represent an input signal.

The TMS320C30 floating-point processor was introduced in the late 1980s. The C31, the C32, and the more recent C33 are all members of the C3x family of floating-point processors [2, 3]. The C4x floating-point processors, introduced subsequently, are code compatible with the C3x processors and are based on the modified Harvard architecture [4].

The TMS320C6201 (C62x), announced in 1997, is the first member of the C6x family of fixed-point digital signal processors. Unlike the previous fixed-point processors, C1x, C2x, and C5x, the C62x is based on a VLIW architecture, still using separate memory spaces for instructions and data, as with the Harvard architecture. The VLIW architecture has simpler instructions, but more are needed for a task than with a conventional DSP architecture.

The C62x is not code compatible with the previous generation of fixed-point processors. Subsequently, the TMS320C6701 (C67x) floating-point processor was introduced as another member of the C6x family of processors. The instruction set of the C62x fixed-point processor is a subset of the instruction set of the C67x processor. Appendix A contains a list of instructions available on the C6x processors. A more recent addition to the family of the C6x fixed-point processors is the C64x. The C64x is introduced in Appendix G.

An application-specific integrated circuit (ASIC) has a DSP core with customized circuitry for a specific application. A C6x processor can be used as a standard general-purpose DSP programmed for a specific application. Specific-purpose digital signal processors are the modem, echo canceler, and others.

A fixed-point processor is better for devices that use batteries, such as cellular phones, since it uses less power than does an equivalent floating-point processor. The fixed-point processors, C1x, C2x, and C5x, are 16-bit processors with limited dynamic range and precision. The C6x fixed-point processor is a 32-bit processor with improved dynamic range and precision. In a fixed-point processor, it is necessary to scale the data. Overflow, which occurs when an operation such as the addition of two numbers produces a result with more bits than can fit within a processor's register, becomes a concern.

A floating-point processor is generally more expensive since it has more "real estate" or is a larger chip because of additional circuitry necessary to handle integer as well as floating-point arithmetic. Several factors, such as cost, power consumption, and speed, come into play when choosing a specific DSP. The C6x processors are particularly useful for applications requiring intensive computations. Family members of the C6x include both fixed-point (e.g., C62x, C64x) and floating-point (e.g., C67x) processors. Other DSPs are also available from companies such as Motorola and Analog Devices [5].

Other architectures include the Super Scalar, which requires special hardware to determine which instructions are executed in parallel. The burden is then on the processor more than on the programmer, as in the VLIW architecture. It does not necessarily execute the same group of instructions, and as a result, it is difficult to time. Thus, it is rarely used in DSPs.

3.2 TMS320C6x ARCHITECTURE

The TMS320C6713 onboard the DSK is a floating-point processor based on the VLIW architecture [6–10]. Internal memory includes a two-level cache architecture with 4 kB of level 1 program cache (L1P), 4 kB of level 1 data cache (L1D), and 256 kB of level 2 memory shared between program and data space. It has a glueless (direct) interface to both synchronous memories (SDRAM and SBSRAM) and asynchronous memories (SRAM and EPROM). Synchronous memory requires clocking but provides a compromise between static SRAM and dynamic DRAM, with SRAM being faster but more expensive than DRAM.

On-chip peripherals include two McBSPs, two timers, a host port interface (HPI), and a 32-bit EMIF. It requires 3.3 V for I/O and 1.26 V for the core (internal). Internal buses include a 32-bit program address bus, a 256-bit program data bus to accommodate eight 32-bit instructions, two 32-bit data address buses, two 64-bit data buses,

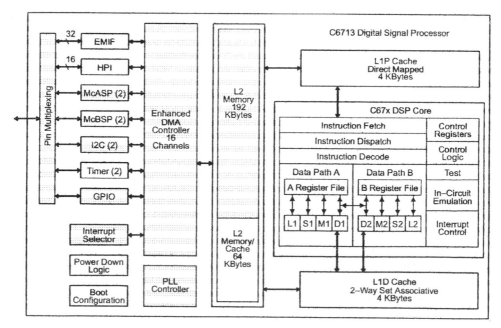

FIGURE 3.1. Functional block diagram of TMS320C6713. (Courtesy of Texas Instruments.)

and two 64-bit store data buses. With a 32-bit address bus, the total memory space is $2^{32} = 4\,GB$, including four external memory spaces: CE0, CE1, CE2, and CE3. Figure 3.1 shows a functional block diagram of the C6713 processor included with CCS.

Independent memory banks on the C6x allow for two memory accesses within one instruction cycle. Two independent memory banks can be accessed using two independent buses. Since internal memory is organized into memory banks, two loads or two stores of instructions can be performed in parallel. No conflict results if the data accessed are in different memory banks. Separate buses for program, data, and direct memory access (DMA) allow the C6x to perform concurrent program fetches, data read and write, and DMA operations. With data and instructions residing in separate memory spaces, concurrent memory accesses are possible. The C6x has a byte-addressable memory space. Internal memory is organized as separate program and data memory spaces, with two 32-bit internal ports (two 64-bit ports with the C64x) to access internal memory.

The C6713 on the DSK includes 264 kB of internal memory, which starts at 0x00000000, and 16 MB of external SDRAM, mapped through CE0 starting at 0x80000000. The DSK also includes 512 kB of Flash memory (256 kB readily available to the user), mapped through CE1 starting at 0x90000000. Figure 3.2 shows the L2 internal memory configuration, included with CCS [7]. Table 3.1 shows the memory map, also included with CCS [7]. A schematic diagram of the DSK is included with CCS (6713dsk_schem.pdf).

With the DSK operating at 225 MHz, one can ideally achieve two multiplies and accumulates per cycle, for a total of 450 million multiplies and accumulates (MACs) per second. With six of the eight functional units in Figure 3.1 (not the .D units described later) capable of handling floating-point operations, it is possible to perform 1350 million floating-point operations per second (MFLOPS). Operating at 225 MHz, this translates into 1800 million instructions per second (MIPS) with a 4.44-ns instruction cycle time.

3.3 FUNCTIONAL UNITS

The CPU consists of eight independent functional units divided into two data paths, A and B, as shown in Figure 3.1. Each path has a unit for multiply operations (.M), for logical and arithmetic operations (.L), for branch, bit manipulation, and arithmetic operations (.S), and for loading/storing and arithmetic operations (.D). The .S and .L units are for arithmetic, logical, and branch instructions. All data transfers make use of the .D units.

The arithmetic operations, such as subtract or add (SUB or ADD), can be performed by all the units, except the .M units (one from each data path). The eight functional units consist of four floating/fixed-point ALUs (two .L and two .S), two fixed-point ALUs (.D units), and two floating/fixed-point multipliers (.M units). Each functional unit can read directly from or write directly to the register file within its own path. Each path includes a set of sixteen 32-bit registers, A0 through A15 and B0 through B15. Units ending in 1 write to register file A, and units ending in 2 write to register file B.

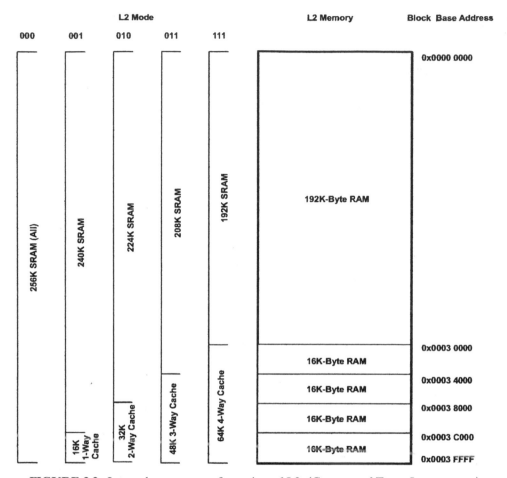

FIGURE 3.2. Internal memory configuration of L2. (Courtesy of Texas Instruments.)

Two cross-paths (`1x` and `2x`) allow functional units from one data path to access a 32-bit operand from the register file on the opposite side. There can be a maximum of two cross-path source reads per cycle. Each functional unit side can access data from the registers on the opposite side using a cross-path (i.e., the functional units on one side can access the register set from the other side). There are 32 general-purpose registers, but some of them are reserved for specific addressing or are used for conditional instructions.

3.4 FETCH AND EXECUTE PACKETS

The architecture VELOCITI, introduced by TI, is derived from the VLIW architecture. An execute packet (EP) consists of a group of instructions that can be executed in parallel within the same cycle time. The number of EPs within a fetch packet (FP)

TABLE 3.1 Memory Map

Memory Block Description	Block Size (Bytes)	Hex Address Range
Internal RAM (L2)	192 K	0000 0000–0002 FFFF
Internal RAM/cache	64 K	0003 0000–0003 FFFF
Reserved	24 M–256 K	0004 0000–017F FFFF
External memory interface (EMIF) registers	256 K	0180 0000–0183 FFFF
L2 registers	128 K	0184 0000–0185 FFFF
Reserved	128 K	0186 0000–0187 FFFF
HPI registers	256 K	0188 0000–018B FFFF
McBSP 0 registers	256 K	018C 0000–018F FFFF
McBSP 1 registers	256 K	0190 0000–0193 FFFF
Timer 0 registers	256 K	0194 0000–0197 FFFF
Timer 1 registers	256 K	0198 0000–019B FFFF
Interrupt selector registers	512	019C 0000–019C 01FF
Device configuration registers	4	019C 0200–019C 0203
Reserved	256 K–516	091C 0204–019F FFFF
EDMA RAM and EDMA registers	256 K	01A0 0000–01A3 FFFF
Reserved	768 K	01A4 0000–01AF FFFF
GPIO registers	16 K	01B0 0000–01B0 3FFF
Reserved	240 K	01B0 4000–01B3 FFFF
I2C0 registers	16 K	01B4 0000–01B4 3FFF
I2C1 registers	16 K	01B4 4000–01B4 7FFF
Reserved	16 K	01B4 8000–01B4 BFFF
McASP0 registers	16 K	01B4 C000–01B4 FFFF
McASP1 registers	16 K	01B5 0000–01B5 3FFF
Reserved	160 K	01B5 4000–01B7 BFFF
PLL registers	8 K	01B7 C000–01B7 DFFF
Reserved	264 K	01B7 E000–01BB FFFF
Emulation registers	256 K	01BC 0000–01BF FFFF
Reserved	4 M	01C0 0000–01FF FFFF
QDMA registers	52	0200 0000–0200 0033
Reserved	16 M–52	0200 0034–02FF FFFF
Reserved	720 M	0300 0000–2FFF FFFF
McBSP0 data port	64 M	3000 0000–33FF FFFF
McBSP1 data port	64 M	3400 0000–37FF FFFF
Reserved	64 M	3800 0000–3BFF FFFF
McASP0 data port	1 M	3C00 0000–3C0F FFFF
McASP1 data port	1 M	3C10 0000–3C1F FFFF
Reserved	1 G + 62 M	3C20 0000–7FFF FFFF
EMIF CE0[a]	256 M	8000 0000–8FFF FFFF
EMIF CE1[a]	256 M	9000 0000–9FFF FFFF
EMIF CE2[a]	256 M	A000 0000–AFFF FFFF
EMIF CE3[a]	256 M	B000 0000–BFFF FFFF
Reserved	1 G	C000 0000–FFFF FFFF

[a] The number of EMIF address pins (EA[21:2]) limits the maximum addressable memory (SDRAM) to 128 MB per CE space.

Source: Courtesy of Texas Instruments.

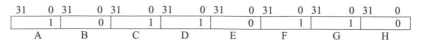

FIGURE 3.3. One FP with three EPs showing the "p" bit of each instruction.

can vary from one (with eight parallel instructions) to eight (with no parallel instructions). The VLIW architecture was modified to allow more than one EP to be included within an FP.

The least significant bit of every 32-bit instruction is used to determine if the next or subsequent instruction belongs in the same EP (if 1) or is part of the next EP (if 0). Consider an FP with three EPs: EP1, with two parallel instructions, and EP2 and EP3, each with three parallel instructions, as follows:

```
   Instruction A
|| Instruction B

   Instruction C
|| Instruction D
|| Instruction E

   Instruction F
|| Instruction G
|| Instruction H
```

EP1 contains the two parallel instructions A and B; EP2 contains the three parallel instructions C, D, and E; and EP3 contains the three parallel instructions F, G, and H. The FP would be as shown in Figure 3.3. Bit 0 (LSB) of each 32-bit instruction contains a "p" bit that signals whether it is in parallel with a subsequent instruction. For example, the "p" bit of instruction B is zero, denoting that it is not within the same EP as the subsequent instruction C. Similarly, instruction E is not within the same EP as instruction F.

3.5 PIPELINING

Pipelining is a key feature in a DSP to get parallel instructions working properly, requiring careful timing. There are three stages of pipelining: program fetch, decode, and execute.

1. The *program fetch stage* is composed of four phases:
   (a) *PG*: program address generate (in the CPU) to fetch an address
   (b) *PS*: program address send (to memory) to send the address
   (c) *PW*: program address ready wait (memory read) to wait for data
   (d) *PR*: program fetch packet receive (at the CPU) to read opcode from memory

TABLE 3.2   Pipeline Phases

Program Fetch				Decode		Execute
PG	PS	PW	PR	DP	DC	E1–E6 (E1–E10 for double precision)

TABLE 3.3   Pipelining Effects

					Clock Cycle						
1	2	3	4	5	6	7	8	9	10	11	12
PG	PS	PW	PR	DP	DC	E1	E2	E3	E4	E5	E6
	PG	PS	PW	PR	DP	DC	E1	E2	E3	E4	E5
		PG	PS	PW	PR	DP	DC	E1	E2	E3	E4
			PG	PS	PW	PR	DP	DC	E1	E2	E3
				PG	PS	PW	PR	DP	DC	E1	E2
					PG	PS	PW	PR	DP	DC	E1
						PG	PS	PW	PR	DP	DC

2. The *decode stage* is composed of two phases:

   (a) *DP*: to dispatch all the instructions within an FP to the appropriate functional units

   (b) *DC*: instruction decode

3. The *execute stage* is composed of 6 phases (with fixed point) to 10 phases (with floating point) due to delays (latencies) associated with the following instructions:

   (a) Multiply instruction, which consists of two phases due to one delay

   (b) Load instruction, which consists of five phases due to four delays

   (c) Branch instruction, which consists of six phases due to five delays

Table 3.2 shows the pipeline phases, and Table 3.3 shows the pipelining effects. The first row in Table 3.3 represents cycle 1, 2, . . . , 12. Each subsequent row represents an FP. The rows represented by PG, PS, . . . illustrate the phases associated with each FP. The program generate (PG) of the first FP starts in cycle 1, and the PG of the second FP starts in cycle 2, and so on. Each FP takes four phases for program fetch and two phases for decoding. However, the execution phase can take from 1 to 10 phases (not all execution phases are shown in Table 3.3). We are assuming that each FP contains one EP.

For example, at cycle 7, while the instructions in the first FP are in the first execution phase E1 (which may be the only one), the instructions in the second FP are in the decoding phase, the instructions in the third FP are in the dispatching phase, and so on. All seven instructions are proceeding through the various phases. Therefore, at cycle 7, "the pipeline is full."

Most instructions have one execute phase. Instructions such as multiply (MPY), load (LDH/LDW), and branch (B) take two, five, and six phases, respectively. Additional execute phases are associated with floating-point and double-precision types of instructions, which can take up to 10 phases. For example, the double-precision multiply operation (MPYDP), available on the C67x, has nine delay slots, so that the execution phase takes a total of 10 phases.

The *functional unit latency*, which represents the number of cycles that an instruction ties up a functional unit, is 1 for all instructions except double-precision instructions, available with the floating-point C67x. Functional unit latency is different from a delay slot. For example, the instruction MPYDP has four functional unit latencies but nine delay slots. This implies that no other instruction can use the associated multiply functional unit for four cycles. A store has no delay slot but finishes its execution in the third execution phase of the pipeline.

If the outcome of a multiply instruction such as MPY is used by a subsequent instruction, a NOP (no operation) must be inserted after the MPY instruction for the pipelining to operate properly. Four or five NOPs are to be inserted in case an instruction uses the outcome of a load or a branch instruction, respectively.

3.6 REGISTERS

Two sets of register files, each set with 16 registers, are available: register file A (A0 through A15) and register file B (B0 through B15). Registers A0, A1, B0, B1, and B2 are used as conditional registers. Registers A4 through A7 and B4 through B7 are used for circular addressing. Registers A0 through A9 and B0 through B9 (except B3) are temporary registers. Any of the registers A10 through A15 and B10 through B15 used are saved and later restored before returning from a subroutine.

A 40-bit data value can be contained across a register pair. The 32 least significant bits (LSBs) are stored in the even register (e.g., A2), and the remaining 8 bits are stored in the 8 LSBs of the next-upper (odd) register (A3). A similar scheme is used to hold a 64-bit double-precision value within a pair of registers (even and odd).

These 32 registers are considered general-purpose registers. Several special-purpose registers are also available for control and interrupts: for example, the address mode register (AMR) used for circular addressing and interrupt control registers, as shown in Appendix B.

3.7 LINEAR AND CIRCULAR ADDRESSING MODES

Addressing modes determine how one accesses memory. They specify how data are accessed, such as retrieving an operand indirectly from a memory location. Both linear and circular modes of addressing are supported. The most commonly used mode is the indirect addressing of memory.

3.7.1 Indirect Addressing

Indirect addressing can be used with or without displacement. Register R represents one of the 32 registers A0 through A15 and B0 through B15 that can specify or point to memory addresses. As such, these registers are pointers. Indirect addressing mode uses a "*" in conjunction with one of the 32 registers. To illustrate, consider R as an address register.

1. *R*. Register R contains the address of a memory location where a data value is stored.
2. *R++(d)*. Register R contains the memory address (location). After the memory address is used, R is postincremented (modified) such that the new address is the current address offset by the displacement value d. If d = 1 (by default), the new address is R + 1, or R is incremented to the next higher address in memory. A double minus (– –) instead of a double plus would update or postdecrement the address to R – d.
3. *++R(d)*. The address is preincremented or offset by d, such that the current address is R + d. A double minus would predecrement the memory address so that the current address is R – d.
4. *+R(d)*. The address is preincremented by d, such that the current address is R + d (as with the preceding case). However, in this case, R preincrements without modification. Unlike the previous case, R is not updated or modified.

3.7.2 Circular Addressing

Circular addressing is used to create a circular buffer. This buffer is created in hardware and is very useful in several DSP algorithms, such as in digital filtering or correlation algorithms where data need to be updated. An example in Chapter 4 illustrates the implementation of a digital filter in assembly code using a circular buffer to update the "delay" samples. Implementing a circular buffer using C code is less efficient.

The C6x has dedicated hardware to allow a circular type of addressing. This addressing mode can be used in conjunction with a circular buffer to update samples by shifting data without the overhead created by shifting data directly. As a pointer reaches the end or "bottom" location of a circular buffer that contains the last element in the buffer, and is then incremented, the pointer is automatically wrapped around or points to the beginning or "top" location of the buffer that contains the first element.

Two independent circular buffers are available using BK0 and BK1 within the AMR. The eight registers A4 through A7 and B4 through B7, in conjunction with the two .D units, can be used as pointers (all registers can be used for linear addressing). The following code segment illustrates the use of a circular buffer using register B2 (only side B can be used) to set the appropriate values within AMR:

```
MVKL   .S2   0x0004,B2   ;lower 16 bits to B2. Select A5 as pointer
MVKH   .S2   0x0005,B2   ;upper 16 bits to B2. Select BK0, set N = 5
MVC    .S2   B2,AMR      ;move 32 bits of B2 to AMR
```

(b) The instruction

```
SUB  .S1  A1,1,A1          ;subtract 1 from A1
```

subtracts 1 from A1 to decrement it using the .s unit.

(c) The parallel instructions

```
   MPY  .M2  A7,B7,B6    ;multiply 16 LSBs of A7, B7 → B6
|| MPYH .M1  A7,B7,A6    ;multiply 16 MSBs of A7, B7 → A6
```

multiplies the lower or least significant 16 bits (LSBs) of both A7 and B7 and places the product in B6, in parallel (concurrently within the same execution packet) with a second instruction that multiplies the higher or most significant 16 bits (MSBs) of A7 and B7 and places the result in A6. In this fashion, two MAC operations can be executed within a single instruction cycle. This can be used to decompose a sum of products into two sets of sum of products: one set using the lower 16 bits to operate on the first, third, fifth, . . . number and another set using the higher 16 bits to operate on the second, fourth, sixth, . . . number. Note that the parallel symbol is not in column 1.

2. *Load/Store*

(a) The instruction

```
   LDH  .D2  *B2++,B7    ;load (B2) → B7, increment B2
|| LDH  .D1  *A2++,A7    ;load (A2) → A7, increment A2
```

loads into B7 the half-word (16 bits) whose address in memory is specified/pointed to by B2. Then register B2 is incremented (postincremented) to point at the next higher memory address. In parallel is another indirect addressing mode instruction to load into A7 the content in memory whose address is specified by A2. Then A2 is incremented to point at the next higher memory address.

The instruction LDW loads a 32-bit word. Two paths using .D1 and .D2 allow for the loading of data from memory to registers A and B using the instruction LDW. The double-word load floating-point instruction LDDW on the C6713 can simultaneously load two 32-bit registers into side A and two 32-bit registers into side B.

(b) The instruction

```
STW  .D2  A1,*+A4[20]     ;store A1→(A4) offset by 20
```

stores the 32-bit word A1 in memory whose address is specified by A4 offset by 20 words (32 bits) or 80 bytes. The address register A4 is preincrcmented with offset, but it is not modified (two plus signs are used if A4 is to be modified).

3. *Branch/Move.* The following code segment illustrates branching and data transfer:

```
Loop  MVKL  .S1  x,A4      ;move 16 LSBs of x address →A4
      MVKH  .S1  x,A4      ;move 16 MSBs of x address →A4
      .
      .
      .
      SUB   .S1  A1,1,A1   ;decrement A1
[A1]  B     .S2  Loop      ;branch to Loop if A1 != 0
      NOP        5         ;five no-operation instructions
      STW   .D1  A3,*A7    ;store A3 into (A7)
```

The first instruction moves the lower 16 bits (LSBs) of address x into register A4. The second instruction moves the higher 16 bits (MSBs) of address x into A4, which now contains the full 32-bit address of x. One must use the instructions MVKL/MVKH in order to get a 32-bit constant into a register.

Register A1 is used as a loop counter. After it is decrcmented with the SUB instruction, it is tested for a conditional branch. Execution branches to the label or address Loop if A1 is not zero. If A1 = 0, execution continues and data in register A3 are stored in memory whose address is specified (pointed to) by A7.

3.9 ASSEMBLER DIRECTIVES

An assembler directive is a message for the assembler (not the compiler) and is not an instruction. It is resolved during the assembling process and does not occupy memory space, as an instruction does. It does not produce executable code. Addresses of different sections can be specified with assembler directives. For example, the assembler directive .sect "my_buffer" defines a section of code or data named my_buffer. The directives .text and .data indicate a section for text and data, respectively. Other assembler directives, such as .ref and .def, are used for undefined and defined symbols, respectively. The assembler creates several sections indicated by directives such as .text for code and .bss for global and static variables.

Other commonly used assembler directives are:

1. .short: to initialize a 16-bit integer.
2. .int: to initialize a 32-bit integer (also .word or .long). The compiler treats a long data value as 40 bits, whereas the C6x assembler treats it as 32 bits.
3. .float: to initialize a 32-bit IEEE single-precision constant.
4. .double: to initialize a 64-bit IEEE double-precision constant.

Initialized values are specified by using the assembler directives .byte, .short, or .int. Uninitialized variables are specified using the directive .usect, which creates an uninitialized section (like the .bss section), whereas the directive .sect creates an initialized section. For example, .usect "variable", 128 designates an uninitialized section named variable with a section size of 128 in bytes.

3.10 LINEAR ASSEMBLY

An alternative to C, or assembly code, is linear assembly. An assembler optimizer (in lieu of a C compiler) is used in conjunction with a linear assembly-coded source program (with extension .sa) to create an assembly source program (with extension .asm) in much the same way that a C compiler optimizer is used in conjunction with a C-coded source program. The resulting assembly-coded program produced by the assembler optimizer is typically more efficient than one resulting from the C compiler optimizer. The assembly-coded program resulting from either a C-coded source program or a linear assembly source program must be assembled to produce an object code.

Linear assembly code programming provides a compromise between coding effort and coding efficiency. The assembler optimizer assigns the functional unit and register to use (optional to be specified by the user), finds instructions that can execute in parallel, and performs software pipelining for optimization (discussed in Chapter 8). Two programming examples at the end of this chapter illustrate a C program calling a linear assembly function. Parallel instructions are not valid in a linear assembly program. Specifying the functional unit is optional in a linear assembly program as well as in an assembly program.

In recent years, the C compiler optimizer has become more and more efficient. Although C code is less efficient (speed performance) than assembly code, it typically involves less coding effort than assembly code, which can be hand-optimized to achieve 100 percent efficiency but with much greater coding effort.

It is interesting to note that the C6x assembly code syntax is not as complex as that of the C2x/C5x or the C3x family of processors. It is actually simpler to "program" the C6x in assembly. For example, the C3x instruction

```
DBNZD AR4,LOOP
```

decrements (due to the first D) a loop counter AR4 and branches (B) conditionally (if AR4 is nonzero) to the address specified by LOOP, with delay (due to the second D). The branch instruction with delay effectively allows the branch instruction to execute in a single cycle (due to pipelining). Such multitask instructions are not available on the C62x and C67x processors, although they were recently introduced on the C64x processor. In fact, C6x types of instructions are simpler. For example, separate instructions are available for decrementing a counter (with a SUB instruction) and branching. The simpler types of instructions are more amenable for a more efficient C compiler.

However, although it is simpler to program in assembly code to perform a desired task, this does not imply or translate into an efficient assembly-coded program. It can be relatively difficult to hand-optimize a program to yield a totally efficient (and meaningful) assembly-coded program.

Linear assembly code is a cross between assembly and C. It uses the syntax of assembly code instructions such as ADD, SUB, and MPY, but with operands/registers as used in C. In some cases this provides a good compromise between C and assembly.

Linear assembler directives include

```
.cproc
.endproc
```

to specify a C-callable procedure or section of code to be optimized by the assembler optimizer. Another directive, .reg, is used to declare variables and use descriptive names for values that will be stored in registers. Programming examples with C calling an assembly function or a linear assembly function are illustrated later in this chapter.

3.11 ASM STATEMENT WITHIN C

Assembly instructions and directives can be incorporated within a C program using the asm statement. The asm statement can provide access to hardware features that cannot be obtained using C code only. The syntax is

```
asm ("assembly code");
```

The assembly line of code within the set of quotation marks has the same format as a valid assembly statement. Note that if the instruction has a label, the first character of the label must start after the first quotation mark so that it is in column 1. The assembly statement should be valid since the compiler does not check it for syntax error but copies it directly into the compiled output file. If the assembly statement has a syntax error, the assembler would detect it.

Avoid using asm statements within a C program, especially within a linear assembly program. This is because the assembler optimizer could rearrange lines of code near the asm statements that may cause undesirable results.

3.12 C-CALLABLE ASSEMBLY FUNCTION

Programming examples are included later in this chapter to illustrate a C program calling an assembly function. Register B3 is preserved and is used to contain the return address of the calling function.

An external declaration of an assembly function called within a C program using `extern` is optional. For example,

```
extern int func();
```

is optional with the assembly function `func` returning an integer value.

3.13 TIMERS

Two 32-bit timers can be used to time and count events or to interrupt the CPU. A timer can direct an external ADC to start conversion or the DMA controller to start a data transfer. A timer includes a time period register, which specifies the timer's frequency; a timer counter register, which contains the value of the incrementing counter; and a timer control register, which monitors the timer's status.

3.14 INTERRUPTS

An interrupt can be issued internally or externally. An interrupt stops the current CPU process so that it can perform a required task initiated by the interrupt. The program flow is redirected to an ISR. The source of the interrupt can be an ADC, a timer, and so on. On an interrupt, the conditions of the current process must be saved so that they can be restored after the interrupt task is performed. On interrupt, registers are saved and processing continues to an ISR. Then the registers are restored.

There are 16 interrupt sources. They include two timer interrupts, four external interrupts, four McBSP interrupts, and four DMA interrupts. Twelve CPU interrupts (INT4–INT15) are available. An interrupt selector is used to choose among the 12 interrupts.

3.14.1 Interrupt Control Registers

The interrupt control registers (Appendix B) are as follows:

1. CSR (control status register): contains the global interrupt enable (GIE) bit and other control/status bits
2. IER (interrupt enable register): enables/disables individual interrupts
3. IFR (interrupt flag register): displays the status of interrupts
4. ISR (interrupt set register): sets pending interrupts
5. ICR (interrupt clear register): clears pending interrupts
6. ISTP (interrupt service table pointer): locates an ISR

7. IRP (interrupt return pointer)

8. NRP (nonmaskable interrupt return pointer)

Interrupts are prioritized, with Reset having the highest priority. The reset interrupt and nonmaskable interrupt (NMI) are external pins that have the first and second highest priority, respectively. The interrupt enable register (IER) is used to set a specific interrupt and can check if and which interrupt has occurred from the interrupt flag register (IFR).

NMI is nonmaskable, along with Reset. NMI can be masked (disabled) by clearing the nonmaskable interrupt enable (NMIE) bit within CSR. It is set to zero only upon reset or upon a nonmaskable interrupt. If NMIE is set to zero, all interrupts INT4 through INT15 are disabled. The interrupt registers are shown in Appendix B.

The reset signal is an active-low signal used to halt the CPU, and the NMI signal alerts the CPU to a potential hardware problem. Twelve CPU interrupts with lower priorities are available, corresponding to the maskable signals INT4 through INT15. The priorities of these interrupts are: INT4, INT5, . . . , INT15, with INT4 having the highest priority and INT15 the lowest priority. For an NMI to occur, the NMIE bit must be 1 (active high). On reset (or after a previously set NMI), the NMIE bit is cleared to zero so that a reset interrupt may occur.

To process a maskable interrupt, the GIE bit within the control status register (CSR) and the NMIE bit within the IER are set to 1. GIE is set to 1 with bit 0 of CSR set to 1, and NMIE is set to 1 with bit 1 of IER set to 1. Note that CSR can be ANDed with −2 (using 2's complement, the LSB is 0, while all other bits are 1's) to set the GIE bit to 0 and disable maskable interrupts globally.

The interrupt enable (IE) bit corresponding to the desired maskable interrupt is also set to 1. When the interrupt occurs, the corresponding IFR bit is set to 1 to show the interrupt status. To process a maskable interrupt, the following apply:

1. The GIE bit is set to 1.

2. The NMIE bit is set to 1.

3. The appropriate IE bit is set to 1.

4. The corresponding IFR bit is set to 1.

For an interrupt to occur, the CPU must not be executing a delay slot associated with a branch instruction.

The interrupt service table (IST) shown in Table 3.5 is used when an interrupt begins. Within each location is an FP associated with each interrupt. The table contains 16 FPs, each with eight instructions. The addresses on the right side correspond to an offset associated with each specific interrupt. For example, the FP for interrupt INT11 is at a base address plus an offset of 160h. Since each FP contains eight 32-bit instructions (256 bits) or 32 bytes, each offset address in the table is incremented by 20h = 32.

TABLE 3.5 Interrupt Service Table

Interrupt	Offset
RESET	000h
NMI	020h
Reserved	040h
Reserved	060h
INT4	080h
INT5	0A0h
INT6	0C0h
INT7	0E0h
INT8	100h
INT9	120h
INT10	140h
INT11	160h
INT12	180h
INT13	1A0h
INT14	1C0h
INT15	1E0h

The reset FP must be at address 0. However, the FPs associated with the other interrupts can be relocated. The relocatable address can be specified by writing this address to the interrupt service table base (ISTB) register of the interrupt service table pointer (ISTP) register, shown in Figure B.7. On reset, ISTB is zero. For relocating the vector table, the ISTP is used; the relocatable address is ISTB plus the offset.

3.14.2 Interrupt Acknowledgment

The signals IACK and INUMx (INUM0 through INUM3) are pins on the C6x that acknowledge that an interrupt has occurred and is being processed. The four INUMx signals indicate the number of the interrupt being processed. For example,

```
INUM3 = 1 (MSB), INUM2 = 0, INUM1 = 1, INUM0 = 1 (LSB)
```

correspond to $(1011)_b = 11$, indicating that INT11 is being processed.

The IE11 bit is set to 1 to enable INT11. The IFR can be read to verify that bit IF11 is set to 1 (INT11 enabled). Writing a 1 to a bit in the interrupt set register (ISR) causes the corresponding interrupt flag to be set in IFR, whereas a 0 to a bit in the interrupt clear register (ICR) causes the corresponding interrupt to be cleared.

All interrupts remain pending while the CPU has a pending branch instruction. Since a branch instruction has five delay slots, a loop smaller than six cycles is noninterruptible. Any pending interrupt will be processed as long as there are no pending branches to be completed. Additional information can be found in Ref. 6.

3.15 MULTICHANNEL BUFFERED SERIAL PORTS

Two McBSPs are available. They provide an interface to inexpensive (industry standard) external peripherals. McBSPs have features such as full-duplex communication, independent clocking and framing for receiving and transmitting, and direct interface to AC97 and IIS compliant devices. They allow several data sizes between 8 and 32 bits. Clocking and framing associated with the McBSPs for input and output are discussed in Ref. 7.

External data communication can occur while data are being moved internally. Figure 3.4 shows an internal block diagram of a McBSP. The data transmit (DX) and data receive (DR) pins are used for data communication. Control information (clocking and frame synchronization) is through CLKX, CLKR, FSX, and FSR. The CPU or DMA controller reads data from the data receive register (DRR) and writes data to be transmitted to the data transmit register (DXR). The transmit shift

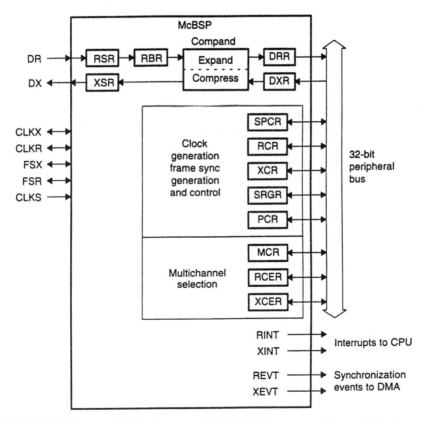

FIGURE 3.4. Internal block diagram of McBSP. (Courtesy of Texas Instruments.)

register (XSR) shifts these data to DX. The receive shift register (RSR) copies the data received on DR to the receive buffer register (RBR). The data in RBR are then copied to DRR to be read by the CPU or the DMA controller.

Other registers—the serial port control register (SPCR), receive/transmit control register (RCR/XCR), receive/transmit channel enable register (RCER/XCER), pin control register (PCR), and sample rate generator register (SRGR)—support further data communication [7].

The two McBSPs are used for input and output through the onboard codec. McBSP0 is used for control and McBSP1 for transmitting and receiving data.

3.16 DIRECT MEMORY ACCESS

Direct memory access (DMA) allows for the transfer of data to and from internal memory or external devices without intervention from the CPU [7]. Sixteen enhanced DMA channels (EDMA) can be configured independently for data transfer. DMA can access on-chip memory and the EMIF, as well as the HPI. Data of different sizes can be transferred: 8-bit bytes, 16-bit half-words, and 32-bit words.

A number of DMA registers are used to configure the DMA: address (source and destination), index, count reload, DMA global data, and control registers. The source and destination addresses can be from internal program memory, internal data memory, an external memory interface, and an internal peripheral bus. DMA transfers can be triggered by interrupts from internal peripherals as well as from external pins.

For each resource, each DMA channel can be programmed for priorities with the CPU, with channel 0 having the highest priority. Each DMA channel can be made to start initiating block transfer of data independently. A block can contain a number of frames. Within each frame can be many elements. Each element is a single data value. The DMA count reload register contains the value to specify the frame count (16 MSBs) and the element count (16 LSBs).

3.17 MEMORY CONSIDERATIONS

3.17.1 Data Allocation

Blocks of code and data can be allocated in memory within sections specified in the linker command file. These sections can be either initialized or uninitialized. The initialized sections are:

1. `.cinit`: for global and static variables
2. `.const`: for global and static constant variables
3. `.switch`: contains jump tables for large switch statements
4. `.text`: for executable code and constants

The uninitialized sections are:

1. .bss: for global and static variables
2. .far: for global and static variables declared far
3. .stack: allocates memory for the system stack
4. .sysmem: reserves space for dynamic memory allocation used by the malloc, calloc, and realloc functions

The linker can be used to place sections such as text in fast internal memory for most efficient operation.

3.17.2 Data Alignment

The C6x always accesses aligned data that allow it to address bytes, half-words, and words (32 bits). The data format consists of four byte boundaries, two half-word boundaries, and one word boundary. For example, to assign a 32-bit load with LDW, the address must be aligned with a word boundary so that the lower 2 bits of the address are zero. Otherwise, incorrect data can be loaded. A double-word (64 bits) also can be accessed. Both .s1 and .s2 can be used to execute the double-word instruction LDDW to load two 64-bit double words, for a total of 128 bits per cycle.

3.17.3 Pragma Directives

The pragma directives tell the compiler to consider certain functions. Pragmas include DATA_ALIGN, DATA_SECTION, and so on. The DATA_ALIGN pragma has the syntax

```
#pragma DATA_ALIGN (symbol,constant);
```

that aligns *symbol* to a boundary. The constant is a power of 2. This pragma directive is used later in several examples (such as in FFT program examples) to align data in memory.

The DATA_SECTION pragma has the following syntax:

```
#pragma DATA_SECTION (symbol, "my_section");
```

which allocates space for *symbol* in the section named *my_section*. This pragma directive is useful to allocate a section in external memory. For example,

```
#pragma DATA_SECTION (buffer, ".extRAM")
```

floating-point double-word load instruction LDDW (with four delay slots, as with the fixed-point LDW) can load 64 bits. Two LDDW instructions can execute in parallel through both units .S1 and .S2 to load a total of 128 bits per cycle.

A single-precision floating-point value can be loaded into a single register, whereas a double-precision floating-point value is a 64-bit value that can be loaded into a register pair such as $A1:A0, A3:A2, \ldots, B1:B0, B3:B2, \ldots$. The least significant 32 bits are loaded into the even register pair, and the most significant 32 bits are loaded into the odd register pair.

One may need to weigh the pros and cons of dynamic range and accuracy with possible degradation in speed when using floating-point types of instructions.

3.18.3 Division

The floating-point C6713 processor has a single-precision reciprocal instruction RCPSP. A division operation can be performed by taking the reciprocal of the denominator and multiplying the result by the numerator [6]. There are no fixed-point instructions for division. Code is available to perform a division operation by using the fixed-point processor to implement a Newton–Raphson equation.

3.19 CODE IMPROVEMENT

Several code optimization schemes are discussed in Chapter 8 using both fixed- and floating-point implementations and ASM code.

3.19.1 Intrinsics

C code can be optimized further by using many of the available *intrinsics* in the run-time library support file. Intrinsic functions are similar to run-time support library functions. *Intrinsics* are available to multiply, to add, to find the reciprocal of a square root, and so on. For example, in lieu of using the asterisk operator to multiply, the intrinsic _mpy can be used. *Intrinsics* are special functions that map directly to inline C6x instructions. For example,

```
int _mpy()
```

is equivalent to the assembly instruction MPY to multiply the 16 LSBs of two numbers. The intrinsic function

```
int _mpyh()
```

is equivalent to the assembly instruction MPYH to multiply the 16 MSBs of two numbers.

3.19.2 Trip Directive for Loop Count

The linear assembly directive `.trip` is used to specify the number of times a loop iterates. If the exact number is known and used, the linear assembler optimizer can produce pipelined code (discussed in Chapter 8) and redundant loops are not generated. This can improve both code size and execution time. A `.trip` count specification, even if it is not the exact value, may improve performance: for example, when the actual number of iterations is a multiple of the specified value. The intrinsic function `_nassert()` can be used in a C program in lieu of `.trip`.

3.19.3 Cross-Paths

Data and address cross-path instructions are used to increase code efficiency. The instruction

```
MPY  .M1x  A2,B2,A4
```

illustrates a data cross-path that multiplies the two sources A2 and B2 from two different sides, A and B, with the result in A4. If the result is in the B register file, a `2x` cross-path is used with the instruction

```
MPY  .M2x  A2,B2,B4
```

with the result in B4. The instruction

```
LDW  .D1T2  *A2,B2
```

illustrates an address cross-path. It loads the content in register A2 (from a register file A) into register B2 (register file B). Only two cross-paths are available on the C6x, so no more than two instructions using cross-paths are allowed within a cycle.

3.19.4 Software Pipelining

Software pipelining uses available resources to obtain efficient pipelining code. The aim is to use all eight functional units within one cycle. However, substantial coding effort can be required when the software pipelining technique is used for more complex programs. There are three stages to a pipelined code:

1. Prolog
2. Loop kernel (or loop cycle)
3. Epilog

```
//Noisegen_casm.c Pseudorandom noise generation calling ASM function

#include "dsk6713_aic23.h"                   //codec-DSK support file
Uint32 fs=DSK6713_AIC23_FREQ_48KHZ;          //set sampling rate
int previous_seed;
short pos = 16000, neg = -16000;             //scaling noise level

interrupt void c_int11()
{
 previous_seed = noisefunc(previous_seed);        //call ASM function
 if(previous_seed & 0x01)    output_left_sample(pos);//positive scaling
 else                        output_left_sample(neg);//negative scaling
}

void main ()
{
 comm_intr();                                //init DSK,codec,McBSP
 previous_seed = noisefunc(0x7E521603);      //call ASM function
 while (1);                                  //infinite loop
}
```

FIGURE 3.10. C program that calls an ASM function to generate a 32-bit noise sequence (noisegen_casm.c).

```
;Noisegen_casmfunc.asm Noise generation C-called function

           .def  _noisefunc  ;ASM function called from C
_noisefunc ZERO  A2          ;init A2 for seed manipulation
           MV    A4,A1       ;seed in A1
           SHR   A1,17,A1    ;shift right 17->bit 17 to LSB
           ADD   A1,A2,A2    ;add A1 to A2 => A2
           SHR   A1,11,A1    ;shift right 11->bit 28 to LSB
           ADD   A1,A2,A2    ;add again
           SHR   A1,2,A1     ;shift right 2->bit 30 to LSB
           ADD   A1,A2,A2    ;
           SHR   A1,1,A1     ;shift right 1->bit 31 to LSB
           ADD   A1,A2,A2    ;
           AND   A2,1,A2     ;Mask LSB of A2
           SHL   A4,1,A4     ;shift seed left 1
           OR    A2,A4,A4    ;Put A2 into LSB of A4
           B     B3          ;return to calling function
           NOP   5           ;5 delays for branch
```

FIGURE 3.11. ASM function called from C to generate a 32-bit noise sequence (noisegen_casmfunc.asm).

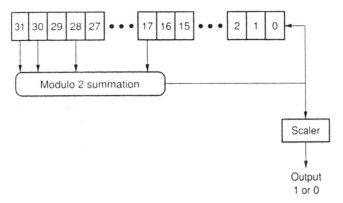

FIGURE 3.12. A 32-bit noise generator diagram.

Build and run this project as **Noisegen_casm**. Sampling at 48 kHz, verify that the noise spectrum is flat, with a bandwidth of approximately 23 kHz. Connect the output to a speaker to verify the generated noise. Change the scaling values to ±8000 and verify that the level of the generated noise is reduced.

Example 3.4: Code Detection Using C Calling an ASM Function (code_casm)

This example detects a four-digit code set initially in the main C source program. Figure 3.13 shows the main C source program *code_casm.c* that calls the *asm* function *code_casmfunc.asm*, shown in Figure 3.14. The code is set with *code1*, . . . , *code4* as 1, 2, 2, 4, respectively. The initial values of *digit1*, . . . , *digit4* set as 1, 1, 1, 1, respectively, are passed to the *asm* function to compare these four digit values with the four code values. Four sliders are used to change the digit values passed to the *asm* function. The C source program, the *asm* function, and the gel file for the sliders are included in the folder **code_casm**.

Build this project example as **code_casm**. Load and run the executable file. Press switch #0 (SW0) and verify that "no match" is continuously being printed (as long as SW0 is pressed). Load the gel file *code_casm.gel* and set the sliders *Digit1*, . . . , *Digit4* to positions 1, 2, 2, 4, respectively. Press SW0 and verify that "correct match" is being printed (with SW0 pressed). Change the slider *Digit2* from position 2 to position 3, and again press SW0 to verify that there is no longer a match. The program is in a continuous loop as long as switch #3 (SW3) is *not* pressed. Note that the initial value for the code (code1, . . . , code4) can readily be changed.

Example 3.5: Dot Product Using Assembly Program Calling an Assembly Function (dotp4a)

This example takes the sum of products of two arrays, each array with four numbers. See also Example 1.3, which implements it using only C code, and Examples 3.1

```
//Code_casm.c Calls ASM function.If code match slider values

#include <stdio.h>
short digit1=1,digit2=1,digit3=1,digit4=1;//init slider values

main()
{
 short code1=1,code2=2,code3=2,code4=4;    //initialize code
 short result;
 DSK6713_init();                           //init BSL
 DSK6713_DIP_init();                       //init dip switches
 while(DSK6713_DIP_get(3) == 1)            //continue til SW #3 pressed
 {
  if(DSK6713_DIP_get(0) == 0)              //if DIP SW #0 is pressed
  {                                        //call ASM function
   result=codefunc(digit1,digit2,digit3,digit4,code1,code2,code3,code4);
   if(result==0) printf("correct match\n");//result from ASM function
   else          printf("no match\n");     //correct match or no match
  }
 }
}
```

FIGURE 3.13. C program that calls an ASM function to detect a four-digit code (code_casm.c).

```
;Code_casmfunc.asm ASM function->if code matches slider values

           .def  _codefunc    ;ASM function called from C
 _codefunc: MV    A8, A2       ;correct code
           MV    B8, B2
           MV    A10, A7
           MV    B10, B7
           CMPEQ A2,A4,A1      ;compare 1st digit(A1=1 if A2=A4)
           CMPEQ A1,0,A1       ;otherwise A1=0
      [A1] B     DONE          ;done if A1=0 since no match
           NOP   5
           MV    B2,A2
           CMPEQ A2,B4,A1      ;compare 2nd digit
           CMPEQ A1,0,A1
      [A1] B     DONE
           NOP   5
           MV    A7,A2
           CMPEQ A2,A6,A1      ;compare 3rd digit
           CMPEQ A1,0,A1
      [A1] B     DONE
           NOP   5
           MV    B7,A2
           CMPEQ A2,B6,A1      ;compare 4th digit
           CMPEQ A1,0,A1
  DONE:    MV    A1,A4         ;return 1 if complete match
           B     B3            ;return to C program
           NOP   5
           .end
```

FIGURE 3.14. ASM function called from C to detect a four-digit code (code_casmfunc.asm).

```
;Dotp4a_init.asm ASM program to init variables.Calls dotp4afunc

                .def        init        ;starting address
                .ref        dotp4afunc  ;called ASM function
                .text                   ;section for code
x_addr          .short      1,2,3,4     ;numbers in x array
y_addr          .short      0,2,4,6     ;numbers in y array
result_addr .short         0           ;initialize sum of products

init            MVK    result_addr,A4   ;result addr -->A4
                MVK    0,A3             ;A3=0
                STH    A3,*A4           ;init result to 0
                MVK    x_addr,A4        ;A4 = address of x
                MVK    y_addr,B4        ;B4 = address of y
                MVK    4,A6             ;A6 = size of array
                B      dotp4afunc       ;B to function dotp4afunc
                MVK    ret_addr,b3      ;B3=return addr from dotp4a
                NOP    3                ;3 more delay slots(branch)
ret_addr        MVK    result_addr,A0   ;A0 = result address
                STW    A4,*A0           ;store result
wait            B      wait             ;wait here
                NOP    5                ;delay slots for branch
```

FIGURE 3.15. ASM program calling an ASM function to find the sum of products (dotp4a_init.asm).

through 3.4, which introduced the syntax of assembly code. Figure 3.15 shows a listing of the assembly program $dotp4a\_init.asm$, which initializes the two arrays of numbers and calls the assembly function $dotp4afunc.asm$, shown in Figure 3.16, which takes the sum of products of the two arrays. It also sets a return address through register B3 and the result address to A0. The addresses of the two arrays and the size of the array are passed to the function $dotp4afunc.asm$ through registers A4, A6, and B4, respectively. The result from the called function is "sent back" through A4. The resulting sum of the products is stored in memory whose address is result_addr. The instruction STW stores the resulting sum of the products in A4 (in memory pointed by A0). Register A0 serves as a pointer with the address result_addr.

The instruction MVK moves the 16 LSBs (equivalent to MVKL). If a 32-bit address (or result) is required, then the pair of instructions MVKL and MVKH can be used to move both the lower and upper 16 bits of the address (or result). The starting address of the calling ASM program is defined as init. The vector file is modified and included in the folder **dotp4a** so that the reference to the entry address is changed from $\_c\_int00$ to the entry address $init$. An alternative vector file $vectors\_dotp4a.asm$, as shown in Figure 3.17, specifies a branch to that entry address. The called asm function $dotp4afunc.asm$ calculates the sum of products. The loop count value was moved to A1 since A6 cannot be used as a conditional register (only A1, A2, B0, B1, and B2 can be used). The two LDH instructions load

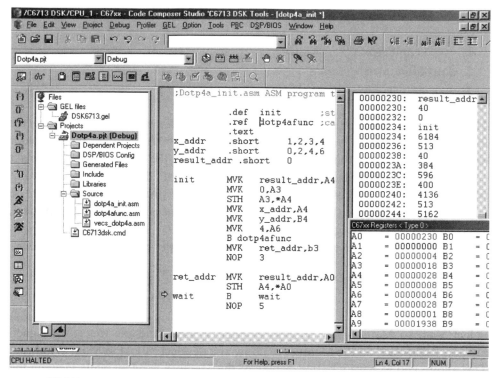

FIGURE 3.18. CCS windows for the sum of products in the project `dotp4a`.

```
//Dotp4clasm.c Multiplies two arrays using C calling linear ASM func

short dotp4clasmfunc(short *a,short *b,short ncount);  //prototype
#include <stdio.h>                              //for printing statement
#include "dotp4.h"                              //arrays of data values
#define   count 4                               //number of data values
short x[count] = {x_array};                     //declare 1st array
short y[count] = {y_array};                     //declare 2nd array
volatile int result = 0;                        //result

main()
{
 result = dotp4clasmfunc(x,y,count);            //call linear ASM func
 printf("result = %d decimal \n", result);      //print result
}
```

FIGURE 3.19. C program calling a linear ASM function to find the sum of products (`dotp4clasm.c`).

```
;Dotp4clasmfunc.sa   Linear assembly function to multiply two arrays
                 .ref    _dotp4clasmfunc  ;ASM func called from C
_dotp4clasmfunc: .cproc  ap,bp,count      ;start section linear ASM
                 .reg    a,b,prod,sum     ;asm optimizer directive
                 zero    sum              ;init sum of products
loop:            ldh     *ap++,a          ;pointer to 1st array->a
                 ldh     *bp++,b          ;pointer to 2nd array->b
                 mpy     a,b,prod         ;product = a*b
                 add     prod,sum,sum     ;sum of products -->sum
                 sub     count,1,count    ;decrement counter
  [count]        b       loop             ;loop back if count # 0
                 .return sum              ;return sum as result
                 .endproc                 ;end linear ASM function
```

FIGURE 3.20. Linear ASM function called from C to find the sum of products (`dotp4clasmfunc.sa`).

of the linear assembly function called is preceded by an underscore since the calling function is in C. The directive `.def` defines the function.

Functional units are optional as in an assembly-coded program. Registers a, b, *prod*, and *sum* are defined by the linear assembler directive .*reg*. The addresses of the two arrays x and y and the size of the array (count) are passed to the linear assembly function through the registers ap, bp, and count. Both ap and bp are registers used as pointers, as in C code. The instruction field is seen to be as in an assembly-coded program, and the subsequent field uses a syntax as in C programming. For example, the instruction

```
loop: ldh *ap++,a
```

(the first time through the loop section of code) loads the content in memory, whose address is specified by register ap, into register a. Then the pointer register ap is postincremented to point to the next higher memory address, pointing at the memory location containing the second value of x within the x array. The value of the sum of the products is accumulated in *sum*, which is returned to the C calling program.

Build and run this project as **dotp4clasm**. Verify that the following is printed: result = 40. You may wish to profile the linear assembly code function and compare its execution time with that of the C-coded version in Example 1.3.

Example 3.7: Factorial Using C Calling a Linear Assembly Function (`factclasm`)

Figure 3.21 shows a listing of the C program `factclasm.c`, which calls the linear asm function `factclasmfunc.sa`, shown in Figure 3.22, to calculate the factorial of a number less than 8. See also Example 3.2, which finds the factorial of a number using a C program that calls an asm function. Example 3.6 illustrates a C program

```
//Factclasm.c Factorial of number. Calls linear ASM function

#include <stdio.h>                  //for print statement
void main()
{
  short number = 7;                 //set value
  short result;                     //result of factorial
  result = factclasmfunc(number);   //call ASM function factlasmfunc
  printf("factorial = %d", result); //result from linear ASM function
}
```

FIGURE 3.21. C program that calls a linear ASM function to find the factorial of a number (`factclasm.c`).

```
;Factclasmfunc.sa Linear ASM function called from C to find factorial

                 .ref    _factclasmfunc ;Linear ASM func called from C
_factclasmfunc:  .cproc  number         ;start of linear ASM function
                 .reg    a,b            ;asm optimizer directive
                 mv      number,b       ;setup loop count in b
                 mv      number,a       ;move number to a
                 sub     b,1,b          ;decrement loop counter
loop:            mpy     a,b,a          ;n(n-1)
                 sub     b,1,b          ;decrement loop counter
    [b]          b       loop           ;loop back to loop if count #0
                 .return a              ;result to calling function
                 .endproc               ;end of linear ASM function
```

FIGURE 3.22. Linear ASM function called from C that finds the factorial of a number (`factclasmfunc.sa`).

calling a linear ASM function to find the sum of products and is instructive for this project. Examples 3.2 and 3.6 cover the essential background for this example.

Support files for this project include `factclasm.c`, `factclasmfunc.sa`, `rts6700.lib`, and `C6713dsk.cmd`. Build and run this project as **factclasm**. Verify that the result of 7! is printed, or `factorial = 5040`.

3.22 ASSIGNMENTS

1. Write a C program that calls an assembly function that takes input values a and b from the C program to calculate the following: $[a^2 + (a + 1)^2 + (a + 2)^2 + \cdots + (2a - 1)^2] - [b^2 + (b + 1)^2 + (b + 2)^2 + \cdots + (2b - 1)^2]$. Set $a = 3$ and $b = 2$ in the C program and verify that the result is printed as 37.

2. Write a C program that calls an assembly function to obtain the determinant of a 3×3 matrix. Set the matrix values in the C program. The first row values are {4, 5, 9}; the second row values are {8, 6, 5}, and the third row values are {2, 1, 2}. Verify that the resulting determinant is printed within CCS as −38.

```
Partial programs C/ASM function to multiply 2 numbers using switches
..
while(m == 100)                         //check for first SW pressed
{
 if(DSK6713_DIP_get(0)== 0)             //true if SW0 is pressed
 {
  m = 1;                                //value if SW0 is pressed
  while(DSK6713_DIP_get(0)==0) DSK6713_LED_on(0);//ON until released
  for(delay=0; delay<5000000; delay++){}          //debounce of SW0
 }
 else if(DSK6713_DIP_get(1)==0)         //true if SW1 is pressed
 {
  m = 2;
  .

  .
  else m = 100;
  .

  .
  while(ii == 0)
   {
   result = values(n, m);              //result from ASM function in A4
   led0 = result0(result);             //returns a 0 or 1 to led0
   if(led0==1)   DSK6713_LED_on(0); //if led0 is 1 turn it on
   .

   .
;ASM function
              ..
_values:    MV     A4,A5        ;setup n as loop counter
            MV     B4,B1
LOOP:       ADD    A5,A4,A4     ;accumulate in A4
            SUB    B1,1,B1      ;decrement loop counter
      [B1]  B      LOOP         ;branch to LOOP if B1#0
            NOP    5            ;five NOPs for delay slots
            SUB    A4,A5,A4     ;answer into A4
            B      B3           ;return to calling routine
            NOP    5            ;five NOPs for delay slots


_result0:   SHL    A4,31,A4     ;shift left 31 bits to keep LSB
            SHRU   A4,31,A4     ;shift right 31 bits to make A4=0 or 1
            B      B3           ;return to calling routine
            NOP    5            ;five NOPs for delay slots
```

FIGURE 3.23. Partial programs (C and ASM function) to multiply two numbers using the dip switches.

3. Write a C program `multi_casm.c` that calls an assembly function `multi_casmfunc.asm` to multiply two numbers using the onboard dip switches. The maximum product is $3 \times 4 = 12$ or $4 \times 3 = 12$. Note that $4 \times 4 = 16$ cannot be represented with the four dip switches. Use delay loops for debouncing the switches. A partial program is included in Figure 3.23. In the main C source

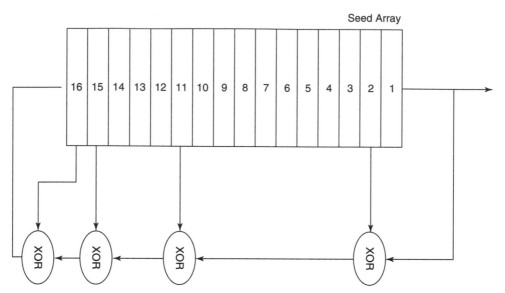

FIGURE 3.24. Pseudorandom noise generation diagram using LFSR.

program, the values of $m = 100$ and $n = 100$ are to check when the first and second switches are pressed. SW0 is tested and, if pressed, $m = 1$, representing the first value. Similarly, $m = 2, 3, 4$ if SW1, SW2, or SW3 is pressed, respectively. Then all LEDs are turned off. This process is repeated while $n = 100$ to check for the second value (when the second switch is pressed).

The function *values* performs the multiplication, adding m (n times) with m and n passed to the asm function through A4 and B4, respectively. Note that led0 is turned on if led0 = 1 (returned from the function result0). Similarly for led1,..., led3. Then, m and n are reset to 100 and ii to 1. The *asm* function multi_casmfunc.asm includes the functions *values, result0, ..., result3*. The functions *result1, result2, result3* are similar to *result0*, but A4 must be shifted first by 1, by 2, and by 3, respectively, in each of these functions. Build and run this project example as **multi_casm**. Press SW2, then SW3 to obtain m = 3 and n = 4, and verify that LED2 and LED3 turn on to represent the result of 12.

4. Write a C program that calls a linear assembly or assembly function to generate a random noise sequence, based on the linear feedback shift register (LFSR) shown in Figure 3.24. In lieu of starting with a 16-bit seed value, 16 integer values are used in an array as the seeds. In this fashion, each 32-bit seed is treated as a theoretical bit. The "tap points" are chosen as shown (bits 1, 2, 11, 15, and 16) to produce a large string of random numbers [11]. Within the *asm* or *linear asm* function, each integer value is taken as a seed, and you can use instructions such as LDW/STW, repeated 15 times, to move each seed "up." XOR bits 1 and 2, the result of which is XORed with bit 11, and so on, as shown in Figure 3.24. The resulting seed generated is placed at the "bottom" of the array, and the process is repeated. The output is a 32-bit value.

Sampling at 8 kHz, verify that the generated noise spectrum is flat until it rolls off at about 3.8 kHz, which is the cutoff frequency of the reconstruction filter on the codec.

REFERENCES

1. R. Chassaing and D. W. Horning, *Digital Signal Processing with the TMS320C25*, Wiley, Hoboken, NJ, 1990.

2. R. Chassaing, *Digital Signal Processing Laboratory Experiments Using C and the TMS320C31 DSK*, Wiley, Hoboken, NJ, 1999.

3. R. Chassaing, *Digital Signal Processing with C and the TMS320C30*, Wiley, Hoboken, NJ, 1992.

4. R. Chassaing and P. Martin, Parallel processing with the TMS320C40, *Proceedings of the 1995 ASEE Annual Conference*, June 1995.

5. R. Chassaing and R. Ayers, Digital signal processing with the SHARC, *Proceedings of the 1996 ASEE Annual Conference*, June 1996.

6. *TMS320C6000 CPU and Instruction Set*, SPRU189F, Texas Instruments, Dallas, TX, 2000.

7. *TMS320C6000 Peripherals*, SPRU190D, Texas Instruments, Dallas, TX, 2001.

8. *TMS320C6000 Programmer's Guide*, SPRU198G, Texas Instruments, Dallas, TX, 2002.

9. *TMS320C6000 Assembly Language Tools User's Guide*, SPRU186K, Texas Instruments, Dallas, TX, 2002.

10. *TMS320C6000 Optimizing C Compiler User's Guide*, SPRU187K, Texas Instruments, Dallas, TX, 2002.

11. Linear Feedback Shift Registers, New Wave Instruments, 2002, www.newwaveinstruments.com/resources.

4

Finite Impulse Response Filters

- Introduction to the z-transform
- Design and implementation of finite impulse response (FIR) filters
- Programming examples using C and TMS320C6x code

The z-transform is introduced in conjunction with discrete-time signals. Mapping from the s-plane, associated with the Laplace transform, to the z-plane, associated with the z-transform, is illustrated. FIR filters are designed with the Fourier series method and implemented by programming a discrete convolution equation. Effects of window functions on the characteristics of FIR filters are covered.

4.1 INTRODUCTION TO THE z-TRANSFORM

The z-transform is utilized for the analysis of discrete-time signals, similar to the Laplace transform for continuous-time signals. We can use the Laplace transform to solve a differential equation that represents an analog filter or the z-transform to solve a difference equation that represents a digital filter. Consider an analog signal $x(t)$ ideally sampled,

$$x_s(t) = \sum_{k=0}^{\infty} x(t)\delta(t - kT) \tag{4.1}$$

Digital Signal Processing and Applications with the TMS320C6713 and TMS320C6416 DSK, Second Edition By Rulph Chassaing and Donald Reay
Copyright © 2008 John Wiley & Sons, Inc.

where $\delta(t - kT)$ is the impulse (delta) function delayed by kT and $T = 1/F_s$ is the sampling period. The function $x_s(t)$ is zero everywhere except at $t = kT$. The Laplace transform of $x_s(t)$ is

$$X_s(s) = \int_0^\infty x_s(t)e^{-st}\,dt$$
$$= \int_0^\infty \{x(t)\,\delta(t) + x(t)\,\delta(t-T) + \cdots\}e^{-st}\,dt \tag{4.2}$$

From the property of the impulse function

$$\int_0^\infty f(t)\delta(t-kT)\,dt = f(kT)$$

$X_s(s)$ in (4.2) becomes

$$X_s(s) = x(0) + x(T)e^{-sT} + x(2T)e^{-2sT} + \cdots = \sum_{n=0}^\infty x(nT)e^{-nsT} \tag{4.3}$$

Let $z = e^{sT}$ in (4.3), which becomes

$$X(z) = \sum_{n=0}^\infty x(nT)z^{-n} \tag{4.4}$$

Let the sampling period T be implied; then $x(nT)$ can be written as $x(n)$, and (4.4) becomes

$$X(z) = \sum_{n=0}^\infty x(n)z^{-n} = ZT\{x(n)\} \tag{4.5}$$

which represents the z-transform (ZT) of $x(n)$. There is a one-to-one correspondence between $x(n)$ and $X(z)$, making the z-transform a unique transformation.

Exercise 4.1: ZT of Exponential Function $x(n) = e^{nk}$

The ZT of $x(n) = e^{nk}$, $n \geq 0$ and k a constant, is

$$X(z) = \sum_{n=0}^\infty e^{nk}z^{-n} = \sum_{n=0}^\infty (e^k z^{-1})^n \tag{4.6}$$

Using the geometric series, obtained from a Taylor series approximation

$$\sum_{n=0}^\infty u^n = \frac{1}{1-u} \qquad |u| < 1$$

(4.6) becomes

$$X(z) = \frac{1}{1 - e^k z^{-1}} = \frac{z}{z - e^k} \tag{4.7}$$

for $|e^k z^{-1}| < 1$ or $|z| > |e^k|$. If $k = 0$, the ZT of $x(n) = 1$ is $X(z) = z/(z - 1)$.

Exercise 4.2: ZT of Sinusoid x(n) = sin nωT

A sinusoidal function can be written in terms of complex exponentials. From Euler's formula $e^{ju} = \cos u + j\sin u$,

$$\sin n\omega T = \frac{e^{jn\omega T} - e^{-jn\omega T}}{2j}$$

Then

$$X(z) = \frac{1}{2j} \sum_{n=0}^{\infty} (e^{jn\omega T} z^{-n} - e^{-jn\omega T} z^{-n}) \tag{4.8}$$

Using the geometric series as in Exercise 4.1, one can solve for $X(z)$; or the results in (4.7) can be used with $k = j\omega T$ in the first summation of (4.8) and $k = -j\omega T$ in the second, to yield

$$\begin{aligned} X(z) &= \frac{1}{2j}\left(\frac{z}{z - e^{j\omega T}} - \frac{z}{z - e^{-j\omega T}}\right) \\ &= \frac{1}{2j} \frac{z^2 - ze^{-j\omega T} - z^2 + ze^{j\omega T}}{z^2 - z(e^{-j\omega T} + e^{j\omega T}) + 1} \\ &= \frac{z \sin \omega T}{z^2 - 2z \cos \omega T + 1} \\ &= \frac{Cz}{z^2 - Az - B} \qquad |z| > 1 \end{aligned} \tag{4.9} \tag{4.10}$$

where $A = 2\cos \omega T$, $B = -1$, and $C = \sin \omega T$. In Chapter 5 we generate a sinusoid based on this result. We can readily generate sinusoidal waveforms of different frequencies by changing the value of ω in (4.9).

Similarly, using Euler's formula for $\cos n\omega T$ as a sum of two complex exponentials, one can find the ZT of $x(n) = \cos n\omega T = (e^{jn\omega T} + e^{-jn\omega T})/2$, as

$$X(z) = \frac{z^2 - z\cos \omega T}{z^2 - 2z \cos \omega T + 1} \qquad |z| > 1 \tag{4.11}$$

4.1.1　Mapping from s-Plane to z-Plane

The Laplace transform can be used to determine the stability of a system. If the poles of a system are on the left side of the $j\omega$ axis on the s-plane, a time-decaying system response will result, yielding a stable system. If the poles are on the right side of the $j\omega$ axis, the response will grow in time, making such a system unstable. Poles located on the $j\omega$ axis, or purely imaginary poles, will yield a sinusoidal response. The sinusoidal frequency is represented by the $j\omega$ axis, and $\omega = 0$ represents dc (direct current).

In a similar fashion, we can determine the stability of a system based on the location of its poles on the z-plane associated with the z-transform, since we can find corresponding regions between the s-plane and the z-plane. Since $z = e^{sT}$ and $s = \sigma + j\omega$,

$$z = e^{\sigma T} e^{j\omega T} \tag{4.12}$$

Hence, the magnitude of z is $|z| = e^{\sigma T}$ with a phase of $\theta = \omega T = 2\pi f/F_s$, where F_s is the sampling frequency. To illustrate the mapping from the s-plane to the z-plane, consider the following regions from Figure 4.1.

$\sigma < 0$

Poles on the left side of the $j\omega$ axis (region 2) in the s-plane represent a stable system, and (4.12) yields a magnitude of $|z| < 1$, because $e^{\sigma T} < 1$. As σ varies from $-\infty$ to 0^-, $|z|$ will vary from 0 to 1^-. Hence, poles *inside* the unit circle within region 2 in the z-plane will yield a stable system. The response of such a system will be a decaying exponential if the poles are real or a decaying sinusoid if the poles are complex.

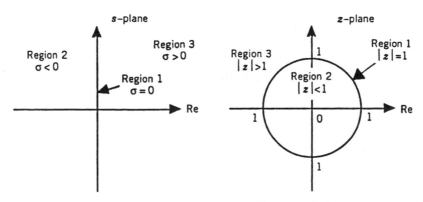

FIGURE 4.1. Mapping from the s-plane to the z-plane.

σ > 0

Poles on the right side of the $j\omega$ axis (region 3) in the s-plane represent an unstable system, and (4.12) yields a magnitude of $|z| > 1$, because $e^{\sigma T} > 1$. As σ varies from 0^+ to ∞, $|z|$ will vary from 1^+ to ∞. Hence, poles *outside* the unit circle within region 3 in the z-plane will yield an unstable system. The response of such a system will be an increasing exponential if the poles are real or a growing sinusoid if the poles are complex.

σ = 0

Poles on the $j\omega$ axis (region 1) in the s-plane represent a marginally stable system, and (4.12) yields a magnitude of $|z| = 1$, which corresponds to region 1. Hence, poles *on* the unit circle in region 1 in the z-plane will yield a sinusoid. In Chapter 5 we implement a sinusoidal signal by programming a difference equation with its poles *on* the unit circle. Note that from Exercise 4.2 the poles of $X(s) = \sin n\omega T$ in (4.9) or $X(s) = \cos n\omega T$ in (4.11) are the roots of $z^2 - 2z\cos \omega T + 1$, or

$$p_{1,2} = \frac{2\cos\omega T \pm \sqrt{4\cos^2\omega T - 4}}{2}$$
$$= \cos\omega T \pm \sqrt{-\sin^2\omega T} = \cos\omega T \pm j\sin\omega T \qquad (4.13)$$

The magnitude of each pole is

$$|p_1| = |p_2| = \sqrt{\cos^2\omega T + \sin^2\omega T} = 1 \qquad (4.14)$$

The phase of z is $\theta = \omega T = 2\pi f/F_s$. As the frequency f varies from zero to $\pm F_s/2$, the phase θ will vary from 0 to π.

4.1.2 Difference Equations

A digital filter is represented by a difference equation in a similar fashion as an analog filter is represented by a differential equation. To solve a difference equation, we need to find the z-transform of expressions such as $x(n - k)$, which corresponds to the kth derivative $d^k x(t)/dt^k$ of an analog signal $x(t)$. The order of the difference equation is determined by the largest value of k. For example, $k = 2$ represents a second-order derivative. From (4.5)

$$X(z) = \sum_{n=0}^{\infty} x(n)z^{-n} = x(0) + x(1)z^{-1} + x(2)z^{-2} + \cdots \qquad (4.15)$$

Then the z-transform of $x(n-1)$, which corresponds to a first-order derivative dx/dt, is

$$
\begin{aligned}
ZT[x(n-1)] &= \sum_{n=0}^{\infty} x(n-1)z^{-n} \\
&= x(-1)+x(0)z^{-1}+x(1)z^{-2}+x(2)z^{-3}+\cdots \\
&= x(-1)+z^{-1}\left[x(0)+x(1)z^{-1}+x(2)z^{-2}+\cdots\right] \\
&= x(-1)+z^{-1}X(z)
\end{aligned}
\tag{4.16}
$$

where we used (4.15), and $x(-1)$ represents the initial condition associated with a first order difference equation. Similarly, the ZT of $x(n-2)$, equivalent to a second derivative $d^2x(t)/dt^2$, is

$$
\begin{aligned}
ZT[x(n-2)] &= \sum_{n=0}^{\infty} x(n-2)z^{-n} \\
&= x(-2)+x(-1)z^{-1}+x(0)z^{-2}+x(1)z^{-3}+\cdots \\
&= x(-2)+x(-1)z^{-1}+z^{-2}\left[x(0)+x(1)z^{-1}+\cdots\right] \\
&= x(-2)+x(-1)z^{-1}+z^{-2}X(z)
\end{aligned}
\tag{4.17}
$$

where $x(-2)$ and $x(-1)$ represent the two initial conditions required to solve a second order difference equation. In general,

$$
ZT[x(n-k)] = z^{-k}\sum_{m=1}^{k} x(-m)z^{m} + z^{k}X(z)
\tag{4.18}
$$

If the initial conditions are all zero, then $x(-m) = 0$ for $m = 1, 2, \ldots, k$, and (4.18) reduces to

$$
ZT[x(n-k)] = z^{-k}X(z)
\tag{4.19}
$$

4.2 DISCRETE SIGNALS

A discrete signal $x(n)$ can be expressed as

$$
x(n) = \sum_{m=-\infty}^{\infty} x(m)\delta(n-m)
\tag{4.20}
$$

where $\delta(n-m)$ is the impulse sequence $\delta(n)$ delayed by m, which is equal to 1 for $n = m$ and is 0 otherwise. It consists of a sequence of values $x(1), x(2), \ldots$, where n is the time, and each sample value of the sequence is taken one sample time apart, determined by the sampling interval or sampling period $T = 1/F_s$.

The signals and systems that we deal with in this book are linear and time invariant, where both superposition and shift invariance apply. Let an input signal $x(n)$ yield an output response $y(n)$, or $x(n) \rightarrow y(n)$. If $a_1x_1(n) \rightarrow a_1y_1(n)$ and $a_2x_2(n) \rightarrow a_2y_2(n)$, then $a_1x_1(n) + a_2x_2(n) \rightarrow a_1y_1(n) + a_2y_2(n)$, where a_1 and a_2 are constants. This is the superposition property, where an overall output response is the sum of the individual responses to each input. Shift invariance implies that if the input is delayed by m samples, the output response will also be delayed by m samples, or $x(n-m) \rightarrow y(n-m)$. If the input is a unit impulse $\delta(n)$, the resulting output response is $h(n)$, or $\delta(n) \rightarrow h(n)$, and $h(n)$ is designated as the impulse response. A delayed impulse $\delta(n-m)$ yields the output response $h(n-m)$ by the shift-invariance property.

Furthermore, if this impulse is multiplied by $x(m)$, then $x(m)\delta(n-m) \rightarrow x(m)h(n-m)$. Using (4.20), the response becomes

$$y(n) = \sum_{m=-\infty}^{\infty} x(m)h(n-m) \tag{4.21}$$

which represents a convolution equation. For a causal system, (4.21) becomes

$$y(n) = \sum_{m=-\infty}^{\infty} x(m)h(n-m) \tag{4.22}$$

Letting $k = n - m$ in (4.22) yields

$$y(n) = \sum_{k=0}^{\infty} h(k)x(n-k) \tag{4.23}$$

4.3 FIR FILTERS

Filtering is one of the most useful signal processing operations [1–47]. DSPs are now available to implement digital filters in real time. The TMS320C6x instruction set and architecture makes it well suited for such filtering operations. An analog filter operates on continuous signals and is typically realized with discrete components such as operational amplifiers, resistors, and capacitors. However, a digital filter, such as an FIR filter, operates on discrete-time signals and can be implemented with a DSP such as the TMS320C6x. This involves use of an ADC to capture an external input signal, processing the input samples, and sending the resulting output through a DAC.

Within the last few years, the cost of DSPs has been reduced significantly, which adds to the numerous advantages that digital filters have over their analog counterparts. These include higher reliability, accuracy, and less sensitivity to temperature and aging. Stringent magnitude and phase characteristics can be achieved with a digital filter. Filter characteristics such as center frequency, bandwidth, and filter type can readily be modified. A number of tools are available to design and implement within a few minutes an FIR filter in real time using the TMS320C6x-based DSK. The filter design consists of the approximation of a transfer function with a resulting set of coefficients.

Different techniques are available for the design of FIR filters, such as a commonly used technique that utilizes the Fourier series, as discussed in Section 4.4. Computer-aided design techniques such as that of Parks and McClellan are also used for the design of FIR filters [5, 6].

The convolution equation (4.23) is very useful for the design of FIR filters, since we can approximate it with a finite number of terms, or

$$y(n) = \sum_{k=0}^{N-1} h(k)x(n-k) \tag{4.24}$$

If the input is a unit impulse $x(n) = \delta(0)$, the output impulse response will be $y(n) = h(n)$. We will see in Section 4.4 how to design an FIR filter with N coefficients $h(0), h(1), \ldots, h(N-1)$, and N input samples $x(n), x(n-1), \ldots, x(n-(N-1))$. The input sample at time n is $x(n)$, and the delayed input samples are $x(n-1), \ldots, x(n-(N-1))$. Equation (4.24) shows that an FIR filter can be implemented with knowledge of the input $x(n)$ at time n and of the delayed inputs $x(n-k)$. It is non-recursive, and no feedback or past outputs are required. Filters with feedback (recursive) that require past outputs are discussed in Chapter 5. Other names used for FIR filters are transversal and tapped-delay filters.

The z-transform of (4.24) with zero initial conditions yields

$$Y(z) = h(0)X(z) + h(1)z^{-1}X(z) + h(2)z^{-2}X(z) + \cdots + h(N-1)z^{-(N-1)}X(z) \tag{4.25}$$

Equation (4.24) represents a convolution in time between the coefficients and the input samples, which is equivalent to a multiplication in the frequency domain, or

$$Y(z) = H(z)X(z) \tag{4.26}$$

where $H(z) = ZT[h(k)]$ is the transfer function, or

$$
\begin{aligned}
H(z) &= \sum_{k=0}^{N-1} h(k)z^{-k} = h(0) + h(1)z^{-1} + h(2)z^{-2} + \cdots + h(N-1)z^{-(N-1)} \\
&= \frac{h(0)z^{(N-1)} + h(1)z^{N-2} + h(2)z^{N-3} + \cdots + h(N-1)}{z^{N-1}}
\end{aligned} \tag{4.27}
$$

which shows that there are $N-1$ poles, all of which are located at the origin. Hence, this FIR filter is inherently stable, with its poles located only inside the unit circle. We usually describe an FIR filter as a filter with "no poles." Figure 4.2 shows an FIR filter structure representing (4.24) and (4.25).

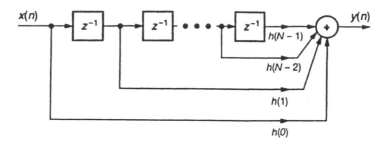

FIGURE 4.2. FIR filter structure showing delays.

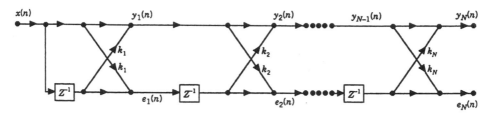

FIGURE 4.3. FIR lattice structure.

A very useful feature of an FIR filter is that it can guarantee *linear phase*. The linear phase feature can be very useful in applications such as speech analysis, where phase distortion can be critical. For example, with linear phase, all input sinusoidal components are delayed by the same amount. Otherwise, harmonic distortion can occur. Linear phase filters are FIR filters; however, not all FIR filters have linear phase.

The Fourier transform of a delayed input sample $x(n - k)$ is $e^{-j\omega kT}X(j\omega)$, yielding a phase of $\theta = -\omega kT$, which is a linear function in terms of ω. Note that the group delay function, defined as the derivative of the phase, is a constant, or $d\theta/d\omega = -kT$.

4.4 FIR LATTICE STRUCTURE

The lattice structure is commonly used for applications in adaptive filtering and speech processing [48, 49], such as in a linear predictive coding (LPC) application. An Nth order lattice structure is shown in Figure 4.3. The coefficients $k_1, k_2, \ldots, k_N$ are commonly referred to as *reflection coefficients* (or *k*-parameters). An advantage of this structure is that the frequency response is not as sensitive as the previous structure to small changes in the coefficients. From the first section in Figure 4.3, with $N = 1$, we have

$$y_1(n) = x(n) + k_1 x(n-1) \tag{4.28}$$

$$e_1(n) = k_1 x(n) + x(n-1) \tag{4.29}$$

From the second section (cascaded with the first), using (4.28) and (4.29),

$$
\begin{aligned}
y_2(n) &= y_1(n) + k_2 e_1(n-1) \\
&= x(n) + k_1 x(n-1) + k_2 k_1 x(n-1) + k_2 x(n-2) \\
&= x(n) + (k_1 + k_1 k_2) x(n-1) + k_2 x(n-2)
\end{aligned}
\tag{4.30}
$$

and

$$
\begin{aligned}
e_2(n) &= k_2 y_1(n) + e_1(n-1) \\
&= k_2 x(n) + k_2 k_1 x(n-1) + k_1 x(n-1) + x(n-2) \\
&= k_2 x(n) + (k_1 + k_1 k_2) x(n-1) + x(n-2)
\end{aligned}
\tag{4.31}
$$

For a specific section i,

$$
y_i(n) = y_{i-1}(n) + k_i e_{i-1}(n-1)
\tag{4.32}
$$

$$
e_i(n) = k_i y_{i-1}(n) + e_{i-1}(n-1)
\tag{4.33}
$$

It is instructive to see that (4.30) and (4.31) have the same coefficients but in reversed order. It can be shown that this property also holds true for a higher order structure. In general, for an Nth order FIR lattice system, (4.30) and (4.31) become

$$
y_N(n) = \sum_{i=0}^{N} a_i x(n-i)
\tag{4.34}
$$

and

$$
e_N(n) = \sum_{i=0}^{N} a_{N-i} x(n-i)
\tag{4.35}
$$

with $a_0 = 1$. If we take the ZT of (4.34) and (4.35) and find their impulse responses,

$$
Y_N(z) = \sum_{i=0}^{N} a_i z^{-i}
\tag{4.36}
$$

$$
E_N(z) = \sum_{i=0}^{N} a_{N-i} z^{-i}
\tag{4.37}
$$

It is interesting to note that

$$
E_N(z) = z^{-N} Y_N(1/z)
\tag{4.38}
$$

Equations (4.36) and (4.37) are referred to as image polynomials. For two sections, $k_2 = a_2$; in general,

$$k_N = a_N \qquad (4.39)$$

For this structure to be useful, it is necessary to find the relationship between the k-parameters and the impulse response coefficients. The lattice network is highly structured, as seen in Figure 4.3 and as demonstrated through the previous difference equations. Starting with k_N in (4.39), we can recursively (with reverse recursion) compute the preceding k-parameters, $k_{N-1}, \ldots, k_1$.

Consider an intermediate section r and, using (4.36) and (4.37),

$$Y_r(z) = Y_{r-1}(z) + k_r z^{-1} E_{r-1}(z) \qquad (4.40)$$

$$E_r(z) = k_r Y_{r-1}(z) + z^{-1} E_{r-1}(z) \qquad (4.41)$$

Solving for $E_{r-1}(z)$ in (4.41) and substituting it into (4.40), $Y_r(z)$ becomes

$$Y_r(z) = Y_{r-1}(z) + k_r z^{-1} \frac{E_r(z) - k_r Y_{r-1}(z)}{z^{-1}} \qquad (4.42)$$

Equation (4.42) now can be solved for $Y_{r-1}(z)$ in terms of $Y_r(z)$, or

$$Y_{r-1}(z) = \frac{Y_r(z) - k_r E_r(z)}{1 - k_r^2}, \quad |k_r| = 1 \qquad (4.43)$$

Using (4.38) with $N = r$, (4.43) becomes

$$Y_{r-1}(z) = \frac{Y_r(z) - k_r z^{-r} Y_r(1/z)}{1 - k_r^2} \qquad (4.44)$$

Equation (4.44) is an important relationship that shows that by using a reverse recursion procedure, we can find Y_{r-1} from Y_r, where $1 \leq r \leq N$. Consequently, we can also find the k-parameters starting with k_r and proceeding to k_1. For r sections, (4.36) can be written

$$Y_r(z) = \sum_{i=0}^{r} a_{ri} z^{-i} \qquad (4.45)$$

Replacing i by $r - i$, and z by $1/z$, (4.45) becomes

$$Y_r\left(\frac{1}{z}\right) = \sum_{i=0}^{r} a_{r(r-i)} z^{r-i} \qquad (4.46)$$

Using (4.45) and (4.46), Equation (4.44) becomes

$$\sum_{i=0}^{r} a_{(r-1)i} z^{-i} = \frac{\sum_{i=0}^{r} a_{ri} z^{-i} - k_r z^{-r} \sum_{i=0}^{r} a_{r(r-i)} z^{r-i}}{1 - k_r^2} \tag{4.47}$$

$$= \frac{\sum_{i=0}^{r} a_{ri} z^{-i} - k_r \sum_{i=0}^{r} a_{r(r-i)} z^{-i}}{1 - k_r^2} \tag{4.48}$$

from which

$$a_{(r-1)i} = \frac{a_{ri} - k_r a_{r(r-i)}}{1 - k_r^2}, \quad i = 0, 1, \ldots, r-1 \tag{4.49}$$

with $r = N, N-1, \ldots, 1$, $|k_r| \neq 1$, $i = 0, 1, \ldots, r-1$, and

$$k_r = a_{rr}, \quad r = N, N-1, \ldots, 1 \tag{4.50}$$

Exercise 4.3: FIR Lattice Structure

This exercise illustrates the use of (4.49) and (4.50) to compute the k-parameters. Given that the impulse response of an FIR filter in the frequency domain is

$$Y_2(z) = 1 + 0.2z^{-1} - 0.5z^{-2}$$

Then, from (4.45), with $r = 2$,

$$Y_2(z) = a_{20} + a_{21} z^{-1} + a_{22} z^{-2}$$

where $a_{20} = 1$, $a_{21} = 0.2$, and $a_{22} = -0.5$. Starting with $r = 2$ in (4.50),

$$k_2 = a_{22} = -0.5$$

Using (4.49), for $i = 0$,

$$a_{10} = \frac{a_{20} - k_2 a_{22}}{1 - k_2^2} = \frac{1 - (-0.5)(-0.5)}{1 - (-0.5)^2} = 1$$

and, for $i = 1$,

$$a_{11} = \frac{a_{21} - k_2 a_{21}}{1 - k_2^2} = \frac{0.2 - (-0.5)(0.2)}{1 - (-0.5)^2} = 0.4$$

From (4.50),

$$k_1 = a_{11} = 0.4$$

Note that the values for the k-parameters $k_2 = -0.5$ and $k_1 = 0.4$ can be verified using (4.30).

4.5 FIR IMPLEMENTATION USING FOURIER SERIES

The design of an FIR filter using a Fourier series method is such that the magnitude response of its transfer function $H(z)$ approximates a desired magnitude response. The transfer function desired is

$$H_d(\omega) = \sum_{n=-\infty}^{\infty} C_n e^{jn\omega T} \qquad |n| < \infty \tag{4.51}$$

where C_n are the Fourier series coefficients. Using a normalized frequency variable v such that $v = f/F_N$, where F_N is the Nyquist frequency, or $F_N = F_s/2$, the desired transfer function in (4.51) can be written

$$H_d(v) = \sum_{n=-\infty}^{\infty} C_n e^{jn\pi v} \tag{4.52}$$

where $\omega T = 2\pi f/F_s = \pi v$ and $|v| < 1$. The coefficients C_n are defined as

$$\begin{aligned} C_n &= \frac{1}{2} \int_{-1}^{1} H_d(v) e^{-jn\pi v} dv \\ &= \frac{1}{2} \int_{-1}^{1} H_d(v)(\cos n\pi v - j\sin n\pi v) dv \end{aligned} \tag{4.53}$$

Assume that $H_d(v)$ is an even function (frequency selective filter); then (4.53) reduces to

$$C_n = \int_{0}^{1} H_d(v) \cos n\pi v \, dv \qquad n \geq 0 \tag{4.54}$$

since $H_d(v) \sin n\pi v$ is an odd function and

$$\int_{-1}^{1} H_d(v) \sin n\pi v \, dv = 0$$

with $C_n = C_{-n}$. The desired transfer function $H_d(v)$ in (4.52) is expressed in terms of an infinite number of coefficients, and to obtain a realizable filter, we must truncate (4.52), which yields the approximated transfer function

$$H_a(v) = \sum_{n=-Q}^{Q} C_n e^{jn\pi v} \tag{4.55}$$

where Q is positive and finite and determines the order of the filter. The larger the value of Q, the higher the order of the FIR filter and the better the approximation in (4.55) of the desired transfer function. The truncation of the infinite series with a finite number of terms results in ignoring the contribution of the terms outside a rectangular window function between $-Q$ and $+Q$. In Section 4.6 we see how the characteristics of a filter can be improved by using window functions other than rectangular.

Let $z = e^{j\pi v}$; then (4.55) becomes

$$H_a(z) = \sum_{n=-Q}^{Q} C_n z^n \tag{4.56}$$

with the impulse response coefficients $C_{-Q}, C_{-Q+1}, \ldots, C_{-1}, C_0, C_1, \ldots, C_{Q-1}, C_Q$. The approximated transfer function in (4.56), with positive powers of z, implies a non-causal or not realizable filter that would produce an output before an input is applied. To remedy this situation, we introduce a delay of Q samples in (4.56) to yield

$$H(z) = z^{-Q} H_a(z) = \sum_{n=-Q}^{Q} C_n z^{n-Q} \tag{4.57}$$

Let $n - Q = -i$; then $H(z)$ in (4.57) becomes

$$H(z) = \sum_{i=0}^{2Q} C_{Q-i} z^{-i} \tag{4.58}$$

Let $h_i = C_{Q-i}$ and $N - 1 = 2Q$; then $H(z)$ becomes

$$H(z) = \sum_{i=0}^{N-1} h_i z^{-i} \tag{4.59}$$

where $H(z)$ is expressed in terms of the impulse response coefficients h_i, and $h_0 = C_Q$, $h_1 = C_{Q-1}, \ldots, h_Q = C_0$, $h_{Q+1} = C_{-1} = C_1, \ldots, h_{2Q} = C_{-Q}$. The impulse response coefficients are symmetric about h_Q, with $C_n = C_{-n}$.

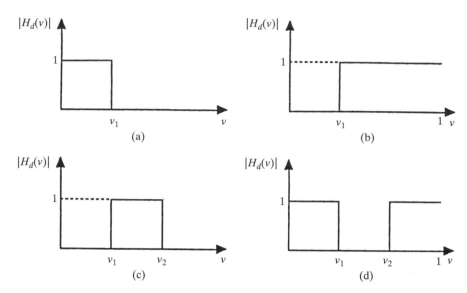

FIGURE 4.4. Desired transfer function: (*a*) lowpass, (*b*) highpass, (*c*) bandpass, and (*d*) bandstop.

The order of the filter is $N = 2Q + 1$. For example, if $Q = 5$, the filter will have 11 coefficients $h_0, h_1, \ldots, h_{10}$, or

$$h_0 = h_{10} = C_5$$
$$h_1 = h_9 = C_4$$
$$h_2 = h_8 = C_3$$
$$h_3 = h_7 = C_2$$
$$h_4 = h_6 = C_1$$
$$h_5 = C_0$$

Figure 4.4 shows the desired transfer functions $H_d(v)$ ideally represented for the frequency selective filters: lowpass, highpass, bandpass, and bandstop for which the coefficients $C_n = C_{-n}$ can be found.

1. *Lowpass:* $C_0 = v_1$

$$C_n = \int_0^{v_1} H_d(v) \cos n\pi v \, dv = \frac{\sin n\pi v_1}{n\pi} \tag{4.60}$$

2. *Highpass:* $C_0 - 1 - v_1$

$$C_n = \sum_{v_1}^{1} H_d(v)\cos n\pi v \, dv = \frac{\sin n\pi v_1}{n\pi} \qquad (4.61)$$

3. *Bandpass:* $C_0 = v_2 - v_1$

$$C_n = \int_{v_1}^{v_2} H_d(v)\cos n\pi v \, dv = \frac{\sin n\pi v_2 - \sin n\pi v_1}{n\pi} \qquad (4.62)$$

4. *Bandstop:* $C_0 = 1 - (v_2 - v_1)$

$$C_n = \int_{0}^{v_1} H_d(v)\cos n\pi v \, dv + \int_{v_2}^{1} H_d(v)\cos n\pi v \, dv = \frac{\sin n\pi v_1 - \sin n\pi v_2}{n\pi} \qquad (4.63)$$

where v_1 and v_2 are the normalized cutoff frequencies shown in Figure 4.4.

Several filter design packages are currently available for the design of FIR filters, as discussed later. When we implement an FIR filter, we develop a generic program such that the specific coefficients will determine the filter type (e.g., whether lowpass or bandpass).

Exercise 4.4: Lowpass FIR Filter

We will find the impulse response coefficients of an FIR filter with $N = 11$, a sampling frequency of 10 kHz, and a cutoff frequency $f_c = 1$ kHz. From (4.60),

$$C_0 = v_1 = \frac{f_c}{F_N} = 0.2$$

where $F_N = F_s/2$ is the Nyquist frequency and

$$C_n = \frac{\sin 0.2n\pi}{n\pi} \qquad n = \pm 1, \pm 2, \ldots, \pm 5 \qquad (4.64)$$

Since the impulse response coefficients $h_i = C_{Q-i}$, $C_n = C_{-n}$, and $Q = 5$, the impulse response coefficients are

$$
\begin{array}{ll}
h_0 = h_{10} = 0 & h_3 = h_7 = 0.1514 \\
h_1 = h_9 = 0.0468 & h_4 = h_6 = 0.1872 \\
h_2 = h_8 = 0.1009 & h_5 = 0.2
\end{array}
\qquad (4.65)
$$

4.6.3 Blackman Window

The Blackman window function is

$$w_B(n) = \begin{cases} 0.42 + 0.5\cos(n\pi/Q) + 0.08\cos(2n\pi/Q) & |n| \le Q \\ 0 & \text{otherwise} \end{cases} \tag{4.73}$$

which has the highest sidelobe level down to approximately −58 dB from the peak of the mainlobe. While the Blackman window produces the largest reduction in the sidelobe compared with the previous window functions, it has the widest mainlobe. As with the previous windows, the width of the mainlobe can be decreased by increasing the width of the window.

4.6.4 Kaiser Window

The design of FIR filters with the Kaiser window has become very popular in recent years. It has a variable parameter to control the size of the sidelobe with respect to the mainlobe. The Kaiser window function is

$$w_K(n) = \begin{cases} I_0(b)/I_0(a) & |n| \le Q \\ 0 & \text{otherwise} \end{cases} \tag{4.74}$$

where a is an empirically determined variable, and $b = a[1 - (n/Q)^2]^{1/2}$. $I_0(x)$ is the modified Bessel function of the first kind defined by

$$I_0(x) = 1 + \frac{0.25x^2}{(1!)^2} + \frac{(0.25x^2)^2}{(2!)^2} + \cdots = 1 + \sum_{n=1}^{\infty} \left[\frac{(x/2)^n}{n!} \right]^2 \tag{4.75}$$

which converges rapidly. A trade-off between the size of the sidelobe and the width of the mainlobe can be achieved by changing the length of the window and the parameter a.

4.6.5 Computer-Aided Approximation

An efficient technique is the computer-aided iterative design based on the Remez exchange algorithm, which produces equiripple approximation of FIR filters [5, 6]. The order of the filter and the edges of both passbands and stopbands are fixed, and the coefficients are varied to provide this equiripple approximation. This minimizes the ripple in both the passbands and the stopbands. The transition regions are left unconstrained and are considered "don't care" regions, where the solution may fail. Several commercial filter design packages include the Parks–McClellan algorithm for the design of an FIR filter.

4.7 PROGRAMMING EXAMPLES USING C AND ASM CODE

The following examples illustrate the implementation of FIR filters. Most of the programs are written in C. A few examples, using a combination of C and assembly language, illustrate the use of a circular buffer as a more efficient way to update delay samples, with the circular buffer in internal or external memory. Several different methods of displaying the magnitude frequency response of a filter are presented.

Example 4.1: Moving Average Filter (average)

The moving average filter is widely used in DSP and arguably is the easiest of all digital filters to understand. It is particularly effective at removing (high frequency) random noise from a signal or at *smoothing* a signal.

The moving average filter operates by taking the arithmetic mean of a number of past input samples in order to produce each output sample. This may be represented by the equation

$$y(n) = \frac{1}{N} \sum_{i=0}^{N-1} x(n-i) \qquad (4.76)$$

where $x(n)$ represents the nth sample of an input signal and $y(n)$ the nth sample of the filter output. The moving average filter is an example of *convolution* using a very simple *filter kernel* or *impulse response* comprising N coefficients each of value $1/N$. Equation (4.76) may be thought of as a particularly simple case of the more general convolution sum implemented by a finite impulse response filter, and introduced in Section 4.3; that is,

$$y(n) = \sum_{i=0}^{N-1} h(i)x(n-i) \qquad (4.77)$$

where the FIR filter coefficients $h(i)$ are samples of the filter impulse response and in the case of the moving average filter each is equal to $1/N$. As far as implementation is concerned, at the nth sampling instant we could either:

1. multiply N past input samples individually by $1/N$ and sum the N products,
2. sum N past input samples and multiply the sum by $1/N$, or
3. maintain a moving average by adding a new input sample (multiplied by $1/N$) to and subtracting the $(n - N + 1)$th input sample (multiplied by $1/N$) from a running total.

The third method of implementation is recursive, that is, calculation of the output $y(n)$ makes use of a previous output value $y(n-1)$. The recursive expression

$$y(n) = \frac{1}{N}x(n) - \frac{1}{N}x(n-N) + y(n-1) \tag{4.78}$$

conforms to the general expression for a recursive or infinite impulse response (IIR) filter:

$$y(n) = \sum_{k=0}^{M} b_k x(n-k) - \sum_{l=1}^{N} a_l y(n-l) \tag{4.79}$$

Program `average.c`, listed in Figure 4.5, uses the first of these options, even though it is not the most computationally efficient. The value of N defined near the start of the source file determines the number of previous input samples to be averaged.

Source file `average.c` is stored in folder `average`, which also contains project file `average.pjt`. Build the project as **average** and run the progam.

Several different methods exist by which the characteristics of the five point moving average filter may be demonstrated. A test file `mefsin.wav`, stored in folder `average`, contains a recording of speech corrupted by the addition of a sinusoidal tone. Listen to this file using *Goldwave, Windows Media Player*, or similar. Then connect the PC soundcard output to the LINE IN socket on the DSK and listen to the filtered test signal (LINE OUT or HEADPHONE). You should find that the sinusoidal tone has been blocked and that the voice sounds muffled. Both observations are consistent with the filter having a lowpass frequency response.

A more rigorous method of assessing the magnitude frequency response of the filter is to use a signal generator and an oscilloscope or spectrum analyzer to measure its gain at different individual frequencies. By using this method, it is straightforward to identify the distinct notches in the magnitude frequency response at 1600 Hz (corresponding to the tone in test file `mefsin.wav`) and at 3200 Hz.

The theoretical frequency response of the filter can be found by taking the discrete time Fourier transform (DTFT) of its coefficients:

$$H(\hat{\omega}) = \frac{1}{2\pi} \sum_{n=0}^{N-1} h[n] e^{-j\hat{\omega}n} \tag{4.80}$$

Evaluated over the frequency range $0 \le \hat{\omega} < 2\pi$, where $\hat{\omega} = \omega T_s$ and T_s is the sampling period.

```
//average.c

#include "DSK6713_AIC23.h"              //codec support
Uint32 fs=DSK6713_AIC23_FREQ_8KHZ;      //set sampling rate
#define DSK6713_AIC23_INPUT_MIC 0x0015
#define DSK6713_AIC23_INPUT_LINE 0x0011
Uint16 inputsource=DSK6713_AIC23_INPUT_LINE; //select input

#define N 5                              //no of points averaged
float x[N];                              //filter input delay line
float h[N];                              //filter coefficients

interrupt void c_int11()                 //interrupt service routine
{
  short i;
  float yn = 0.0;

  x[0]=(float)(input_left_sample()); //get new input sample
  for (i=0 ; i<N ; i++)                  //calculate filter output
    yn += h[i]*x[i];
  for (i=(N-1) ; i>0 ; i--)              //shift delay line contents
    x[i] = x[i-1];
  output_left_sample((short)(yn));       //output to codec
  return;
}

void main()
{
  short i;                               //index variable

  for (i=0 ; i<N ; i++)                  //initialise coefficients
    h[i] = 1.0/N;
  comm_intr();                           //initialise DSK
  while(1);                              //infinite loop
}
```

FIGURE 4.5. Five point moving average filter program (`average.c`).

In this case,

$$
\begin{aligned}
|H(\hat{\omega})| &= \left| \sum_{n=0}^{4} 0.2 e^{-j\hat{\omega}n} \right| \\
&= \left| \sum_{n=-2}^{2} 0.2 e^{-j\hat{\omega}n} \right| \\
&= \left| 0.2 \left(e^{j2\hat{\omega}} + e^{j\hat{\omega}} + 1 + e^{-j\hat{\omega}} + e^{-j2\hat{\omega}} \right) \right| \\
&= \left| 0.2 (1 + 2\cos(\hat{\omega}) + 2\cos(2\hat{\omega})) \right|
\end{aligned}
\tag{4.81}
$$

Changing the summation limits from $0 \le n \le 4$ to $-2 \le n \le 2$ changes the phase but not the magnitude of the frequency response of the filter. The theoretical magnitude frequency response of the filter is illustrated in Figure 4.6.

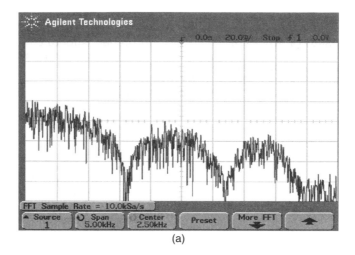

(a)

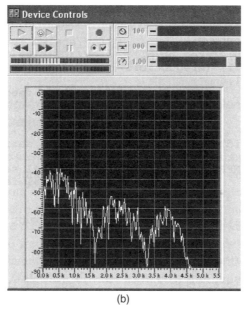

(b)

FIGURE 4.8. Magnitude frequency response of five point moving average filter illustrated using program `averagen.c` and displayed (a) using FFT function on oscilloscope and (b) using *GoldWave*.

points A and B in Figure 4.9, including the codec DAC between point A and the LINE OUT socket and the codec ADC between the LINE IN socket and point B. In broad terms, it identifies the system connected between LINE OUT and LINE IN sockets. After program `sysid.c` has run for a few seconds, halt the program and select *View → Graph*. The *Graph Property* settings required are shown in Figure 4.10. You should see something similar to that shown in Figure 4.11. Figure 4.12 shows the data illustrated in Figure 4.11 (exported from Code Composer as a text

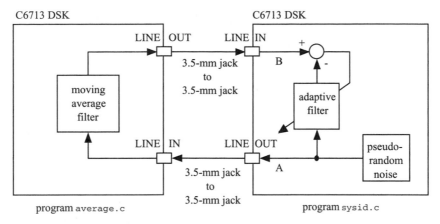

C6713 DSK   C6713 DSK

FIGURE 4.9. Connection diagram for use of program `sysid.c` to identify characteristics of the moving average filter.

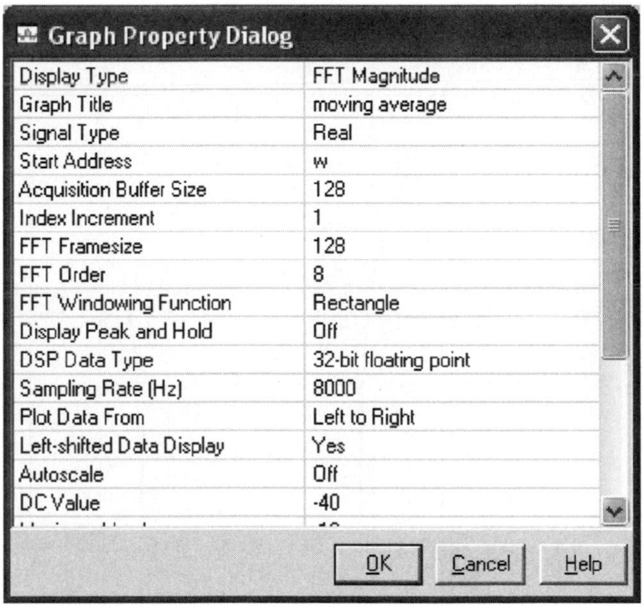

FIGURE 4.10. *Graph Property* settings for use with program `sysid.c` to identify characteristics of the moving average filter.

file and imported to MATLAB) plotted on the same axes as the theoretical magnitude frequency response of the five point moving average filter. Program `sysid.c` gives a reasonably accurate indication of the magnitude frequency response of the filter. The discrepancy between theoretical and identified responses at frequencies greater than 3.5 kHz is due to the characteristics of the antialiasing and reconstruction filters in the AIC23 codec.

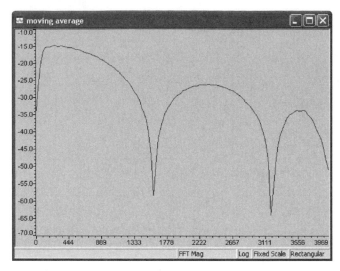

FIGURE 4.11. Magnitude frequency response of five point moving average filter identified by program `sysid.c`.

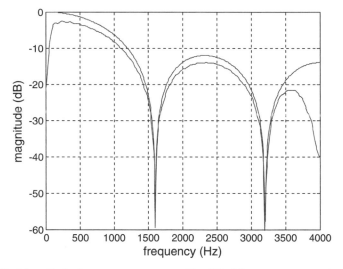

FIGURE 4.12. Magnitude frequency response identified by program `sysid.c` plotted on same axes as theoretical magnitude frequency response (dotted). Experimentally identified response has been multiplied by 4 to take into account resistor networks on codec inputs.

Altering the Coefficients of the Moving Average Filter

The frequency response of the moving average filter can be changed by altering the number of previous input samples that are averaged. Modify program `averagen.c` so that it implements an eleven point moving average filter; that is, change the line that reads

```
#define N 5
```

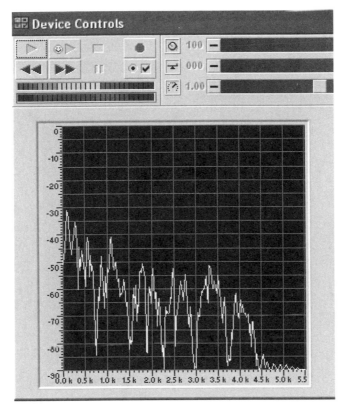

FIGURE 4.13. Magnitude frequency response of eleven point moving average filter implemented using program `averagen.c` and displayed using *GoldWave*.

to read

```
#define N 11
```

Build and run the project and verify that the frequency response of the filter has changed to that shown in Figure 4.13.

The frequency response of the eleven point moving average filter has the same basic form as that of the five point moving average filter but the notches in the frequency response occur at integer multiples of (8000/11) Hz, that is, at 727, 1455, 2182, and 2909 Hz.

The frequency response of the filter can also be changed by altering the values of the coefficients. Modify program `averagen.c` again, changing the lines that read

```
#define N 11
float h[N];
```

to read

```
#define N 5
float h[N] = {0.0833, 0.2500, 0.3333. 0.2500, 0.0833};
```

```
//firprn.c FIR with internally generated input noise sequence

#include "DSK6713_AIC23.h"              //codec support
Uint32 fs=DSK6713_AIC23_FREQ_8KHZ;     //set sampling rate
#define DSK6713_AIC23_INPUT_MIC 0x0015
#define DSK6713_AIC23_INPUT_LINE 0x0011
Uint16 inputsource=DSK6713_AIC23_INPUT_LINE; //select line in

#include "bs2700f.cof"                  //filter coefficient file
#include "noise_gen.h"                  //support file for noise
int fb;                                 //feedback variable
shift_reg sreg;                         //shift register
#define NOISELEVEL 8000                 //scale factor for noise
float x[N];                             //filter delay line

int prand(void)                         //pseudo-random noise
{
  int prnseq;
  if(sreg.bt.b0)
    prnseq = -NOISELEVEL;               //scaled -ve noise level
  else
    prnseq = NOISELEVEL;                //scaled +ve noise level
  fb =(sreg.bt.b0)^(sreg.bt.b1);        //XOR bits 0,1
  fb^=(sreg.bt.b11)^(sreg.bt.b13);      //with bits 11,13 -> fb
  sreg.regval<<=1;                      //shift register 1 bit left
  sreg.bt.b0=fb;                        //close feedback path
  return prnseq;
}

void resetreg(void)                     //reset shift register
{
  sreg.regval=0xFFFF;                   //initial seed value
  fb = 1;                              //initial feedback value
}

interrupt void c_int11()                //interrupt service routine
{
  short i;                              //declare index variable
  float yn = 0.0;

  x[0] = (float)(prand());              //get new input sample
  for (i=0 ; i<N ; i++)                 //calculate filter output
    yn += h[i]*x[i];
  for (i=(N-1) ; i>0 ; i--)             //shift delay line contents
    x[i] = x[i-1];
  output_left_sample((short)(yn));      //output to codec
  return;                              //return from interrupt
}

void main()
{
  resetreg();                          //reset shift register
  comm_intr();                         //initialise DSK
  while (1);                           //infinite loop
}
```

FIGURE 4.20. FIR filter with internally generated pseudorandom noise as input (firprn.c).

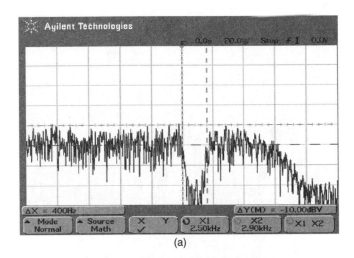

(a)

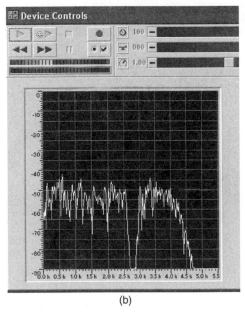

(b)

FIGURE 4.21. Output generated using program `firprn.c` and coefficient file `bs2700f.cof` displayed using (a) an oscilloscope and (b) *Goldwave*.

1. `bp55f.cof`: bandpass with center frequency $F_s/4$
2. `bs55f.cof`: bandstop with center frequency $F_s/4$
3. `lp55f.cof`: lowpass with cutoff frequency $F_s/4$
4. `hp55f.cof`: highpass with bandwidth $F_s/4$
5. `pass2bf.cof`: with two passbands
6. `pass3bf.cof`: with three passbands
7. `pass4bf.cof`: with four passbands
8. `comb14f.cof`: with multiple notches (comb filter)

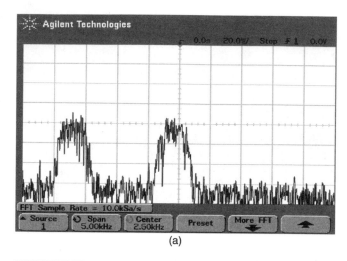

(a)

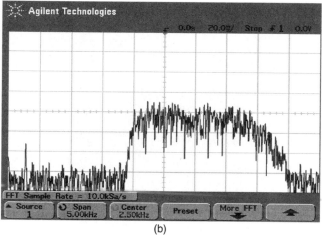

(b)

FIGURE 4.22. Output generated using program firprn.c and coefficient files (a) pass2bf.cof and (b) hp55f.cof diplayed using an oscilloscope.

These filters were designed using MATLAB (see Appendix D). Figure 4.22a shows the FFT of the output of an FIR filter with two passbands, using the coefficient file pass2bf.cof. Figure 4.22b shows the FFT of the output of a highpass FIR filter using the coefficient file hp55f.cof. These plots were obtained using the FFT function of an *Agilent 54621A* oscilloscope.

Example 4.6: FIR Filter with Internally Generated Pseudorandom Noise as Input to a Filter and Output Stored in Memory (*firprnbuf*)

This example builds on the previous one that generates a pseudorandom noise sequence as the input to an FIR filter, with the filter output also stored in a memory buffer. Figure 4.23 shows a listing of the program firprnbuf.c, which implements this example.

```
//firprnbuf.c

#include "DSK6713_AIC23.h"                //codec support
Uint32 fs=DSK6713_AIC23_FREQ_8KHZ;       //set sampling rate
#define DSK6713_AIC23_INPUT_MIC 0x0015
#define DSK6713_AIC23_INPUT_LINE 0x0011
Uint16 inputsource=DSK6713_AIC23_INPUT_LINE; //select line in

#include "pass4bf.cof"               //filter coefficient file
#include "noise_gen.h"               //support file for noise
int fb;                             //feedback variable
shift_reg sreg;                     //shift register
#define NOISELEVEL 8000             //scale factor for  noise
float x[N];                         //filter delay line
#define YNBUFLENGTH 1024
float yn_buffer[YNBUFLENGTH];
short ynbufindex = 0;

int prand(void)                     //pseudo-random noise
{
  int prnseq;
  if(sreg.bt.b0)
    prnseq = -NOISELEVEL;           //scaled -ve noise level
  else
    prnseq = NOISELEVEL;            //scaled +ve noise level
  fb =(sreg.bt.b0)^(sreg.bt.b1);    //XOR bits 0,1
  fb^=(sreg.bt.b11)^(sreg.bt.b13);  //with bits 11,13 -> fb
  sreg.regval<<=1;                  //shift register 1 bit left
  sreg.bt.b0=fb;                    //close feedback path
  return prnseq;
}

void resetreg(void)                 //reset shift register
{
  sreg.regval=0xFFFF;               //initial seed value
  fb = 1;                           //initial feedback value
}

interrupt void c_int11()            //interrupt service routine
{
  short i;                          //declare index variable
  float yn = 0.0;

  x[0] = (float)(prand());          //get new input sample
  for (i=0 ; i<N ; i++)             //calculate filter output
    yn += h[i]*x[i];
  for (i=(N-1) ; i>0 ; i--)         //shift delay line contents
    x[i] = x[i-1];
  output_left_sample((short)(yn));  //output to codec
  yn_buffer[ynbufindex++] = yn;
  if(ynbufindex >= YNBUFLENGTH) ynbufindex = 0;
  return;                           //return from interrupt
}

void main()
{
  resetreg();                       //reset shift register
  comm_intr();                      //initialise DSK
  while (1);                        //infinite loop
}
```

FIGURE 4.23. FIR filter with internally generated pseudorandom noise as input and output stored in memory (firprnbuf.c).

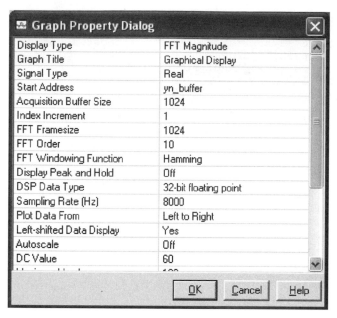

FIGURE 4.24. *Graph Properties* for use with program firprnbuf.c.

The coefficient file bp41f.cof represents a 41-coefficient FIR bandpass filter centered at 1000 Hz.

Build and run this project as **firprnbuf**. Verify that the output signal is band-limited noise. Then halt the program, select *View → Graph*, and set the *Graph Properties* as shown in Figure 4.24 in order to look at the frequency content of 1024 stored output samples.

Figure 4.25 shows several Code Composer windows, including plots of the filter coefficients in the time and frequency domains and the magnitude FFT of the buffer contents.

Edit the program (firprnbuf.c), changing the lines that read

```
output_left_sample((short)(yn)); //output to codec
yn_buffer[ynbufindex++] = yn;
```

to read

```
output_left_sample((short)(x[0])); //output to codec
yn_buffer[ynbufindex++] = x[0];
```

This effectively disables the FIR filter, passes the pseudorandom binary input signal directly to the DAC, and stores it in the circular buffer implemented using array yn_buffer.

Run the program again and plot the FFT magnitude of the noise sequence; that is, use the same *Graphical Display* window as before. It does not appear perfectly flat since the resulting plot is not averaged. An example is shown in Figure 4.26. With the output to an oscilloscope with FFT function, verify that the noise spectrum

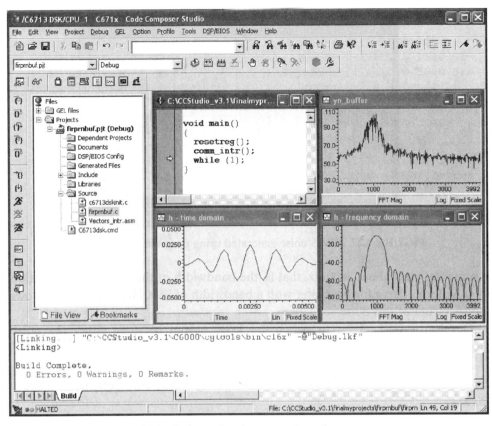

FIGURE 4.25. CCS windows showing operation of program `firprnbuf.c`.

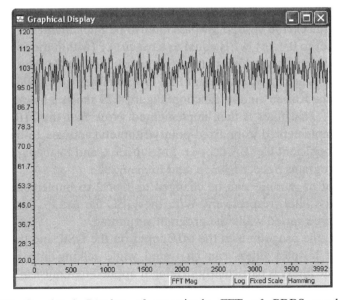

FIGURE 4.26. *Graphical Display* of magnitude FFT of PRBS used in program `firprnbuf.c`.

```
//LP600.cof FIR lowpass filter coefficients using Kaiser window

#define N 81                    //length of filter

short hlp600[N] = {0,-6,-14,-22,-26,-24,-13,8,34,61,80,83,63,
19,-43,-113,-171,-201,-185,-117,0,146,292,398,428,355,174,-99,
-416,-712,-905,-921,-700,-218,511,1424,2425,3391,4196,4729,
4915,4729,4196,3391,2425,1424,511,-218,-700,-921,-905,-712,
-416,-99,174,355,428,398,292,146,0,-117,-185,-201,-171,-113,
-43,19,63,83,80,61,34,8,-13,-24,-26,-22,-14,-6,0};
```

FIGURE 4.29. Coefficient file for FIR lowpass filter with 600-Hz cutoff frequency (LP600.cof).

to read

```
Uint16 inputsource=DSK6713_AIC23_INPUT_LINE; // select LINE IN
```

and *Rebuild* the project. Use a CD or MP3 player as a source connected to the LINE IN socket on the DSK. With the lower bandwidth of 600 Hz, using the first set of coefficients, the frequency components of the input signal above 600 Hz are suppressed. Connect the output to a speaker or a spectrum analyzer to verify such results, and listen to the effect of the different bandwidths of the three FIR lowpass filters. Alternatively, the effects of the filters can be illustrated using an oscilloscope and a signal generator set to input a 200-Hz square wave to the LINE IN socket. Figure 4.30 shows a 200-Hz square wave that has been passed through the three lowpass filters.

Example 4.8: Implementation of Four Different Filters: Lowpass, Highpass, Bandpass, and Bandstop (fir4types)

This example illustrates the use of a GEL slider to step through four different types of FIR filters (Figure 4.31). Each filter has 81 coefficients, designed using MATLAB. The four coefficient files (on the accompanying CD) are:

1. lp1500.cof: lowpass with bandwidth of 1500 Hz
2. hp2200.cof: highpass with bandwidth of 2200 Hz
3. bp1750.cof: bandpass with center frequency at 1750 Hz
4. bs790.cof: bandstop with center frequency at 790 Hz

Program fir4types.c implements this project. Build and run this project as **fir4types**. Load the GEL file fir4types.gel and select *GEL → Filter Characteristics → Filter* to bring up a GEL slider to switch between the four different FIR filters. This example could readily be expanded to implement more FIR

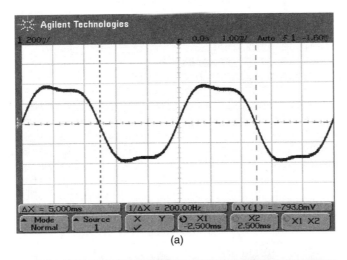

(a)

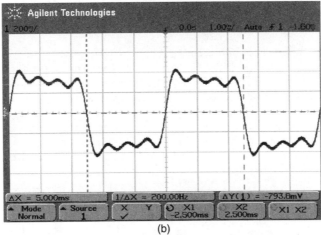

(b)

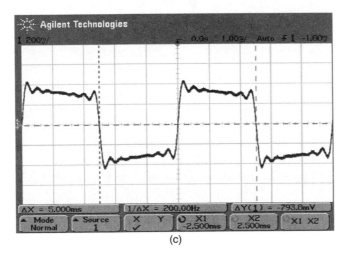

(c)

FIGURE 4.30. A 200-Hz square wave passed through three different lowpass filters implemented using program fir3lp.c.

```
//fir4types.c Lowpass, Highpass, bandpass, Bandstop FIR filters

#include "DSK6713_AIC23.h"
Uint32 fs=DSK6713_AIC23_FREQ_8KHZ;
#define DSK6713_AIC23_INPUT_MIC 0x0015
#define DSK6713_AIC23_INPUT_LINE 0x0011
Uint16 inputsource=DSK6713_AIC23_INPUT_LINE; //select line in
#include "lp1500.cof"               //coeff file LP @ 1500 Hz
#include "hp2200.cof"               //coeff file HP @ 2200 Hz
#include "bp1750.cof"               //coeff file BP @ 1750 Hz
#include "bs790.cof"                //coeff file BS @  790 Hz
short FIR_number = 0;               //start with 1st LP filter
int yn = 0;                         //initialize filter output
short dly[N];                       //delay samples
short h[4][N];                      //filter characteristics

interrupt void c_int11()            //ISR
{
  short i;

  dly[0] = input_left_sample();     //new input @ top of buffer
  yn = 0;                           //initialize filter output
  for (i = 0; i< N; i++)
    yn +=(h[FIR_number][i]*dly[i]); //y(n) += h(LP#,i)*x(n-i)
  for (i = N-1; i > 0; i--)         //start @ bottom of buffer
    dly[i] = dly[i-1];              //update delays
  output_left_sample(yn >> 15);     //output filter
  return;                           //return from interrupt
}

void main()
{
  short i;
  for (i=0; i<N; i++)
  {
    dly[i] = 0;                     //init buffer
    h[0][i] = hlp[i];               //start of lp1500 coeffs
    h[1][i] = hhp[i];               //start of hp2200 coeffs
    h[2][i] = hbp[i];               //start of bp1750 coeffs
    h[3][i] = hbs[i];               //start of bs790 coeffs
  }
  comm_intr();                      //init DSK, codec, McBSP
  while(1);                         //infinite loop
}
```

FIGURE 4.31. FIR program to implement four different types of filter (fir4types.c).

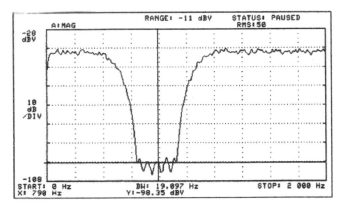

FIGURE 4.32. Magnitude frequency response of bandstop filter implemented using program `fir4types.c`.

filters. The effects of the four different filters on musical input are particularly striking. Figure 4.32 shows the magnitude frequency response of the FIR bandstop filter centered at 790 Hz, implemented using the coefficient file `bs790.cof`.

Example 4.9: Two Notch Filters to Recover a Corrupted Speech Recording (notch2)

This example illustrates the use of two notch (bandstop) FIR filters in series to recover a speech recording corrupted by the addition of two sinusoidal signals at frequencies of 900 and 2700 Hz. Program `notch2.c` is listed in Figure 4.33. Two coefficient files, `bs900.cof` and `bs2700.cof`, each containing 89 coefficients and designed using MATLAB, are used by the program. They implement two FIR notch filters, centered at 900 and 2700 Hz, respectively. The output of the first notch filter, centered at 900 Hz, is used as the input to the second notch filter, centered at 2700 Hz.

Build this project as **notch2**. The file `corrupt.wav`, stored in folder `notch2`, contains a recording of speech corrupted by the addition of 900- and 2700-Hz sinusoidal tones. Listen to this file using *Goldwave, Windows Media Player*, or similar. Then connect the PC soundcard output to the LINE IN socket on the DSK and listen to the filtered test signal (LINE OUT or HEADPHONE).

A GEL slider (`notch2.gel`) can be used to select either the output of the two cascaded notch filters (default) or the output of the first notch filter.

Compare the results of this example with those obtained in Example 4.1, in which a notch in the magnitude frequency response of a moving average filter was exploited in order to filter out an unwanted sinusoidal tone. In this case, the filtered speech sounds brighter because the notch filters do not have a lowpass characteristic.

```
//fir2ways.c FIR with alternative ways of storing/updating samples

#include "DSK6713_AIC23.h"              //codec support
Uint32 fs=DSK6713_AIC23_FREQ_8KHZ;     //set sampling rate
#define DSK6713_AIC23_INPUT_MIC 0x0015
#define DSK6713_AIC23_INPUT_LINE 0x0011
Uint16 inputsource=DSK6713_AIC23_INPUT_LINE; //select input
#include "BP41.cof"                    //BP coeff centered at Fs/8
#define METHOD 'A'                     //or change to B
int yn = 0;                            //initialize filter's output
short dly[N];                          //delay samples array
#if METHOD == 'B'
short *start_ptr;
short *end_ptr;
short *h_ptr;
short *dly_ptr;
#endif
interrupt void c_int11()               //ISR
{
  short i;

  yn = 0;                              //initialize filter's output

#if METHOD == 'A'
  dly[0] = input_left_sample();
  for (i = 0; i< N; i++) yn += (h[i] * dly[i]);
  for (i = N-1; i > 0; i--) dly[i] = dly[i-1];
#elif METHOD == 'B'
  *dly_ptr = input_left_sample();
  if (++dly_ptr > end_ptr) dly_ptr = start_ptr;
  for (i = 0; i < N ; i++)
  {
    dly_ptr++;
    if (dly_ptr > end_ptr) dly_ptr = start_ptr;
    yn += *(h_ptr + i)* *dly_ptr;
  }
#endif
  output_left_sample((short)(yn>>15));
  return;
}

void main()
{
#if METHOD == 'B'
  dly_ptr = dly;
  start_ptr = dly;
  end_ptr = dly + N - 1;
  h_ptr = h;
#endif
  comm_intr();
  while(1);
}
```

FIGURE 4.34. FIR program using two alternative methods for convolution and updating of delay samples (fir2ways.c).

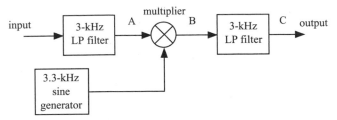

FIGURE 4.35. Block diagram of scrambler system.

with suppressed carrier. At point C the upper sideband and the section of the lower sideband between 3 and 3.3 kHz are filtered out. The scheme is attractive because of its simplicity. Only simple DSP algorithms—namely, filtering, sine wave generation, and amplitude modulation—are required for its implementation.

Figure 4.36 shows a listing of program `scrambler.c`, which operates at a sampling rate, `fs`, of 16 kHz. The input signal is first lowpass filtered using an FIR filter with 65 coefficients, `h`, defined in the file `lp3k64.cof`. The filtering algorithm used is identical to that used in, for example, program `fir.c`. The filter delay line is implemented using array `x1` and the output is assigned to variable `yn1`. The filter output (at point A in Figure 4.36) is multiplied (modulated) by a 3.3-kHz sinusoid stored as 160 samples (exactly 33 cycles) in array `sine160` (file `sine160.h`). Finally, the modulated signal (at point B) is lowpass filtered again, using the same set of filter coefficients `h` (`lp3k64.cof`) but a different filter delay line implemented using array `x2` and the output variable `yn2`. The output is a scrambled signal (at point C). Using this scrambled signal as the input to a second DSK running the same algorithm, the original descrambled input is recovered as the output of the second DSK.

Build and run this project as **scrambler**. First, test the program using a 2-kHz sine wave as input. The resulting output is a lower sideband signal at 1.3 kHz. The upper sideband signal at 5.3 kHz is filtered out by the second lowpass filter. By varying the frequency of the sinusoidal input, you should be able to verify that input frequencies in the range 300–3000 Hz appear as output frequencies in the inverted range 3000 to 300 Hz.

A second DSK running the same program can be used to recover the original signal (simulating the receiving end). Use the output of the first DSK as the input to the second DSK.

Change the input source used by the program from LINE IN to MIC IN and test the scrambler and descrambler using speech from a microphone as the input. Run exactly the same program on each DSK, that is, including the line

```
Uint16 inputsource=DSK6713_AIC23_INPUT_MIC
```

and connect LINE OUT on the first DSK (scrambler) to MIC IN on the second DSK (descrambler).

//**scrambler.c**

```
#include "DSK6713_AIC23.h"              // codec support
Uint32 fs=DSK6713_AIC23_FREQ_16KHZ;   //set sampling rate
#define DSK6713_AIC23_INPUT_MIC 0x0015
#define DSK6713_AIC23_INPUT_LINE 0x0011
Uint16 inputsource=DSK6713_AIC23_INPUT_LINE; //select line in
#include "sine160.h"
#include "lp3k64.cof"                   //filter coefficient file
float yn1, yn2;                         //filter outputs
float x1[N],x2[N];                      //filter delay lines
int index = 0;

interrupt void c_int11()
{
 short i;
                                        // first filter input
 x1[0]=(float)(input_left_sample()); //get input into delay line
 yn1 = 0.0;                             //initialise filter output
 for (i=0 ; i<N ; i++) yn1 += h[i]*x1[i];
 for (i=(N-1) ; i>0 ; i--) x1[i] = x1[i-1];
                                        // next mix with 3300Hz
 yn1 *= sine160[index++];
 if (index >= NSINE) index = 0;
                                        // now filter again
 x2[0] = yn1;                           //get input into delay line
 yn2 = 0.0;                             //initialise filter output
 for (i=0 ; i<N ; i++) yn2 += h[i]*x2[i];
 for (i=(N-1) ; i>0 ; i--) x2[i] = x2[i-1];
 output_left_sample((short)(yn2));   //output to codec
 return;
}

void main()
{
 comm_intr();                           //initialise McBSP, AD535
 while(1);                              //infinite loop
}
```

FIGURE 4.36. Scrambler program scrambler.c.

Interception of the speech signal could be made more difficult by changing the modulation frequency dynamically and by including (or omitting) the carrier frequency according to a predefined sequence: for example, a code for no modulation, another for modulating at frequency fc1, and a third code for modulating at frequency fc2.

This project was first implemented using the TMS320C25 [50] and also the TMS320C31 DSK.

Example 4.12: FIR Implementation Using C Calling an ASM Function (`FIRcasm`)

The C program `FIRcasm.c` (Figure 4.37) calls the assembly language function `_fircasmfunc` defined in file `FIRcasmfunc.asm` (Figure 4.38), and which implements an FIR filter.

Build and run this project as **FIRcasm**. Verify that the program implements a 1-kHz FIR bandpass filter. Two buffers are used by program `FIRcasm.c`. Array `dly` is used to store N previous input samples and array `h` stores N filter coefficients. The value of constant N is defined in the filter coefficient (`.cof`) file. On each interrupt, a new input sample is acquired and stored at the end (higher memory address) of the buffer `dly`. The delay samples and the filter coefficients are arranged in memory as shown in Table 4.1. The delay samples are stored in memory starting with the oldest sample stored at the lowest memory address. The newest sample is at the end of the buffer. The coefficients are arranged in memory with $h(0)$ at the beginning of the coefficient buffer and $h(N-1)$ at the end.

```
//FIRcasm.c FIR C program calling ASM function fircasmfunc.asm

#include "DSK6713_AIC23.h"      //codec-DSK support file
Uint32 fs=DSK6713_AIC23_FREQ_8KHZ;   //set sampling rate
#include "bp41.cof"             //BP @ Fs/8 coefficient file
#define DSK6713_AIC23_INPUT_MIC 0x0015
#define DSK6713_AIC23_INPUT_LINE 0x0011
Uint16 inputsource=DSK6713_AIC23_INPUT_LINE; // select mic in
int yn = 0;                             //initialize filter's output
short dly[N];                           //delay samples

interrupt void c_int11()                //ISR
{
    dly[N-1] = input_left_sample(); //newest sample @bottom buffer
    yn = fircasmfunc(dly,h,N); //to ASM func through A4,B4,A6
    output_left_sample((short)(yn>>15));   //filter's output
    return;                         //return from ISR
}

void main()
{
    short i;

    for (i = 0; i<N; i++)
    dly[i] = 0;                 //init buffer for delays
    comm_intr();                //init DSK, codec, McBSP
    while(1);                   //infinite loop
}
```

FIGURE 4.37. C program calling an ASM function for FIR implementation (`FIRcasm.c`).

buffer. This value is then moved up the buffer to a lower address. Observe after a while that the samples are being updated, with each value in the buffer moving up in memory. You can also observe the register (pointer) A4 incrementing by 2 (two bytes) and B4 decrementing by 2.

Example 4.13: FIR Implementation Using C Calling a Faster ASM Function (FIRcasmfast)

The same C calling program, FIRcasm.c, is used in this example as in Example 4.12. It calls the ASM function _fircasmfunc within the file FIRcasmfuncfast.asm, as shown in Figure 4.39. This ASM function executes faster than the function in the previous example by having parallel instructions and rearranging the sequence of instructions. There are two parallel instructions: LDH/LDH and SUB/LDH.

1. The number of NOPs is reduced from 19 to 11.
2. The SUB instruction to decrement the loop count is moved up the program.
3. The sequence of some instructions is changed to fill some of the NOP slots. For example, the conditional branch instruction executes after the ADD instruction

```
;FIRcasmfuncfast.asm C-called faster function to implement FIR
            .def  _fircasmfunc
_fircasmfunc:                    ;ASM function called from C
            MV    A6,A1          ;setup loop count
            MPY   A6,2,A6        ;since dly buffer data as byte
            ZERO  A8             ;init A8 for accumulation
            ADD   A6,B4,B4       ;since coeff buffer data as byte
            SUB   B4,1,B4        ;B4=bottom coeff array h[N-1]
loop:                           ;start of FIR loop
            LDH   *A4++,A2       ;A2=x[n-(N-1)+i] i=0,1,...,N-1
    ||      LDH   *B4--,B2       ;B2=h[N-1-i] i=0,1,...,N-1
            SUB   A1,1,A1        ;decrement loop count
    ||      LDH   *A4,A7         ;A7=x[(n-(N-1)+i+1]update delays
            NOP   4
            STH   A7,*-A4[1]     ;-->x[(n-(N-1)+i] update sample
    [A1]    B     loop           ;branch to loop if count # 0
            NOP   2
            MPY   A2,B2,A6       ;A6=x[n-(N-1)+i]*h[N-1-i]
            NOP
            ADD   A6,A8,A8       ;accumlate in A8

            B     B3             ;return addr to calling routine
            MV    A8,A4          ;result returned in A4
            NOP   4
```

FIGURE 4.39. FIR ASM function with parallel instructions for faster execution (FIRcasmfuncfast.asm).

to accumulate in A8, since branching has five delay slots. Additional changes to make it faster would also make it less comprehensible due to further resequencing of the instructions.

Build this project as **FIRcasmfast**, so that the linker option names the output executable file FIRcasmfast.out. The resulting output is the same 1-kHz bandpass filter as in the previous example.

Example 4.14: FIR Implementation Using C Calling an ASM Function with a Circular Buffer (FIRcirc)

The C program FIRcirc.c (Figure 4.40) calls the ASM function FIRcircfunc.asm (Figure 4.41). This example expands Example 4.12 to implement an FIR filter using a circular buffer. The coefficients within the file bp1750.cof were designed with MATLAB using a Kaiser window and represent a 128-coefficient FIR bandpass filter with a center frequency of 1750 Hz. Figure 4.42 shows the characteristics of this filter, obtained using MATLAB's filter designer fdatool (described in Appendix D).

```
//FIRcirc.c   C program calling ASM function using circular buffer

#include "DSK6713_AIC23.h"     //codec-DSK support file
Uint32 fs=DSK6713_AIC23_FREQ_8KHZ;  //set sampling rate
#define DSK6713_AIC23_INPUT_MIC 0x0015
#define DSK6713_AIC23_INPUT_LINE 0x0011
Uint16 inputsource=DSK6713_AIC23_INPUT_LINE; // select input source
#include "bp1750.cof"               //BP at 1750 Hz coeff file
int yn = 0;                         //init filter's output

interrupt void c_int11()            //ISR
{
  short sample_data;

  sample_data = (input_sample());      //newest input sample data
  yn = fircircfunc(sample_data,h,N); //ASM func passing to A4,B4,A6
  output_sample((short)(yn>>15));      //filter's output
  return;                            //return to calling function
}

void main()
{
  comm_intr();                      //init DSK, codec, McBSP
  while(1);                         //infinite loop
}
```

FIGURE 4.40. C program calling an ASM function using a circular buffer (FIRcirc.c).

```
;FIRcircfunc.asm ASM function called from C using circular addressing
;A4=newest sample, B4=coefficient address, A6=filter order
;Delay samples organized: x[n-(N-1)]...x[n]; coeff as h(0)...h[N-1]

                .def    _fircircfunc
                .def    last_addr
                .def    delays
                .sect   "circdata"      ;circular data section
                .align  256             ;align delay buffer 256-byte boundary
delays          .space  256             ;init 256-byte buffer with 0's
last_addr       .int    last_addr-1     ;point to bottom of delays buffer
                .text                   ;code section
_fircircfunc:                           ;FIR function using circ addr
                MV      A6,A1           ;setup loop count
                MPY     A6,2,A6         ;since dly buffer data as byte
                ZERO    A8              ;init A8 for accumulation

                ADD     A6,B4,B4        ;since coeff buffer data as bytes
                SUB     B4,1,B4         ;B4=bottom coeff array h[N-1]

                MVKL 0x00070040,B6      ;select A7 as pointer and BK0
                MVKH 0x00070040,B6      ;BK0 for 256 bytes (128 shorts)

                MVC     B6,AMR          ;set address mode register AMR

                MVKL    last_addr,A9    ;A9=last circ addr(lower 16 bits)
                MVKH last_addr,A9       ;last circ addr (higher 16 bits)

                LDW     *A9,A7          ;A7=last circ addr
                NOP     4
                STH     A4,*A7++        ;newest sample-->last address

loop:                                   ;begin FIR loop
                LDH     *A7++,A2        ;A2=x[n-(N-1)+i] i=0,1,...,N-1
     ||         LDH     *B4--,B2        ;B2=h[N-1-i] i=0,1,...,N-1
                SUB     A1,1,A1         ;decrement count
     [A1]       B       loop            ;branch to loop if count # 0
                NOP     2
                MPY     A2,B2,A6        ;A6=x[n-(N-1)+i]*h[N-1+i]
                NOP
                ADD     A6,A8,A8        ;accumulate in A8

                STW     A7,*A9          ;store last circ addr to last_addr
                B       B3              ;return addr to calling routine
                MV      A8,A4           ;result returned in A4
                NOP     4
```

FIGURE 4.41. FIR ASM function using a circular buffer for updating samples (FIRcircfunc.asm).

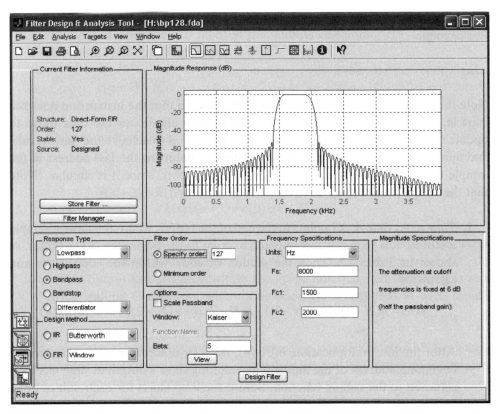

FIGURE 4.42. Frequency characteristics of a 128-coefficient FIR bandpass filter centered at 1750 Hz, designed using MATLAB filter design and analysis tool `fdatool`.

In lieu of moving the data to update the delay samples, a pointer is used. The 16 LSBs of the address mode register are set with a value of

```
0x0040 = 0000 0000 0100 0000
```

This selects A7 mode as the circular buffer pointer register. The 16 MSBs of AMR are set with N = 0x0007 to select the block BK0 as a circular buffer. The buffer size is $2^{N+1} = 256$. A circular buffer is used in this example only for the delay samples.

It is also possible to use a second circular buffer for the coefficients. For example, using

```
0x0140 = 0000 0001 0100 0000
```

would select two pointers, B4 and A7.

Example 4.15: FIR Implementation Using C Calling an ASM Function Using a Circular Buffer in External Memory (FIRcirc_ext)

This example implements an FIR filter using a circular buffer in external memory. The same C source program FIRcirc.c and ASM function FIRcircfunc.asm as in the previous example are used, but with a modified linker command file. This linker command file FIRcirc_ext.cmd is listed in Figure 4.43. The section circdata designates the memory section buffer_ext, which starts in external memory at 0x80000000.

Build this project as **FIRcirc_ext**. Load the executable file and view the memory at the address delays. This should display the external memory section that starts at 0x80000000. Verify that the circular buffer is in external memory, where all the delay samples are initialized to zero. Place a breakpoint as in Example 4.14, run the program up to the breakpoint, and verify that the newest input sample is stored at the end of the circular buffer at 0x800000FE and 0x800000FF. Register A9 contains the last address, and register A7 contains the address where the subsequent 16-bit input sample is to be stored (0x80000001). Run the program again (to the set

```
/*FIRcirc_ext.cmd Linker command file for external memory*/

MEMORY
{
   IVECS:       org =          0h,  len =       0x220
   IRAM:        org = 0x00000220,  len = 0x0002FFFF
   SRAM_EXT1:   org = 0x80000000,  len = 0x00000110
   SRAM_EXT2:   org = 0x80000110,  len = 0x00100000
   FLASH:       org = 0x90000000,  len = 0x00020000
}

SECTIONS
{
   circdata  :> SRAM_EXT1 /*buffer in external mem*/
   .vecs     :> IVECS     /*Created in vectors file*/
   .text     :> IRAM      /*Created by C Compiler*/
   .bss      :> IRAM
   .cinit    :> IRAM
   .stack    :> IRAM
   .sysmem   :> IRAM
   .const    :> IRAM
   .switch   :> IRAM
   .far      :> IRAM
   .cio      :> IRAM
   .csldata  :> IRAM
}
```

FIGURE 4.43. Linker command file for a circular buffer in external memory (FIRcirc_ext.cmd).

breakpoint) and verify that the subsequent acquired sample is stored at the beginning of the buffer at the address 0x80000001. Remove the breakpoint, restart/run, and verify that the output is the same FIR bandpass filter centered at 1750 Hz, as in Example 4.14.

4.8 ASSIGNMENTS

1. **(a)** Design a 65-coefficient FIR lowpass filter, using a Hamming window, with a cut-off frequency of 2500 Hz and a sampling frequency of 8 kHz. Implement it in real time using program firprnbuf.c.

   (b) Compare the characteristics of the filter designed using a Hamming window with those of filters designed using Hann and Kaiser windows.

2. The coefficient file LP1500_256.cof (stored in folder fir) contains the 256 coefficients of an FIR lowpass filter, with a bandwidth of 1500 Hz when sampling at 48 kHz. Implement this filter in real time. The C-coded examples in this chapter may not be efficient enough to implement this filter at a sampling rate of 48 kHz. Consider using an ASM-coded FIR function with a circular buffer.

3. Design and implement an FIR filter with two passbands, one centered at 2500 and the other at 3500 Hz. Use a sampling frequency of 16 kHz.

4. Rather than using an internal noise generator coded in C as input to a C-coded FIR function (see program firprn.c), generate the noise using ASM code (see program noisegen_casm.asm in Chapter 3).

REFERENCES

1. W. J. Gomes III and R. Chassaing, Filter design and implementation using the TMS320C6x interfaced with MATLAB, *Proceedings of the 2000 ASEE Annual Conference*, 2000.

2. A. V. Oppenheim and R. Schafer, *Discrete-Time Signal Processing*, Prentice Hall, Upper Saddle River, NJ, 1989.

3. B. Gold and C. M. Rader, *Digital Signal Processing of Signals*, McGraw-Hill, New York, 1969.

4. L. R. Rabiner and B. Gold, *Theory and Application of Digital Signal Processing*, Prentice Hall, Upper Saddle River, NJ, 1975.

5. T. W. Parks and J. H. McClellan, Chebychev approximation for nonrecursive digital filter with linear phase, *IEEE Transactions on Circuit Theory*, Vol. CT-19, pp. 189–194, 1972.

6. J. H. McClellan and T. W. Parks, A unified approach to the design of optimum linear phase digital filters, *IEEE Transactions on Circuit Theory*, Vol. CT-20, pp. 697–701, 1973.

7. J. F. Kaiser, Nonrecursive digital filter design using the I0-sinh window function, *Proceedings of the IEEE International Symposium on Circuits and Systems*, 1974.

8. J. F. Kaiser, Some practical considerations in the realization of linear digital filters, *Proceedings of the 3rd Allerton Conference on Circuit System Theory*, Oct. 1965, pp. 621–633.

9. L. B. Jackson, *Digital Filters and Signal Processing*, Kluwer Academic, Norwell, MA, 1996.

10. J. G. Proakis and D. G. Manolakis, *Digital Signal Processing: Principles, Algorithms, and Applications*, Prentice Hall, Upper Saddle River, NJ, 1996.

11. R. G. Lyons, *Understanding Digital Signal Processing*, Addison-Wesley, Reading, MA, 1997.

12. F. J. Harris, On the use of windows for harmonic analysis with the discrete Fourier transform, *Proceedings of the IEEE*, Vol. 66, pp. 51–83, 1978.

13. I. F. Progri, W. R. Michalson, and R. Chassaing, Fast and efficient filter design and implementation on the TMS320C6711 digital signal processor, *International Conference on Acoustics, Speech, and Signal Processing Student Forum*, May 2001.

14. B. Porat, *A Course in Digital Signal Processing*, Wiley, Hoboken, NJ, 1997.

15. T. W. Parks and C. S. Burrus, *Digital Filter Design*, Wiley, Hoboken, NJ, 1987.

16. S. D. Stearns and R. A. David, *Signal Processing In Fortran and C*, Prentice Hall, Upper Saddle River, NJ, 1993.

17. N. Ahmed and T. Natarajan, *Discrete-Time Signals and Systems*, Reston Publishing, Reston, VA, 1983.

18. S. J. Orfanidis, *Introduction to Signal Processing*, Prentice Hall, Upper Saddle River, NJ, 1996.

19. A. Antoniou, *Digital Filters: Analysis, Design, and Applications*, McGraw-Hill, New York, 1993.

20. E. C. Ifeachor and B. W. Jervis, *Digital Signal Processing: A Practical Approach*, Addison-Wesley, Reading, MA, 1993.

21. P. A. Lynn and W. Fuerst, *Introductory Digital Signal Processing with Computer Applications*, Wiley, Hoboken, NJ, 1994.

22. R. D. Strum and D. E. Kirk, *First Principles of Discrete Systems and Digital Signal Processing*, Addison-Wesley, Reading, MA, 1988.

23. D. J. DeFatta, J. G. Lucas, and W. S. Hodgkiss, *Digital Signal Processing: A System Approach*, Wiley, Hoboken, NJ, 1988.

24. C. S. Williams, *Designing Digital Filters*, Prentice Hall, Upper Saddle River, NJ, 1986.

25. R. W. Hamming, *Digital Filters*, Prentice Hall, Upper Saddle River, NJ, 1983.

26. S. K. Mitra and J. F. Kaiser, Eds., *Handbook for Digital Signal Processing*, Wiley, Hoboken, NJ, 1993.

27. S. K. Mitra, *Digital Signal Processing: A Computer-Based Approach*, McGraw-Hill, New York, 2001.

28. R. Chassaing, B. Bitler, and D. W. Horning, Real-time digital filters in C, *Proceedings of the 1991 ASEE Annual Conference*, June 1991.

29. R. Chassaing and P. Martin, Digital filtering with the floating-point TMS320C30 digital signal processor, *Proceedings of the 21st Annual Pittsburgh Conference on Modeling and Simulation*, May 1990.

30. S. D. Stearns and R. A. David, *Signal Processing In Fortran and C*, Prentice Hall, Upper Saddle River, NJ, 1993.

31. R. A. Roberts and C. T. Mullis, *Digital Signal Processing*, Addison-Wesley, Reading, MA, 1987.

32. E. P. Cunningham, *Digital Filtering: An Introduction*, Houghton Mifflin, Boston, 1992.

33. N. J. Loy, *An Engineer's Guide to FIR Digital Filters*, Prentice Hall, Upper Saddle River, NJ, 1988.

34. H. Nuttall, Some windows with very good sidelobe behavior, *IEEE Transactions on Acoustics, Speech, and Signal Processing*, Vol. ASSP-29, No. 1, Feb. 1981.

35. L. C. Ludemen, *Fundamentals of Digital Signal Processing*, Harper & Row, New York, 1986.

36. M. Bellanger, *Digital Processing of Signals: Theory and Practice*, Wiley, Hoboken, NJ, 1989.

37. M. G. Bellanger, *Digital Filters and Signal Analysis*, Prentice Hall, Upper Saddle River, NJ, 1986.

38. F. J. Taylor, *Principles of Signals and Systems*, McGraw-Hill, New York, 1994.

39. F. J. Taylor, *Digital Filter Design Handbook*, Marcel Dekker, New York, 1983.

40. W. D. Stanley, G. R. Dougherty, and R. Dougherty, *Digital Signal Processing*, Reston Publishing, Reston, VA, 1984.

41. R. Kuc, *Introduction to Digital Signal Processing*, McGraw-Hill, New York, 1988.

42. H. Baher, *Analog and Digital Signal Processing*, Wiley, Hoboken, NJ, 1990.

43. J. R. Johnson, *Introduction to Digital Signal Processing*, Prentice Hall, Upper Saddle River, NJ, 1989.

44. S. Haykin, *Modern Filters*, Macmillan, New York, 1989.

45. T. Young, *Linear Systems and Digital Signal Processing*, Prentice Hall, Upper Saddle River, NJ, 1985.

46. A. Ambardar, *Analog and Digital Signal Processing*, PWS, Boston, MA, 1995.

47. A. W. M. van den Enden and N. A. M. Verhoeckx, *Discrete-Time Signal Processing*, Prentice-Hall International, Hemel Hempstead, Hertfordshire, England, 1989.

48. A. H. Gray and J. D. Markel, Digital lattice and ladder filter synthesis, *IEEE Transactions on Acoustics, Speech, and Signal Processing*, Vol. ASSP-21, pp. 491–500, Dec. 1973.

49. A. H. Gray and J. D. Markel, A normalized digital filter structure, *IEEE Transactions on Acoustics, Speech, and Signal Processing*, Vol. ASSP-23, pp. 258–277, June 1975.

50. R. Chassaing and D. W. Horning, *Digital Signal Processing with the TMS320C25*, Wiley, Hoboken, NJ, 1990.

5.2.1 Direct Form I Structure

With the direct form I structure shown in Figure 5.1, the filter in (5.2) can be realized. For an Nth order filter, this structure has $2N$ delay elements, represented by z^{-1}. For example, a second order filter with $N = 2$ will have four delay elements.

5.2.2 Direct Form II Structure

The direct form II structure shown in Figure 5.2 is one of the most commonly used structures. It requires half as many delay elements as the direct form I. For example, a second order filter requires two delay elements z^{-1}, as opposed to four with the direct form I.

From the block diagram of Figure 5.2 it can be seen that

$$w(n) = x(n) - a_1 w(n-1) - a_2 w(n-2) - \cdots - a_N w(n-N) \tag{5.6}$$

and that

$$y(n) = b_0 w(n) + b_1 w(n-1) + b_2 w(n-2) + \cdots + b_N w(n-N) \tag{5.7}$$

Taking z-transforms of equations (5.6) and (5.7), we find

$$W(z) = X(z) - a_1 z^{-1} W(z) - a_2 z^{-2} W(z) - \cdots - a_N z^{-N} W(z) \tag{5.8}$$

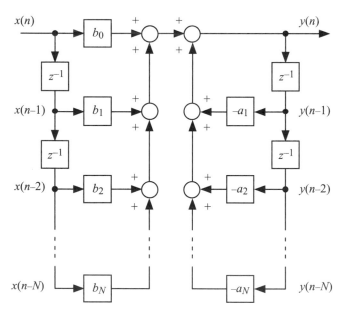

FIGURE 5.1. Direct form I IIR filter structure.

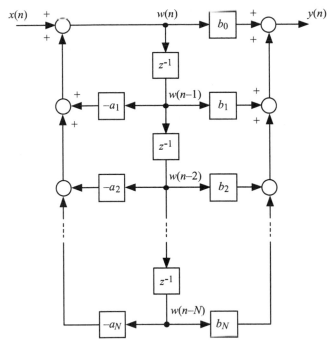

FIGURE 5.2. Direct form II IIR filter structure.

and hence

$$X(z) = (1 + a_1 z^{-1} + a_2 z^{-2} + \cdots + a_N z^{-N}) W(z) \tag{5.9}$$

and

$$Y(z) = (b_0 + b_1 z^{-1} + b_2 z^{-2} + \cdots + b_N z^{-N}) W(z) \tag{5.10}$$

Thus,

$$H(z) = \frac{Y(z)}{X(z)} = \frac{b_0 + b_1 z^{-1} + b_2 z^{-2} + \cdots + b_N z^{-N}}{1 + a_1 z^{-1} + a_2 z^{-2} + \cdots + a_N z^{-N}} \tag{5.11}$$

that is, the same as equation (5.4).

The direct form II structure can be represented by difference equations (5.6) and (5.7) taking the place of equation (5.2).

Equations (5.6) and (5.7) are used to program an IIR filter. Initially, $w(n - 1)$, $w(n - 2), \ldots$ are set to zero. At time n, a new sample $x(n)$ is acquired, and (5.6) is used to solve for $w(n)$; then the output $y(n)$ is calculated using (5.7).

5.2.3 Direct Form II Transpose

The direct form II transpose structure shown in Figure 5.3 is a modified version of the direct form II and requires the same number of delay elements.

From inspection of the block diagram, the filter output can be computed using

$$y(n) = b_0 x(n) + w_0(n-1) \tag{5.12}$$

Subsequently, the contents of the delay line can be updated using

$$w_0(n) = b_1 x(n) + w_1(n-1) - a_1 y(n) \tag{5.13}$$

$$w_1(n) = b_2 x(n) + w_2(n-1) - a_2 y(n) \tag{5.14}$$

and so on until finally

$$w_{N-1}(n) = b_N x(n) - a_N y(n) \tag{5.15}$$

Using equation (5.13) to find $w_0(n-1)$,

$$w_0(n-1) = b_1 x(n-1) + w_1(n-2) - a_1 y(n-1)$$

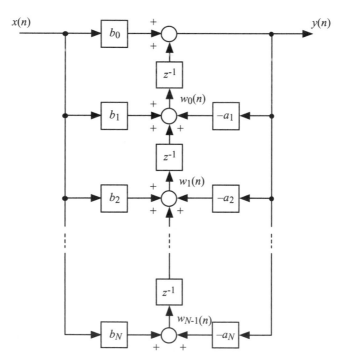

FIGURE 5.3. Direct form II transpose IIR filter structure.

equation (5.12) becomes

$$y(n) = b_0 x(n) + [b_1 x(n-1) + w_1(n-2) - a_1 y(n-1)]$$

Similarly, using equation (5.14) to find $w_1(n-2)$,

$$w_1(n-2) = b_2 x(n-2) + w_2(n-3) - a_2 y(n-2)$$

equation (5.12) becomes

$$y(n) = b_0 x(n) + [b_1 x(n-1) + [b_2 x(n-2) + w_2(n-3) - a_2 y(n-2)] - a_1 y(n-1)] \quad (5.16)$$

Continuing this procedure until equation (5.15) has been used, it can be shown that equation (5.12) is equivalent to equation (5.2) and hence that the block diagram of Figure 5.3 is equivalent to that of Figures 5.1 and 5.2. The transposed structure implements the zeros first and then the poles, whereas the direct form II structure implements the poles first.

5.2.4 Cascade Structure

The transfer function in (5.5) can be factorized as

$$H(z) = CH_1(z)H_2(z) \cdots H_r(z) \quad (5.17)$$

in terms of first or second order transfer functions. The cascade (or series) structure is shown in Figure 5.4. An overall transfer function can be represented with cascaded transfer functions. For each section, the direct form II structure or its transpose version can be used. Figure 5.5 shows a fourth order IIR structure in terms of two direct form II second order sections in cascade. The transfer function $H(z)$, in terms of cascaded second order transfer functions, can be written

$$H(z) = \prod_{i=1}^{N/2} \frac{b_{0i} + b_{1i}z^{-1} + b_{2i}z^{-2}}{1 + a_{1i}z^{-1} + a_{2i}z^{-2}} \quad (5.18)$$

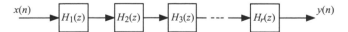

FIGURE 5.4. Cascade form IIR filter structure.

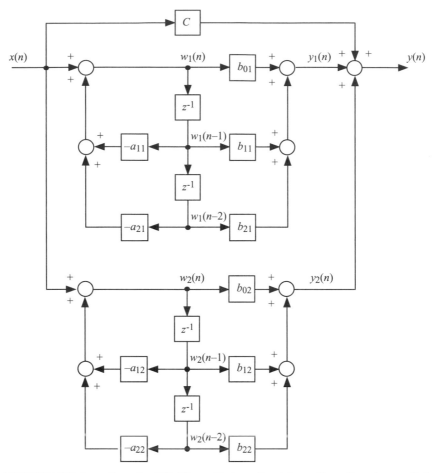

FIGURE 5.7. Fourth order IIR filter with two direct form II sections in parallel.

$$s = K\frac{(z-1)}{(z+1)} \tag{5.24}$$

The constant K in (5.24) is commonly chosen as $K = 2/T$, where T represents the sampling period, in seconds, of the digital filter. Other values for K can be selected, as described in Section 5.3.1.

This transformation allows the following:

1. The left region in the s-plane, corresponding to $\sigma < 0$, maps *inside* the unit circle in the z-plane.
2. The right region in the s-plane, corresponding to $\sigma > 0$, maps *outside* the unit circle in the z-plane.
3. The imaginary $j\omega$ axis in the s-plane maps *on* the unit circle in the z-plane.

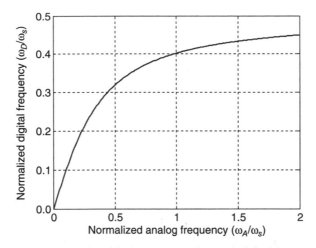

FIGURE 5.8. Relationship between analog and digital frequencies.

Let ω_A and ω_D represent analog and digital frequencies, respectively. With $s = j\omega_A$ and $z = e^{j\omega_D T}$, (5.24) becomes

$$j\omega_A = K\frac{(e^{j\omega_D T} - 1)}{(e^{j\omega_D T} + 1)} = K\frac{e^{j\omega_D T/2}(e^{j\omega_D T/2} - e^{-j\omega_D T/2})}{e^{j\omega_D T/2}(e^{j\omega_D T/2} + e^{-j\omega_D T/2})} \qquad (5.25)$$

Using Euler's expressions for sine and cosine in terms of complex exponential functions, ω_A from (5.25) becomes

$$\omega_A = K\tan\left(\frac{\omega_D T}{2}\right) \qquad (5.26)$$

which relates the analog frequency ω_A to the digital frequency ω_D. This relationship is plotted in Figure 5.8 for positive values of ω_A. The nonlinear compression of the entire analog frequency range into the digital frequency range from zero to $\omega_s/2$ is referred to as frequency warping ($\omega_s = 2\pi/T$).

5.3.1 BLT Design Procedure

The BLT design procedure for transforming an analog filter design expressed as a transfer function $H(s)$ into a z-transfer function $H(z)$ representing a discrete-time IIR filter is described by

$$H(z) = H(s)\big|_{s=2(z-1)/T(z+1)} \qquad (5.27)$$

$H(s)$ can be chosen according to well-documented analog filter design theory (e.g., Butterworth, Chebyshev, Bessel, or elliptic).

It is common to choose $K = 2/T$. Alternatively, it is possible to prewarp the analog filter frequency response in such a way that the bilinear transform maps an analog frequency $\omega_A = \omega_c$, in the range 0 to $\omega_s/2$, to exactly the same digital frequency $\omega_D = \omega_c$. This is achieved by choosing

$$K = \frac{\omega_c}{\tan(\pi\omega_c/\omega_s)} \tag{5.28}$$

5.4 PROGRAMMING EXAMPLES USING C AND ASM CODE

The examples in this section introduce and illustrate the implementation of infinite impulse response (IIR) filtering. Many different approaches to IIR filter design are possible and most often IIR filters are designed with the aid of software tools. Before using such a design package, and in order to appreciate better what such design packages do, a simple example will be used to illustrate some of the basic principles of IIR filter design.

Design of a Simple IIR Lowpass Filter

Traditionally, IIR filter design is based on the concept of transforming a continuous-time, or analog, design into the discrete-time domain. Butterworth, Chebyshev, Bessel, and elliptical classes of analog filter are widely used. For our example we will choose a *second order, type 1 Chebyshev, lowpass filter with 2 dB of passband ripple and a cutoff frequency of 1500 Hz* (9425 rad/s).

The continuous-time transfer function of such a filter is

$$H(s) = \frac{58072962}{s^2 + 7576s + 73109527} \tag{5.29}$$

and its frequency response is shown in Figure 5.9.

This transfer function can be generated by typing

```
>>  [b,a]  =  cheby1(2,2,2*pi*1500,'s');
```

at the MATLAB command line.

Our task is to transform this design into the discrete-time domain. One method of achieving this is the *impulse invariance* method.

Impulse Invariance Method

This method is based on the concept of mapping each s-plane pole of the continuous-time filter to a corresponding z-plane pole using the substitution $(1 - e^{-p_k t_s} z^{-1})$ for $(s + p_k)$ in $H(s)$. This can be achieved by several different means. Partial fraction expansion of $H(s)$ and substitution of $(1 - e^{-p_k t_s} z^{-1})$ for $(s + p_k)$ can involve a lot of

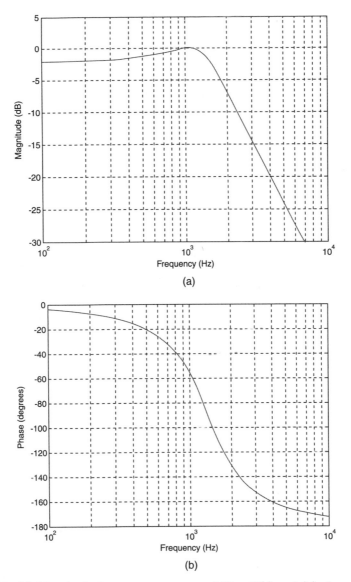

FIGURE 5.9. (a) Magnitude frequency response of filter $H(s)$ and (b) phase response of filter $H(s)$.

algebra. An equivalent method of making the transformation is to use tables of Laplace and z-transforms as follows.

In our example, starting with the filter transfer function (5.29), we can make use of the Laplace transform pair

$$L\{Ae^{-\alpha t}\sin(\omega t)\} = \frac{A\omega}{s^2 + 2\alpha s + (\alpha^2 + \omega^2)} \qquad (5.30)$$

(the filter's transfer function is equal to the Laplace transform of its impulse response) and use the values

$$\alpha = 7576/2 = 3787.9$$

$$\omega = \sqrt{73109527 - 3787.9^2} = 7665.6$$

$$A = 58072962/7665.6 = 7575.8$$

Hence, the impulse response of the filter in this example is given by

$$h(t) = 7575.8 e^{-3788t} \sin(7665.6t) \tag{5.31}$$

The z-transform pair

$$Z\{Ae^{-\alpha t} \sin(\omega t)\} = \frac{Ae^{-\alpha t_s} \sin(\omega t_s) z^{-1}}{1 - 2(e^{-\alpha t_s} \cos(\omega t_s)) z^{-1} + e^{-2\alpha t_s} z^{-2}} \tag{5.32}$$

yields the following discrete-time transfer function when we substitute for ω, A, and α and set $t_s = 0.000125$ in equation (5.32):

$$H(z) = \frac{Y(z)}{X(z)} = \frac{0.48255 z^{-1}}{1 - 0.71624315 z^{-1} + 0.38791310 z^{-2}} \tag{5.33}$$

From $H(z)$, the following difference equation may be derived:

$$y(n) = 0.48255 x(n-1) + 0.71624315 y(n-1) - 0.38791310 y(n-2) \tag{5.34}$$

In terms of equation (5.1), we can see that $a_1 = 0.71624315$, $a_2 = -0.38791310$, $b_0 = 0.0000$, and $b_1 = 0.48255$.

In order to apply the impulse invariant method using MATLAB, type

```
>> [bz,az] = impinvar(b,a,8000);
```

This discrete-time filter has the property that its discrete-time impulse response, $h(n)$, is exactly equal to samples of the continuous-time impulse response, $h(t)$, as shown in Figure 5.10.

Although it is evident from Figure 5.10 that the discrete-time impulse response $h(n)$ decays almost to zero, this sequence is not finite. Whereas the impulse response of an FIR filter is given explicitly by its finite set of coefficients, the coefficients of an IIR filter are used in a recursive equation (5.1) to determine its impulse response $h(n)$.

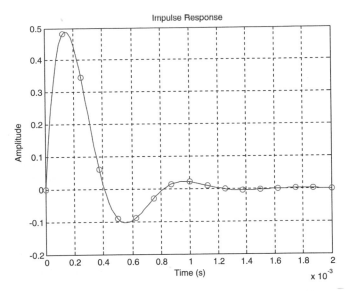

FIGURE 5.10. Impulse responses $h(t)$ and $h(n)$ of continuous-time filter and its impulse invariant digital implementation.

Example 5.1: Implementation of IIR Filter Using Cascaded Second Order Direct Form II Sections (`iirsos`)

Program `iirsos.c`, stored in folder `iirsos` and listed in Figure 5.11, implements a generic IIR filter using cascaded direct form II second order stages (sections) and coefficient values stored in a separate file. The program uses the following two expressions:

$$w(n) = x(n) - a_1 w(n-1) - a_2 w(n-2)$$
$$y(n) = b_0 w(n) + b_1 w(n-1) + b_2 y(n-2)$$

implemented by the lines

```
wn = input - a[section][0]*w[section][0] - [section][1]
     *w[section][1];
yn = b[section][0]*wn + b[section][1]*w[section][0] +
     a[section][2]*w[section][1];
```

With reference to Figure 5.5 and to equation (5.18), the coefficients a_{1i}, a_{2i}, b_{0i}, b_{1i}, and b_{2i} are stored as

```
a[i][0], a[i][1], b[i][0], b[i][1] and b[i][2] respectively.
w[i][0] and w[i][1] correspond to w_i(n - 1) and w_i(n - 2).
```

The impulse invariant filter is implemented using program `iirsos.c` by including the coefficient file `impinv.cof`, listed in Figure 5.12. The number of cascaded second order sections is defined as NUM_SECTIONS in that file.

```
//iirsos.c iir filter using cascaded second order sections

#include "DSK6713_AIC23.h"              //codec support
Uint32 fs=DSK6713_AIC23_FREQ_8KHZ;     //set sampling rate

#define DSK6713_AIC23_INPUT_MIC 0x0015
#define DSK6713_AIC23_INPUT_LINE 0x0011
Uint16 inputsource=DSK6713_AIC23_INPUT_LINE;
#include "impinv.cof"

float w[NUM_SECTIONS][2] = {0};

interrupt void c_int11()                //interrupt service routine
{
  int section;                          //index for section number
  float input;                          //input to each section
  float wn,yn;                          //intermediate and output
                                        //values in each stage

  input = ((float)input_left_sample());

  for (section=0 ; section< NUM_SECTIONS ; section++)
  {
    wn = input - a[section][0]*w[section][0]
         - a[section][1]*w[section][1];
    yn = b[section][0]*wn + b[section][1]*w[section][0]
         + b[section][2]*w[section][1];
    w[section][1] = w[section][0];
    w[section][0] = wn;
    input = yn;                         //output of current section
                                        //will be input to next
  }
  output_left_sample((short)(yn));      //before writing to codec
  return;                               //return from ISR
}

void main()
{
  comm_intr();                          //init DSK, codec, McBSP
  while(1);                             //infinite loop
}
```

FIGURE 5.11. IIR filter program using second order stages in cascade (iirsos.c).

Build and run the project as **iirsos**.

You can use a signal generator and oscilloscope to measure the magnitude frequency response of the filter and you will find that the attenuation of frequencies above 2500 Hz is not very pronounced. That is due to the low order of the filter and to inherent shortcomings of the impulse invariant transformation method. A number

```
// impinv.cof
// second order type 1 Chebyshev LPF with 2dB passband ripple
// and cutoff frequency 1500Hz

#define NUM_SECTIONS 1

float b[NUM_SECTIONS][3]={ {0.0, 0.48255, 0.0} };
float a[NUM_SECTIONS][2]={ {-0.71624, 0.387913} };
```

FIGURE 5.12. Listing of coefficient file `impinv.cof`.

of alternative methods of assessing the magnitude frequency response of the filter will be described in the next few examples.

Example 5.2: Implementation of IIR Filter Using Cascaded Second Order Transposed Direct Form II Sections (`iirsostr`)

A transposed direct form II structure can be implemented using program `iirsos.c` by replacing the lines that read

```
 wn = input - a[section][0]*w[section][0] -
a[section][1]*w[section][1];
 yn = b[section][0]*wn + b[section][1]*w[section][0] +
b[section][2]*w[section][1];
 w[section][1] = w[section][0];
 w[section][0] = wn;
```

with the following:

```
 yn = b[section][0]*input + w[section][0]; w[section][0] =
b[section][1]*input + w[section][1] - a[section][0]*yn;
w[section][1] = b[section][2]*input - a[section][1]*yn;
```

(variable `wn` is not required in the latter case). This substitution has been made in program `iirsostr.c`, stored in folder `iirsos`. You should not notice any difference in the filter characteristics implemented using program `iirsostr.c`.

Example 5.3: Estimating the Frequency Response of an IIR Filter Using Pseudorandom Noise as Input (`iirsosprn`)

Program `iirsosprn.c` is closely related to program `firprn.c`, described in Chapter 4. In real time, it generates a pseudorandom binary sequence and uses this wideband noise signal as the input to an IIR filter (Figure 5.13). The output of the filter is written to the DAC in the AIC23 codec and the resulting analog signal (filtered noise) can be analyzed using an oscilloscope, spectrum analyzer, *Goldwave* (Figure

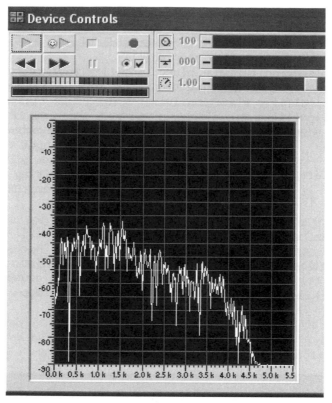

FIGURE 5.15. Output from program `iirsosprn.c`, using coefficient file `impinv.cof`, viewed using *Goldwave*.

of the figure between 0 and 2.5 kHz. The steeper roll-off beyond 3.5 kHz, in the right-hand third of the figure, is due to the reconstruction filter in the AIC23 codec.

Example 5.4: Estimating the Frequency Response of an IIR Filter Using a Sequence of Impulses as Input (`iirsosdelta`)

Instead of a pseudorandom binary sequence, program `iirsosdelta.c` generates a sequence of discrete-time impulses as the input to an IIR filter. The resultant output is an approximation to a repetitive sequence of filter impulse responses. This relies on the filter impulse response decaying practically to zero within the period between successive input impulses. The filter output is written to the DAC in the AIC23 codec and the resulting analog signal can be analyzed using an oscilloscope, spectrum analyzer, *Goldwave*, or other instrument. In addition, program `iirsosdelta.c` stores BUFSIZE samples of the filter output, $y(n)$, in buffer `response` and we can use the *View→Graph* facility in Code Composer to view that data in both time and frequency domains (Figure 5.16).

```
// iirsosdelta.c iir filter using cascaded second order sections
// input internally generated delta sequence, output to line out
// and save in buffer
// float coefficients read from included .cof file

#include "DSK6713_AIC23.h"            //codec support
Uint32 fs=DSK6713_AIC23_FREQ_8KHZ;   //set sampling rate

#define DSK6713_AIC23_INPUT_MIC 0x0015
#define DSK6713_AIC23_INPUT_LINE 0x0011
Uint16 inputsource=DSK6713_AIC23_INPUT_LINE;
#define BUFSIZE 256
#define AMPLITUDE 20000
#include "impinv.cof"

float w[NUM_SECTIONS][2] = {0};

float dimpulse[BUFSIZE];
float response[BUFSIZE];
int index = 0;

float w[NUM_SECTIONS][2] = {0};

interrupt void c_int11()               //interrupt service routine
{
  int section;                         //index for section number
  float input;                         //input to each section
  float wn,yn;                         //intermediate and output
                                       //values in each stage

  input = dimpulse[index];             //input to first section is
                                       //read from impulse sequence

  for (section=0 ; section< NUM_SECTIONS ; section++)
  {
    wn = input - a[section][0]*w[section][0]
         - a[section][1]*w[section][1];
    yn = b[section][0]*wn + b[section][1]*w[section][0]
         + b[section][2]*w[section][1];
    w[section][1] = w[section][0];
    w[section][0] = wn;
    input = yn;                        //output of current section
                                       //will be input to next
  }
  output_left_sample((short)(yn));     //before writing to codec
  return;                              //return from ISR
}

void main()
{
  int i;

  for (i=0 ; i< BUFSIZE ; i++) dimpulse[i] = 0.0;
  dimpulse[0] = 1.0;
  comm_intr();                         //init DSK, codec, McBSP
  while(1);                            //infinite loop
}
```

FIGURE 5.16. IIR filter program using second order stages in cascade and internally generated impulses as input (iirsosdelta.c).

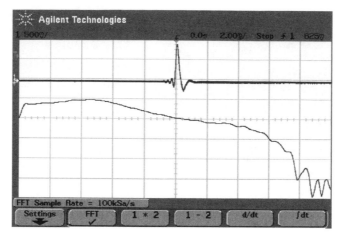

FIGURE 5.17. Output from program `iirsosdelta.c`, using coefficient file `impinv.cof`, viewed using the FFT function of an *Agilent 54621A* oscilloscope.

Build and run the project as **iirsosdelta**. The necessary files are stored in folder `iirsosdelta`.

Figure 5.17 shows the analog output signal generated by the program, captured using an *Agilent 54621A* oscilloscope. The upper trace shows the time domain impulse response of the filter (2 ms per division) and the lower trace shows the FFT of that impulse response over a frequency range of 0–5 kHz. The output waveform is shaped both by the IIR filter and by the AIC23 codec reconstruction filter. In the frequency domain, the codec reconstruction filter is responsible for the steep roll-off of gain at frequencies above 3500 Hz and the ac coupling of the codec output is responsible for the steep roll-off of gain at frequencies below 100 Hz. In the time domain, the characteristics of the codec reconstruction filter are evident in the ringing that precedes the greater part of the impulse response waveform.

Halt the program and select *View→Graph*. Set the Graph Properties as indicated in Figure 5.18 and you should see something similar to the right-hand graph shown in Figure 5.19.

Aliasing in the Impulse Invariant Method

There are significant differences between the magnitude frequency response of the analog prototype filter used in this example (Figure 5.9) and that of its impulse invariant digital implementation (Figure 5.19). The gain of the analog prototype has a magnitude of −15 dB at 3000 Hz, whereas, according to Figure 5.19, the gain of the digital filter at that frequency has a magnitude closer to −11 dB. This difference is due to *aliasing*. Whenever a signal is sampled, the problem of aliasing should be addressed and in order to avoid aliasing, the signal to be sampled should not contain any frequency components at frequencies greater than or equal to half the sampling frequency. In this case, **the impulse invariant transformation is equivalent to sam-**

Graph Property Dialog

Display Type	FFT Magnitude
Graph Title	response
Signal Type	Real
Start Address	response
Acquisition Buffer Size	256
Index Increment	1
FFT Framesize	256
FFT Order	8
FFT Windowing Function	Rectangle
Display Peak and Hold	Off
DSP Data Type	32-bit floating point
Sampling Rate (Hz)	8000
Plot Data From	Left to Right
Left-shifted Data Display	Yes
Autoscale	On
DC Value	0
Axes Display	On
Frequency Display Unit	Hz

[OK] [Cancel] [Help]

FIGURE 5.18. *Graph Property* settings for use with program `iirsosdelta.c`.

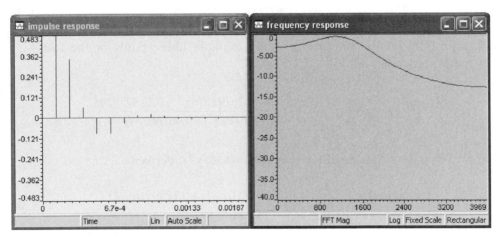

FIGURE 5.19. Impulse and magnitude frequency response of example filter captured using Code Composer and program `iirsosdelta.c`.

pling the continuous-time impulse response of the analog prototype $h(t)$**. However, this transformation does not in itself consider the frequency content of** $h(t)$**. The** impulse invariant method will be completely free of aliasing effects *only* if the impulse response $h(t)$ contains no frequency components at frequencies greater than or equal to half the sampling frequency.

In our example, the magnitude frequency response of the analog prototype filter will be folded back on itself about the 4000-Hz point.

This can be verified using MATLAB function `freqz()`, which assesses the frequency response of a digital filter. Type

```
>> freqz(bz,az);
```

An alternative method of transforming an analog filter design to a discrete-time implementation that eliminates this effect is the use of the *bilinear transform*.

Bilinear Transform Method of Digital Filter Implementation

The bilinear transform method of converting an analog filter design to discrete time is relatively straightforward, often involving less algebraic manipulation than the impulse invariant method. It is achieved by making the substitution

$$s = \frac{2(z-1)}{T(z+1)} \tag{5.35}$$

in $H(s)$, where T is the sampling period of the digital filter; that is,

$$H(z) = H(s)|_{s=2(z-1)/T(z+1)} \tag{5.36}$$

Applying the bilinear transform to the example filter results in the following z-transfer function:

$$H(z) = \frac{Y(z)}{X(z)} = \frac{0.12895869 + 0.25791738z^{-1} + 0.12895869z^{-2}}{1 - 0.81226498z^{-1} + 0.46166249z^{-2}} \tag{5.37}$$

From $H(z)$, the following difference equation may be derived:

$$\begin{aligned} y(n) = &\, 0.1290x(n) + 0.2579x(n-1) + 0.1290x(n-2) \\ &+ 0.8123y(n-1) - 0.4617y(n-2) \end{aligned} \tag{5.38}$$

This can be achieved in MATLAB by typing

```
>> [bd,ad]  = bilinear(b,a,8000);
```

The characteristics of the filter can be examined by changing the coefficient file used by programs `iirsos.c`, `iirsosprn.c`, and `iirsosdelta.c` from `impinv.cof` to `bilinear.cof`. In each case, change the line that reads

```
#include "impinv.cof"
```

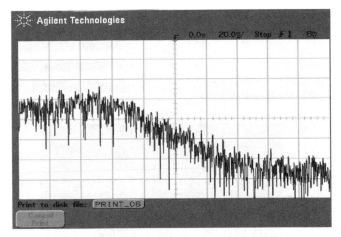

FIGURE 5.20. Output from program `iirsosprn.c`, using coefficient file `bilinear.cof`, viewed using the FFT function of an *Agilent 54621A* oscilloscope.

to read

```
#include "bilinear.cof"
```

before building, loading, and running the programs.

Figures 5.20–5.23 show results obtained using programs `iirsosprn.c` and `iirsosdelta.c` with coefficient file `bilinear.cof`. The attenuation provided by this filter at high frequencies is much greater than in the impulse invariant case. In fact, the attenuation at frequencies higher than 2 kHz is significantly greater than that of the analog prototype filter.

Frequency Warping in the Bilinear Transform

The concept behind the bilinear transform is that of compressing the frequency response of an analog filter design such that its response over the entire range of frequencies from zero to infinity is mapped into the frequency range zero to half the sampling frequency of the digital filter. This may be represented by

$$f_D = \frac{\arctan(\pi f_A T_S)}{\pi T_S} \quad \text{or} \quad \omega_D = \frac{2}{T_S}\arctan\left(\frac{\omega_A T_S}{2}\right) \tag{5.39a}$$

and

$$f_A = \frac{\tan(\pi f_D T_S)}{\pi T_S} \quad \text{or} \quad \omega_A = \frac{2}{T_S}\tan\left(\frac{\omega_D T_S}{2}\right) \tag{5.39b}$$

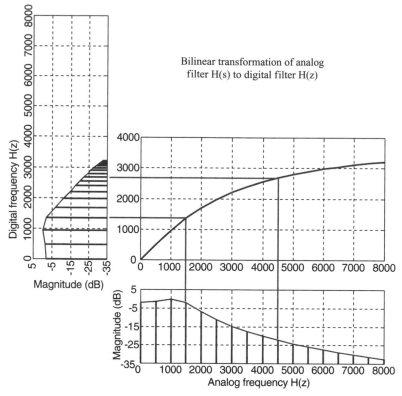

FIGURE 5.24. Effect of bilinear transform on magnitude frequency response of example filter.

The analog filter represented by equation (5.40) can be produced using the MATLAB command

```
>> [bb, aa] = cheby1(2,2,2*pi*1702,'s');
```

and the bilinear transformation applied by typing

```
>> [bbd, aad] = bilinear(bb,aa,8000);
```

to yield the result given by equation (5.41).

Alternatively, prewarping of the analog filter design considered previously can be combined with application of the bilinear transform by typing

```
>> [bbd, aad]=bilinear(b, a, 8000,1500);
```

at the MATLAB command line.

Coefficient file `bilinearw.cof`, stored in folder `iirsos`, contains the coefficients obtained as described above.

Using MATLAB's Filter Design and Analysis Tool

MATLAB provides a filter design and analysis tool, `fdatool`, that makes the design of IIR filter coefficients simple. Coefficients can be exported in direct form II, second order section format and a MATLAB function `dsk_sos_iir67()`, supplied on the CD as file `dsk_sos_iir67.m`, can be used to generate coefficient files compatible with the programs in this chapter.

Example 5.5 Fourth Order Elliptical Lowpass IIR Filter Designed Using `fdatool`

To invoke the *Filter Design and Analysis Tool* window, type

```
>> fdatool
```

in the MATLAB command window. Enter the parameters for a *fourth order elliptical lowpass IIR filter with a cutoff frequency of 800 Hz and 1 dB of ripple in the passband and 50 dB of stopband attenuation.* Click on *Design Filter* and then look at the characteristics of the filter using options from the *Analysis* menu (Figure 5.25).

This example illustrates the steep transition from passband to stopband possible even with relatively few filter coefficients.

Select *Filter Coefficients* from the *Analysis* menu. `fdatool` automatically designs filters as cascaded second order sections. Each section is similar to those shown in block diagram form in Figure 5.5 and each section is characterised by six parameter values a_0, a_1, a_2, b_0, b_1, and b_2.

By default, `fdatool` uses the bilinear transform method of designing a digital filter starting from an analog prototype. Figure 5.26 shows the use of `fdatool` to design the Chebyshev filter considered in the preceding examples. Note that the magnitude frequency response decreases more and more rapidly with frequency, approaching half the sampling frequency. This is characteristic of filters designed using the bilinear transform. Compare this with Figure 5.23.

Implementing a Filter Designed Using `fdatool` on the C6713 DSK

In order to implement a filter designed using `fdatool` on the C6713 DSK, carry out the following steps.

1. Design the IIR filter using `fdatool`.
2. Click on *Export* in the `fdatool` *File* menu.
3. Select *Workspace, Coefficients, SOS*, and *G* and click *Export*.
4. At the MATLAB command line, type `dsk_sos_iir67(SOS,G)` and enter a filename (e.g., `elliptic.cof`).

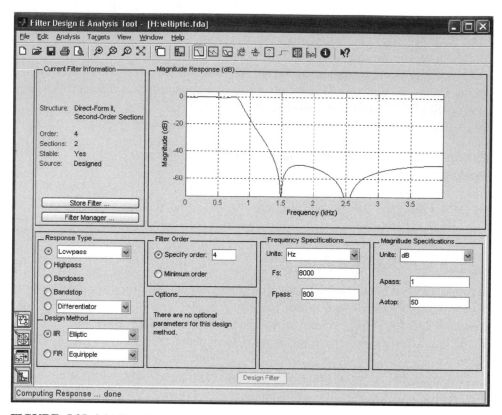

FIGURE 5.25. MATLAB `fdatool` window showing magnitude frequency response of fourth order elliptical lowpass filter.

Figure 5.27 shows an example of a coefficient file produced using `dsk_sos_iir67()` (Figure 5.28).

Program `iirsos.c` is a generic IIR filter program that uses cascaded second order sections and reads the filter coefficients from a separate `.cof` file. In order to implement your filter, edit the line in program `iirsos.c` that reads

```
#include "bilinear.cof"
```

to read

```
#include "elliptic.cof"
```

and *Build, Load Program*, and *Run*.

Figures 5.29 and 5.30 show results obtained with programs `iirsosdelta.c` and `iirsosprn.c` using coefficient file `elliptic.cof`.

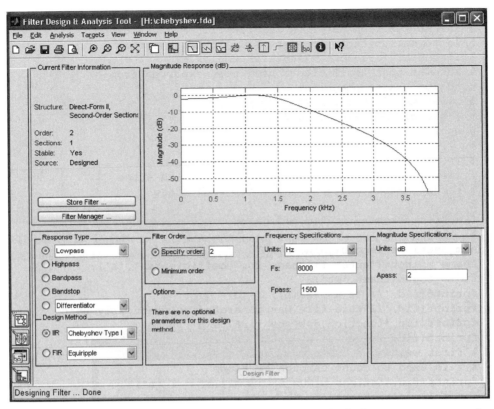

FIGURE 5.26. MATLAB `fdatool` window showing magnitude frequency response of a second order Chebyshev lowpass filter.

```
// elliptic.cof
// this file was generated automatically using function
dsk_sos_iir67.m

#define NUM_SECTIONS 2

float b[NUM_SECTIONS][3] = {
{1.00494714E-002, 7.90748088E-003, 1.00494714E-002},
{1.00000000E+000, -7.76817178E-001, 1.00000000E+000} };

float a[NUM_SECTIONS][2] = {
{-1.52873456E+000, 6.37031997E-001},
{-1.51375640E+000, 8.68676718E-001} };
```

FIGURE 5.27. Listing of coefficient file `elliptic.cof`.

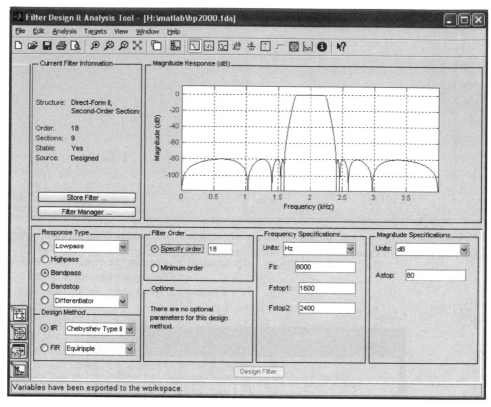

FIGURE 5.31. MATLAB `fdatool` window showing magnitude frequency response of 18th order bandpass pass filter centered on 2000 Hz.

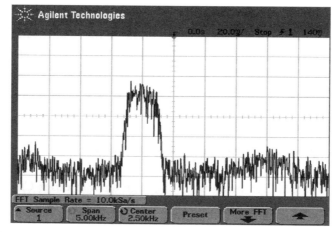

FIGURE 5.32. Output from program `iirsosprn.c`, using coefficient file `bp2000.cof`, viewed using the FFT function of an *Agilent 54621A* oscilloscope.

Example 5.7: Fixed-Point Implementation of IIR Filter (iir)

Program iir.c, listed in Figure 5.33, implements a generic IIR filter using cascaded second order stages (sections) and fixed-point (integer) coefficients. The program implements each second order stage as a direct form II structure using the following two expressions:

```
// iir.c filter using cascaded second order sections
// 16-bit integer coefficients read from .cof file

#include "DSK6713_AIC23.h"                //codec support
Uint32 fs=DSK6713_AIC23_FREQ_8KHZ; //set sampling rate
#define DSK6713_AIC23_INPUT_MIC 0x0015
#define DSK6713_AIC23_INPUT_LINE 0x0011
Uint16 inputsource=DSK6713_AIC23_INPUT_LINE;

#include "bs1800int.cof"
short w[NUM_SECTIONS][2] = {0};

interrupt void c_int11()                  //interrupt service routine
{
  short section;                          //index for section number
  short input;                            //input to each section
  int wn,yn;                              //intermediate and output
                                          //values in each stage
  input = input_left_sample();

  for (section=0 ; section< NUM_SECTIONS ; section++)
  {
    wn = input - ((a[section][0]*w[section][0])>>15)
        - ((a[section][1]*w[section][1])>>15);
    yn = ((b[section][0]*wn)>>15)
        + ((b[section][1]*w[section][0])>>15)
        + ((b[section][2]*w[section][1])>>15);
    w[section][1] = w[section][0];
    w[section][0] = wn;
    input = yn;                           //output of current section
                                          //will be input to next
  }

  output_left_sample((short)(yn)); //before writing to codec
  return;                                 //return from ISR
}

void main()
{
  comm_intr();                            //init DSK, codec, McBSP
  while(1);                               //infinite loop
}
```

FIGURE 5.33. IIR filter program using second order sections in cascade (iir.c).

```
// bs1800int.cof
// this file was generated automatically using function
dsk_sos_iir67int.m

#define NUM_SECTIONS 3

int b[NUM_SECTIONS][3] = {
{11538, -4052, 11538},
{32768, 599, 32768},
{32768, -22852, 32768} };

int a[NUM_SECTIONS][2] = {
{-5832, 450},
{17837, 26830},
{-35076, 27634} };
```

FIGURE 5.34. Coefficient file for a sixth order IIR bandstop filter designed using MATLAB as described in Appendix D (bs1800int.cof).

$$w(n) = x(n) - a_1 w(n-1) - a_2 w(n-2)$$

$$y(n) = b_0 w(n) + b_1 w(n-1) + b_2 y(n-2)$$

implemented by the lines

```
wn = input - ((a[section][0]*w[section][0])>>15) -
((a[section][1]*w[section][1])>>15);
yn = ((b[section][0]*wn)>>15) + ((b[section][1]*w[section][0])
>>15) + ((b[section][2]*w[section][1])>>15);
```

The values of the coefficients in the files bs1800int.cof (Figure 5.34) and ellipint.cof were calculated using MATLAB's fdatool and function dsk_sos_iir67int(), an integer version of function dsk_sos_iir67(), which multiplies the filter coefficients generated using fdatool by 32768 and casts them as integers. Build and run this project as **iir**. Verify that an IIR bandstop filter centered at 1800 Hz is implemented if coefficient file bs1800int.cof is used.

Example 5.8: Generation of a Sine Wave Using a Difference Equation (*sinegenDE*)

In Chapter 4 it was shown that the z-transform of a sinusoidal sequence $y(n) = \sin(n\omega T)$ is given by

$$Y(z) = \frac{z\sin(\omega T)}{z^2 - 2\cos(\omega T)z + 1} \qquad (5.42)$$

Comparing this with the z-transfer function of the second order filter of Example 5.1,

$$H(z) = \frac{Y(z)}{X(z)} = \frac{b_1 z}{z^2 + a_1 z + a_2} \tag{5.43}$$

It is apparent that by appropriate choice of filter coefficients we can design that filter to act as a sine wave generator, that is, to have a sinusoidal impulse response.

Choosing $a_2 = 1.0$ and $a_1 = -2\cos(\omega T)$, the denominator of the transfer function becomes $z^2 - 2\cos(\omega T)z - 1$, which corresponds to a pair of complex conjugate poles located *on* the unit circle in the z-plane. The filter can be set oscillating by applying an impulse to its input. Rearranging equation (5.43) and setting $x(n) = \delta(n)(X(z) = 1.0)$ and $b_1 = \sin(\omega T)$,

$$Y(z) = \frac{X(z)b_1 z}{z^2 - 2\cos(\omega T)z + 1} = \frac{\sin(\omega T)z}{z^2 - 2\cos(\omega T)z + 1} \tag{5.44}$$

Equation (5.44) is equivalent to equation (5.42), implying that the filter impulse response is $y(n) = \sin(n\omega T)$. Equation (5.44) corresponds to the difference equation

$$y(n) = \sin(\omega T)x(n-1) + 2\cos(\omega T)y(n-1) - y(n-2) \tag{5.45}$$

which is illustrated in Figure 5.35.

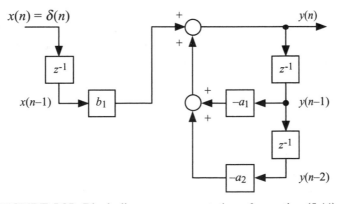

FIGURE 5.35. Block diagram representation of equation (5.44).

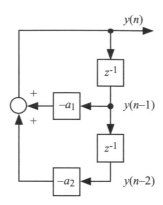

FIGURE 5.36. Block diagram representation of equation (5.45).

Since the input, $x(n) = \delta(n)$, to the filter is nonzero only at sampling instant $n = 0$, for all other n, the difference equation is

$$y(n) = 2\cos(\omega T)y(n-1) - y(n-2) \tag{5.46}$$

and the sine wave generator may be implemented as shown in Figure 5.36, using no input signal but using nonzero initial values for $y(n-1)$ and $y(n-2)$. The initial values used determine the amplitude of the sinusoidal output.

Since the frequency of oscillation, ω, is fixed by the choice of $a_1 = -2\cos(\omega T)$ and $a_2 = 1$, the initial values chosen for $y(n-1)$ and $y(n-2)$ represent two samples of a sinusoid of frequency ω that are one sampling period, or T seconds, apart in time; that is,

$$y(n-1) = A\sin(\omega t + \phi)$$
$$y(n-2) = A\sin(\omega(t+T) + \phi)$$

The initial values of $y(n-1)$ and $y(n-2)$ determine the amplitude, A, of the sine wave generated. Assuming that an output amplitude $A = 1$ is desired, a simple solution to the equations implemented in program sinegenDE.c is

$$y(n-1) = 0$$
$$y(n-2) = \sin(\omega T)$$

Build and run this project as **sinegenDE**. Verify that the output is a 2-kHz tone. Change the value of the constant FREQ; build and run and verify the generation of a tone of the frequency selected (Figure 5.37).

```
//sinegenDE.c generates sinusoid using difference equations

#include "DSK6713_AIC23.h"              //codec support
Uint32 fs=DSK6713_AIC23_FREQ_8KHZ;   //set sampling rate
#define DSK6713_AIC23_INPUT_MIC 0x0015
#define DSK6713_AIC23_INPUT_LINE 0x0011
Uint16 inputsource=DSK6713_AIC23_INPUT_LINE; // select input
#include <math.h>
#define FREQ 2000
#define SAMPLING_FREQ 8000
#define AMPLITUDE 10000
#define PI 3.14159265358979

float y[3] = {0.0, 0.0, 0.0};
float a1;

interrupt void c_int11()                //ISR
{
  y[0] =(y[1]*a1)-y[2];
  y[2] = y[1];                                     //update y1(n-2)
  y[1] = y[0];                                     //update y1(n-1)
  output_left_sample((short)(y[0]*AMPLITUDE));  //output result
  return;                                //return to main
}

void main()
{
  y[1] = sin(2.0*PI*FREQ/SAMPLING_FREQ);
  a1 = 2.0*cos(2.0*PI*FREQ/SAMPLING_FREQ);
  comm_intr();                          //init DSK, codec, McBSP
  while(1);                             //infinite loop
}
```

FIGURE 5.37. Program to generate a sine wave using a difference equation (sinegenDE.c).

Example 5.9: Generation of DTMF Signal Using Difference Equations (sinegenDTMF)

Program sinegenDTMF.c, listed in Figure 5.39, uses the same difference equation method as program sinegenDE.c to generate two sinusoidal signals of different frequencies, which, added together, form a DTMF tone (see also Example 2.12, which used a table lookup method). Build this example as **sinegenDTMF**. The DTMF tone is output via the codec only while DIP switch #0 is pressed down. The program incorporates a buffer that is used to store the 256 most recent output samples. Figure 5.38 shows the contents of that buffer in time and frequency domains, plotted using Code Composer.

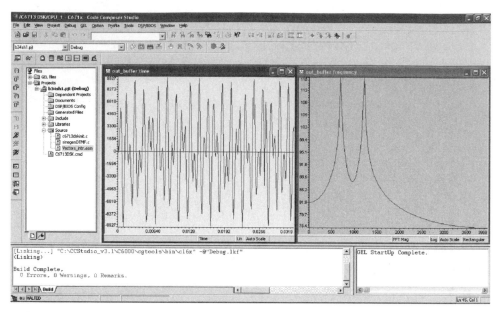

FIGURE 5.38. The 256 samples of waveform generated by program `sinegenDTMF.c` displayed using Code Composer.

Example 5.10: Generation of a Swept Sinusoid Using a Difference Equation (sweepDE)

Figure 5.40 shows a listing of the program `sweepDE.c`, which generates a sinusoidal signal, sweeping in frequency. The program implements the difference equation

$$y(n) = 2\cos(\omega T)y(n-1) - y(n-2)$$

where $A = 2\cos(\omega T)$ and the initial conditions are $y(n-1) = \sin(\omega T)$ and $y(n-2) = 0$. Example 5.8 illustrated the generation of a sine wave using this difference equation.

Compared with the lookup table method of Example 2.11, making step changes in the frequency of the output signal generated using a difference equation is slightly more problematic. Each time program `sweepDE.c` changes its output frequency it reinitializes the stored values of previous output samples $y(n-1)$ and $y(n-2)$. These values determine the amplitude of the sinusoidal output at the new frequency and must be chosen appropriately. Using the existing values, left over from the generation of a sinusoid at the previous frequency, might cause the amplitude of the output sinusoid to change. In order to avoid discontinuities, or glitches, in the output waveform, a further constraint on the parameters of the program must be observed. Since at each change in frequency the output waveform starts at the same phase in its cycle, it is necessary to ensure that each different frequency segment is output for an integer number of cycles. This can be achieved by making the number of samples

```
//sinegenDTMF.c generates DTMF tone using difference equations

#include "DSK6713_AIC23.h"              //codec support
Uint32 fs=DSK6713_AIC23_FREQ_8KHZ;   //set sampling rate
#define DSK6713_AIC23_INPUT_MIC 0x0015
#define DSK6713_AIC23_INPUT_LINE 0x0011
Uint16 inputsource=DSK6713_AIC23_INPUT_LINE; // select input
#include <math.h>
#define FREQLO 770
#define FREQHI 1336
#define SAMPLING_FREQ 8000
#define AMPLITUDE 5000
#define BUFSIZE 256
#define PI 3.14159265358979

float ylo[3] = {0.0, 0.0, 0.0};
float yhi[3] = {0.0, 0.0, 0.0};
float a1lo, a1hi;
float out_buffer[BUFSIZE];
int bufindex = 0;
float output;
float DIP0pressed = 0;

interrupt void c_int11()               //ISR
{
  ylo[0] =(ylo[1]*a1lo)-ylo[2];
  ylo[2] = ylo[1];                                  //update y1(n-2)
  ylo[1] = ylo[0];                                  //update y1(n-1)
  yhi[0] =(yhi[1]*a1hi)-yhi[2];
  yhi[2] = yhi[1];                                  //update y1(n-2)
  yhi[1] = yhi[0];                                  //update y1(n-1)
  output = (yhi[0]+ylo[0])*AMPLITUDE;
  out_buffer[bufindex++] = output;
  if (bufindex >= BUFSIZE) bufindex = 0;
  if (DIP0pressed) output_left_sample((short)(output));
  else output_left_sample(0);                  //output result
  return;                              //return to main
}

void main()
{
  ylo[1] = sin(2.0*PI*FREQLO/SAMPLING_FREQ);
  a1lo = 2.0*cos(2.0*PI*FREQLO/SAMPLING_FREQ);
  yhi[1] = sin(2.0*PI*FREQHI/SAMPLING_FREQ);
  a1hi = 2.0*cos(2.0*PI*FREQHI/SAMPLING_FREQ);
  DSK6713_DIP_init();
  comm_intr();                         //init DSK, codec, McBSP
  while(1)                             //infinite loop
  {
    if (DSK6713_DIP_get(0) == 0) DIP0pressed = 1;
    else DIP0pressed = 0;
  }
}
```

FIGURE 5.39. Program to generate DTMF tone using difference equations (sinegenDTMF.cof).

//**sweepDE.c** generates sweeping sinusoid using difference equations

```
#include "DSK6713_AIC23.h"              //codec support
Uint32 fs=DSK6713_AIC23_FREQ_8KHZ;   //set sampling rate
#define DSK6713_AIC23_INPUT_MIC 0x0015
#define DSK6713_AIC23_INPUT_LINE 0x0011
Uint16 inputsource=DSK6713_AIC23_INPUT_LINE; // select input
#include <math.h>
#define MIN_FREQ 200
#define MAX_FREQ 3800
#define STEP_FREQ 20
#define SWEEP_PERIOD 400
#define SAMPLING_FREQ 8000
#define AMPLITUDE 5000
#define PI 3.14159265358979

float y[3] = {0.0, 0.0, 0.0};
float a1;
float freq = MIN_FREQ;
short sweep_count = 0;

void coeff_gen(float freq)
{
  a1 = 2.0*cos(2.0*PI*freq/SAMPLING_FREQ);
  y[0] = 0.0;
  y[1] = sin(2.0*PI*freq/SAMPLING_FREQ);
  y[2] = 0.0;
  return;
}

interrupt void c_int11()                //ISR
{
  sweep_count++;
  if (sweep_count >= SWEEP_PERIOD)
  {
    if (freq >= MAX_FREQ) freq = MIN_FREQ;
    else freq += STEP_FREQ;
    coeff_gen(freq);
    sweep_count = 0;
  }
  y[0] =(y[1]*a1)-y[2];
  y[2] = y[1];                               //update y1(n-2)
  y[1] = y[0];                               //update y1(n-1)
  output_left_sample((short)(y[0]*AMPLITUDE));  //output result
  return;                               //return to main
}

void main()
{
  y[1] = sin(2.0*PI*freq/SAMPLING_FREQ);
  a1 = 2.0*cos(2.0*PI*freq/SAMPLING_FREQ);
  comm_intr();                              //init DSK, codec, McBSP
  while(1);                                 //infinite loop
}
```

FIGURE 5.40. Program to generate a sweeping sinusoid using a difference equation (sweepDE.c).

output between step changes in frequency equal to the sampling frequency divided by the frequency increment. As listed in Figure 5.40, the frequency increment is 20 Hz and the sampling frequency is 8000 Hz. Hence, the number of samples output at each different frequency is equal to 8000/20 = 400. Different choices for the values of the constants STEP_FREQ and SWEEP_PERIOD are possible.

Build and run this project as **sweepDE**. Verify that the output is a swept sinusoidal signal starting at frequency 200 Hz and taking (SWEEP_PERIOD/SAMPLING_FREQ)*(MAX_FREQ-MIN_FREQ)/STEP_FREQ seconds to increase in frequency to 3800 Hz. Change the values of MIN_FREQ and MAX_FREQ to 2000 and 3000, respectively. Build the project again, load and run program sweepDE.out, and verify that the frequency sweep is from 2000 to 3000 Hz.

Example 5.11: Sine Wave Generation Using a Difference Equation with C Calling an ASM Function (*sinegencasm*)

This example is based on Example 5.8 but uses an assembly language function to generate a sine wave using a difference equation. Program sinegencasm.c, listed in Figure 5.41, calls the assembly language function sinegencasmfunc, defined in

```
//Sinegencasm.c Sine gen using DE with asm function

#include "dsk6713_aic23.h"                     //codec-DSK support file
Uint32 fs=DSK6713_AIC23_FREQ_8KHZ;  //set sampling rate
#define DSK6713_AIC23_INPUT_MIC 0x0015
#define DSK6713_AIC23_INPUT_LINE 0x0011
Uint16 inputsource=DSK6713_AIC23_INPUT_LINE; // select input source

short y[3] = {0, 15137, 11585};              //y(1)=sinwT (f=1.5kHz)
short A = 12540;                     //A=2*coswT * 2^14
short n = 2;

interrupt void c_int11()             //interrupt service routine
{
      sinegencasmfunc(&y[0], A);     //calls ASM function
      output_sample(y[n]);
      return;
}

void main()
{
  comm_intr();                          //init DSK, codec, McBSP
  while(1);                       //infinite loop
}
```

FIGURE 5.41. C source program that calls an ASM function to generate a sine wave using a difference equation (sinegencasm.c).

```
;Sinegencasmfunc.asm ASM func to generate sine using DE
;A4 = address of y array, B4 = A

            .def  _sinegencasmfunc    ;ASM function called from C
_sinegencasmfunc:
            LDH       *+A4[0], A5      ;y[n-2]-->A5
            LDH       *+A4[1], A2      ;y[n-1]-->A2
            LDH       *+A4[2], A3      ;y[n]-->A3
            NOP       3                ;NOP due to LDH
            MPY       B4, A2, A8       ;A*y[n-1]
            NOP       1                ;NOP due to MPY
            SHR       A8, 14, A8       ;shift right by 14
            SUB       A8, A5, A8       ;A*y[n-1]-y[n-2]
            STH       A8, *+A4[2]      ;y[n]=A*y[n-1]-y[n-2]
            STH       A2, *+A4[0]      ;y[n-2]=y[n-1]
            STH       A8, *+A4[1]      ;y[n-1] = y[n]
            B         B3               ;return addr to call routine
            NOP       5                ;delays to to branching
            .end
```

FIGURE 5.42. ASM function called from C to generate a sine wave using a difference equation (sinegencasmfunc.asm).

file sinegencasmfunc.asm (Figure 5.42). The C source program shows the array y[3], which contains the values $y(0)$, $y(1)$, and $y(2)$ and the coefficient $A = 2\cos(\omega T)$, calculated to generate a 1.5-kHz sine wave. The address of the array y[3] and the value of the coefficient A are passed to the ASM function using registers A4 and B4, respectively. The values in the array y[3] and the coefficient A were scaled by 2^{14} to allow for a fixed-point implementation. As a result, within the ASM function, A8 initially containing $Ay(n-1)$ is scaled back (shifted right) by 2^{14}.

Build this project as **sinegencasm**. Verify that a 1.5-kHz sine wave is generated. Verify that changing the initial contents of the array to y[3] ={0, 16384, 0} and setting A = 0 yields a 2-kHz sine wave.

5.5 ASSIGNMENTS

1. Design and implement in real time a 12th order Chebyshev type 2 lowpass IIR filter with a cutoff frequency of 1700 Hz, using a sampling frequency of 8 kHz. Compare the characteristics of this filter with those of comparable elliptical and Butterworth designs.

2. Modify program sweepDE.c to generate a swept sine wave that decreases in frequency, starting at 3200 Hz and resetting when it reaches 400 Hz.

3. Three sets of coefficients corresponding to fourth, sixth, and eighth order IIR filters implemented as cascaded second order stages are shown below. Use programs iirsos.c, iirsosprn.c, and iirsosdelta.c (use a sampling frequency of 8 kHz) in order to determine the characteristics of these three filters.

Filter (a)

	First Stage	Second Stage
b0	0.894858606	1.00000000
b1	0.687012957	0.767733531
b2	0.894858606	1.00000000
a1	0.626940111	0.823551047
a2	0.892574561	0.897182915

Filter (b)

	First Stage	Second Stage	Third Stage
b0	4.22434573E–003	1.00000000	1.00000000
b1	–7.40347363E–003	–7.51020138E–001	–2.42042682E–001
b2	4.22434573E–003	1.00000000	1.00000000
a1	1.38530785E+000	1.08202283	8.72945011E–001
a2	5.49723350E–001	7.24171197E–001	9.12022866E–001

Filter (c)

	First Stage	Second Stage	Third Stage	Fourth Stage
b0	0.0799986548	1.00000000	1.00000000	1.00000000
b1	0.159997310	–2.00000000	2.00000000	–2.00000000
b2	0.0799986548	1.00000000	1.00000000	1.00000000
a1	0.131667585	–1.11285289	0.568937617	–1.56515908
a2	0.112608874	0.365045225	0.582994098	0.767928609

REFERENCES

1. L. B. Jackson, *Digital Filters and Signal Processing*, Kluwer Academic, Norwell, MA, 1996.

2. L. B. Jackson, Roundoff noise analysis for fixed-point digital filters realized in cascade or parallel form, *IEEE Transactions on Audio and Electroacoustics*, Vol. AU-18, pp. 107–122, June 1970.

3. L. B. Jackson, An analysis of limit cycles due to multiplicative rounding in recursive digital filters, *Proceedings of the 7th Allerton Conference on Circuit and System Theory*, 1969, pp. 69–78.

4. L. B. Lawrence and K. V. Mirna, A new and interesting class of limit cycles in recursive digital filters, *Proceedings of the IEEE International Symposium on Circuit and Systems*, Apr. 1977, pp. 191–194.

5. R. Chassaing and D. W. Horning, *Digital Signal Processing with the TMS320C25*, Wiley, Hoboken, NJ, 1990.

6

Fast Fourier Transform

- The fast Fourier transform using radix-2 and radix-4
- Decimation or decomposition in frequency and in time
- Programming examples

The fast Fourier transform (FFT) is an efficient algorithm that is used for converting a time-domain signal into an equivalent frequency-domain signal, based on the discrete Fourier transform (DFT). Several real-time programming examples on FFT are included.

6.1 INTRODUCTION

The DFT converts a time-domain sequence into an equivalent frequency-domain sequence. The inverse DFT performs the reverse operation and converts a frequency-domain sequence into an equivalent time-domain sequence. The FFT is a very efficient algorithm technique based on the DFT but with fewer computations required. The FFT is one of the most commonly used operations in digital signal processing to provide a frequency spectrum analysis [1–6]. Two different procedures are introduced to compute an FFT: the decimation-in-frequency and the decimation-in-time. Several variants of the FFT have been used, such as the Winograd transform [7, 8], the discrete cosine transform (DCT) [9], and the discrete Hartley transform [10–12]. The fast Hartley transform (FHT) is described in Appendix E. Transform methods such as the DCT have become increasingly

Digital Signal Processing and Applications with the TMS320C6713 and TMS320C6416 DSK,
Second Edition   By Rulph Chassaing and Donald Reay
Copyright © 2008 John Wiley & Sons, Inc.

popular in recent years, especially for real-time systems. They provide a large compression ratio.

6.2 DEVELOPMENT OF THE FFT ALGORITHM WITH RADIX-2

The FFT reduces considerably the computational requirements of the DFT. The DFT of a discrete-time signal $x(nT)$ is

$$X(k) = \sum_{n=0}^{N-1} x(n)W^{nk} \qquad k = 0, 1, \ldots, N-1 \qquad (6.1)$$

where the sampling period T is implied in $x(n)$ and N is the frame length. The constants W are referred to as *twiddle constants* or *factors*, which represent the phase, or

$$W = e^{-j2\pi/N} \qquad (6.2)$$

and are a function of the length N. Equation (6.1) can be written for $k = 0, 1, \ldots, N-1$, as

$$X(k) = x(0) + x(1)W^k + x(2)W^{2k} + \cdots + x(N-1)W^{(N-1)k} \qquad (6.3)$$

This represents a matrix of $N \times N$ terms, since $X(k)$ needs to be calculated for N values for k. Since (6.3) is an equation in terms of a complex exponential, for each specific k there are $(N-1)$ complex additions and N complex multiplications. This results in a total of $(N^2 - N)$ complex additions and N^2 complex multiplications. Hence, the computational requirements of the DFT can be very intensive, especially for large values of N. FFT reduces computational complexity from N^2 to N $\log N$.

The FFT algorithm takes advantage of the periodicity and symmetry of the twiddle constants to reduce the computational requirements of the FFT. From the periodicity of W,

$$W^{k+N} = W^k \qquad (6.4)$$

and from the symmetry of W,

$$W^{k+N/2} = -W^k \qquad (6.5)$$

Figure 6.1 illustrates the properties of the twiddle constants W for $N = 8$. For example, let $k = 2$, and note that from (6.4), $W^{10} = W^2$, and from (6.5), $W^6 = -W^2$.

For a radix-2 (base 2), the FFT decomposes an N-point DFT into two $(N/2)$-point or smaller DFTs. Each $(N/2)$-point DFT is further decomposed into two

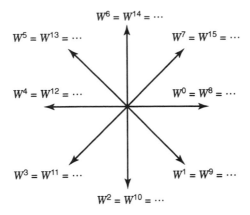

FIGURE 6.1. Periodicity and symmetry of twiddle constant W.

($N/4$)-point DFTs, and so on. The last decomposition consists of ($N/2$) two-point DFTs. The smallest transform is determined by the radix of the FFT. For a radix-2 FFT, N must be a power or base of 2, and the smallest transform or the last decomposition is the two-point DFT. For a radix-4, the last decomposition is a four-point DFT.

6.3 DECIMATION-IN-FREQUENCY FFT ALGORITHM WITH RADIX-2

Let a time-domain input sequence $x(n)$ be separated into two halves:

$$x(0), x(1), \ldots, x\left(\frac{N}{2}-1\right) \tag{6.6}$$

and

$$x\left(\frac{N}{2}\right), x\left(\frac{N}{2}+1\right), \ldots, x(N-1) \tag{6.7}$$

Taking the DFT of each set of the sequence in (6.6) and (6.7) gives us

$$X(k) = \sum_{n=0}^{(N/2)-1} x(n)W^{nk} + \sum_{n=N/2}^{N-1} x(n)W^{nk} \tag{6.8}$$

Let $n = n + N/2$ in the second summation of (6.8); $X(k)$ becomes

$$X(k) = \sum_{n=0}^{(N/2)-1} x(n)W^{nk} + W^{kN/2} \sum_{n=0}^{(N/2)-1} x\left(n+\frac{N}{2}\right)W^{nk} \tag{6.9}$$

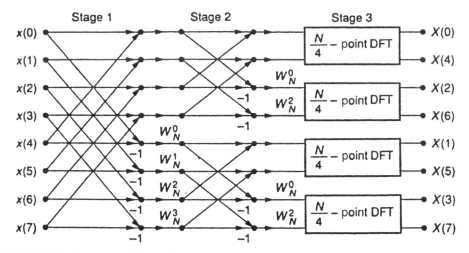

FIGURE 6.3. Decomposition of two ($N/2$)-point DFTs into four ($N/4$)-point DFTs for $N = 8$.

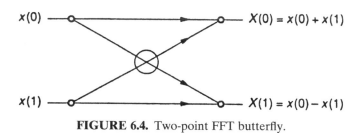

FIGURE 6.4. Two-point FFT butterfly.

This is the last decomposition, since we now have a set of ($N/2$) two-point DFTs, the lowest decomposition for a radix-2. For the two-point DFT, $X(k)$ in (6.1) can be written

$$X(k) = \sum_{n=0}^{1} x(n)W^{nk} \qquad k = 0, 1 \tag{6.19}$$

or

$$X(0) = x(0)W^0 + x(1)W^0 = x(0) + x(1) \tag{6.20}$$

$$X(1) = x(0)W^0 - x(1)W^0 = x(0) - x(1) \tag{6.21}$$

since $W^1 = e^{-j2\pi/2} = -1$. Equations (6.20) and (6.21) can be represented by the flow graph in Figure 6.4, usually referred to as a *butterfly*. The final flow graph of an eight-point FFT algorithm is shown in Figure 6.5. This algorithm is referred to as *decimation-in-frequency* (DIF) because the output sequence $X(k)$ is decomposed (decimated) into smaller subsequences, and this process continues through M stages

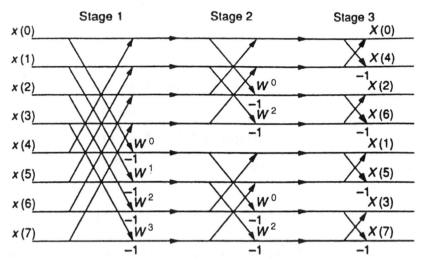

FIGURE 6.5. Eight-point FFT flow graph using DIF.

or iterations, where $N = 2^M$. The output $X(k)$ is complex with both real and imaginary components, and the FFT algorithm can accommodate either complex or real input values.

The FFT is not an approximation of the DFT. It yields the same result as the DFT with fewer computations required. This reduction becomes more and more important with higher-order FFT.

There are other FFT structures that have been used to illustrate the FFT. An alternative flow graph to that in Figure 6.5 can be obtained with ordered output and scrambled input.

An eight-point FFT is illustrated through the following exercise. We will see that flow graphs for higher order FFT (larger N) can readily be obtained.

Exercise 6.1: Eight-Point FFT Using DIF

Let the input $x(n)$ represent a rectangular waveform, or $x(0) = x(1) = x(2) = x(3) = 1$ and $x(4) = x(5) = x(6) = x(7) = 0$. The eight-point FFT flow graph in Figure 6.5 can be used to find the output sequence $X(k), k = 0, 1, \ldots, 7$. With $N = 8$, four twiddle constants need to be calculated, or

$$W^0 = 1$$

$$W^1 = e^{-j2\pi/8} = \cos(\pi/4) - j\sin(\pi/4) = 0.707 - j0.707$$

$$W^2 = e^{-j4\pi/8} = -j$$

$$W^3 = e^{-j6\pi/8} = -0.707 - j0.707$$

The intermediate output sequence can be found after each stage.

Stage 1

$$x(0) + x(4) = 1 \rightarrow x'(0)$$

$$x(1) + x(5) = 1 \rightarrow x'(1)$$

$$x(2) + x(6) = 1 \rightarrow x'(2)$$

$$x(3) + x(7) = 1 \rightarrow x'(3)$$

$$[x(0) - x(4)]W^0 = 1 \rightarrow x'(4)$$

$$[x(1) - x(5)]W^1 = 0.707 - j0.707 \rightarrow x'(5)$$

$$[x(2) - x(6)]W^2 = -j \rightarrow x'(6)$$

$$[x(3) - x(7)]W^3 = -0.707 - j0.707 \rightarrow x'(7)$$

where $x'(0), x'(1), \ldots, x'(7)$ represent the intermediate output sequence after the first iteration, which becomes the input to the second stage.

Stage 2

$$x'(0) + x'(2) = 2 \rightarrow x''(0)$$

$$x'(1) + x'(3) = 2 \rightarrow x''(1)$$

$$[x'(0) - x'(2)]W^0 = 0 \rightarrow x''(2)$$

$$[x'(1) - x'(3)]W^2 = 0 \rightarrow x''(3)$$

$$x'(4) + x'(6) = 1 - j \rightarrow x''(4)$$

$$x'(5) + x'(7) = (0.707 - j0.707) + (-0.707 - j0.707) = -j1.41 \rightarrow x''(5)$$

$$[x'(4) - x'(6)]W^0 = 1 + j \rightarrow x''(6)$$

$$[x'(5) - x'(7)]W^2 = -j1.41 \rightarrow x''(7)$$

The resulting intermediate, second-stage output sequence $x''(0), x''(1), \ldots, x''(7)$ becomes the input sequence to the third stage.

Stage 3

$$X(0) = x''(0) + x''(1) = 4$$

$$X(4) = x''(0) - x''(1) = 0$$

$$X(2) = x''(2) + x''(3) = 0$$

$$X(6) = x''(2) - x''(3) = 0$$

$$X(1) = x''(4) + x''(5) = (1 - j) + (-j1.41) = 1 - j2.41$$

$$X(5) = x''(4) - x''(5) = 1 + j0.41$$

$$X(3) = x''(6) + x''(7) = (1 + j) + (-j1.41) = 1 - j0.41$$

$$X(7) = x''(6) - x''(7) = 1 + j2.41$$

We now use the notation of X's to represent the final output sequence. The values $X(0), X(1), \ldots, X(7)$ form the scrambled output sequence. We show later how to reorder the output sequence and plot the output magnitude.

Exercise 6.2: Sixteen-Point FFT

Given $x(0) = x(1) = \cdots = x(7) = 1$, and $x(8) = x(9) = \cdots = x(15) = 0$, which represents a rectangular input sequence, the output sequence can be found using the 16-point flow graph shown in Figure 6.6. The intermediate output results after each stage are found in a manner similar to that in Exercise 6.1. Eight twiddle constants $W^0, W^1, \ldots,$ W^7 need to be calculated for $N = 16$.

Verify the scrambled output sequence X's as shown in Figure 6.6. Reorder this output sequence and take its magnitude. Verify the plot in Figure 6.7, which represents a sinc function. The output $X(8)$ represents the magnitude at the Nyquist frequency.

6.4 DECIMATION-IN-TIME FFT ALGORITHM WITH RADIX-2

Whereas the DIF process decomposes an output sequence into smaller subsequences, *decimation-in-time* (DIT) is a process that decomposes the input sequence into smaller subsequences. Let the input sequence be decomposed into an even sequence and an odd sequence, or

$$x(0), x(2), x(4), \ldots, x(2n)$$

and

$$x(1), x(3), x(5), \ldots, x(2n+1)$$

We can apply (6.1) to these two sequences to obtain

$$X(k) = \sum_{n=0}^{(N/2)-1} x(2n)W^{2nk} + \sum_{n=0}^{(N/2)-1} x(2n+1)W^{(2n+1)k} \tag{6.22}$$

Using $W_N^2 = W_{N/2}$ in (6.22) yields

$$X(k) = \sum_{n=0}^{(N/2)-1} x(2n)W_{N/2}^{nk} + W_N^k \sum_{n=0}^{(N/2)-1} x(2n+1)W_{N/2}^{nk} \tag{6.23}$$

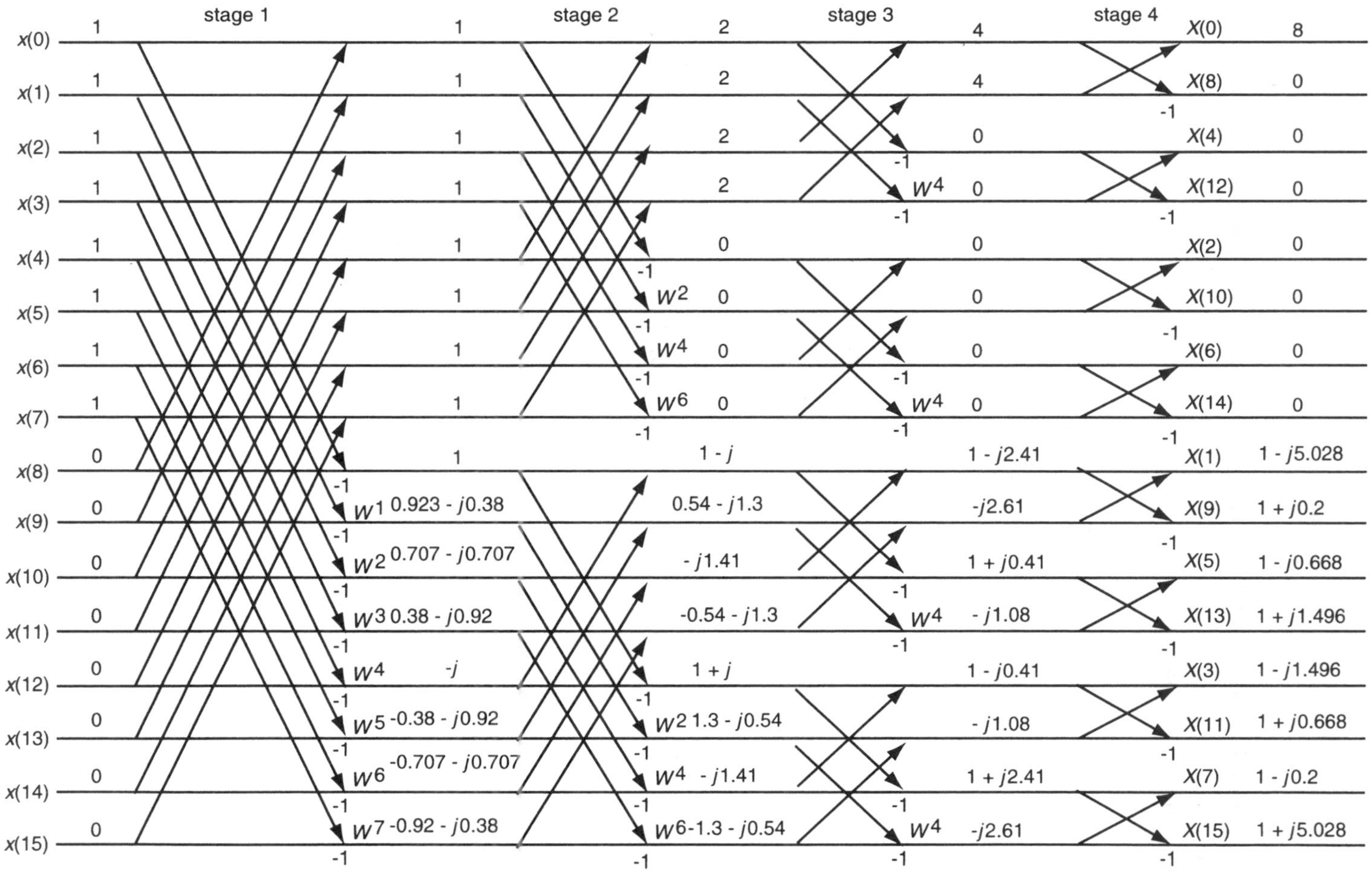

FIGURE 6.6. Sixteen-point FFT flow graph using DIF.

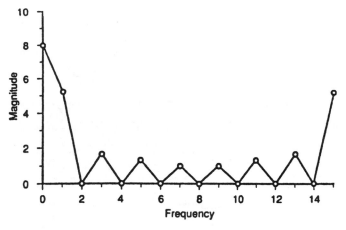

FIGURE 6.7. Output magnitude for 16-point FFT.

which represents two $(N/2)$-point DFTs. Let

$$C(k) = \sum_{n=0}^{(N/2)-1} x(2n) W_{N/2}^{nk} \tag{6.24}$$

$$D(k) = \sum_{n=0}^{(N/2)-1} X(2n+1) W_{N/2}^{nk} \tag{6.25}$$

Then $X(k)$ in (6.23) can be written

$$X(k) = C(k) + W_N^k D(k) \tag{6.26}$$

Equation (6.26) needs to be interpreted for $k > (N/2) - 1$. Using the symmetry property (6.5) of the twiddle constant, $W^{k+N/2} = -W^k$,

$$X(k + N/2) = C(k) - W^k D(k) \qquad k = 0, 1, \ldots, (N/2) - 1 \tag{6.27}$$

For example, for $N = 8$, (6.26) and (6.27) become

$$X(k) = C(k) + W^k D(k) \qquad k = 0, 1, 2, 3 \tag{6.28}$$

$$X(k + 4) = C(k) - W^k D(k) \qquad k = 0, 1, 2, 3 \tag{6.29}$$

Figure 6.8 shows the decomposition of an eight-point DFT into two four-point DFTs with the DIT procedure. This decomposition or decimation process is repeated so that each four-point DFT is further decomposed into two two-point DFTs, as shown in Figure 6.9. Since the last decomposition is $(N/2)$ two-point DFTs, this is as far as this process goes.

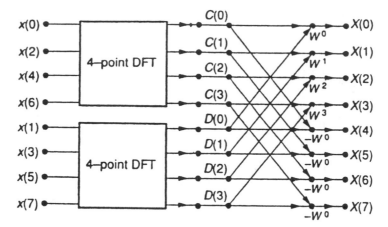

FIGURE 6.8. Decomposition of eight-point DFT into four-point DFTs using DIT.

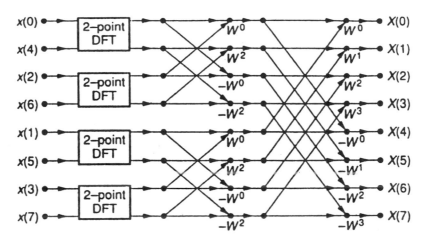

FIGURE 6.9. Decomposition of two four-point DFTs into four two-point DFTs using DIT.

Figure 6.10 shows the final flow graph for an eight-point FFT using a DIT process. The input sequence is shown to be scrambled in Figure 6.10 in the same manner as the output sequence $X(k)$ was scrambled during the DIF process. With the input sequence $x(n)$ scrambled, the resulting output sequence $X(k)$ becomes properly ordered. Identical results are obtained with an FFT using either the DIF or the DIT process. An alternative DIT flow graph to the one shown in Figure 6.10, with ordered input and scrambled output, can also be obtained.

The following exercise shows that the same results are obtained for an eight-point FFT with the DIT process as in Exercise 6.1 with the DIF process.

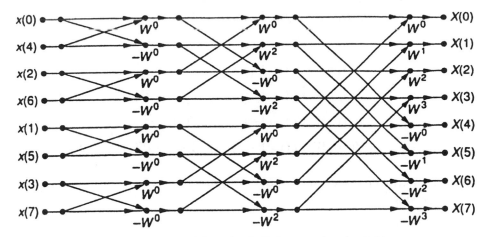

FIGURE 6.10. Eight-point FFT flow graph using DIT.

Exercise 6.3: Eight-Point FFT Using DIT

Given the input sequence $x(n)$ representing a rectangular waveform as in Exercise 6.1, the output sequence $X(k)$, using the DIT flow graph in Figure 6.10, is the same as in Exercise 6.1. The twiddle constants are the same as in Exercise 6.1. Note that the twiddle constant W is multiplied with the second term only (not with the first).

Stage 1

$$x(0) + W^0 x(4) = 1 + 0 = 1 \rightarrow x'(0)$$

$$x(0) - W^0 x(4) = 1 - 0 = 1 \rightarrow x'(4)$$

$$x(2) + W^0 x(6) = 1 + 0 = 1 \rightarrow x'(2)$$

$$x(2) - W^0 x(6) = 1 - 0 = 1 \rightarrow x'(6)$$

$$x(1) + W^0 x(5) = 1 + 0 = 1 \rightarrow x'(1)$$

$$x(1) - W^0 x(5) = 1 - 0 = 1 \rightarrow x'(5)$$

$$x(3) + W^0 x(7) = 1 + 0 = 1 \rightarrow x'(3)$$

$$x(3) - W^0 x(7) = 1 - 0 = 1 \rightarrow x'(7)$$

where the sequence x' represents the intermediate output after the first iteration and becomes the input to the subsequent stage.

Stage 2

$$x'(0) + W^0 x'(2) = 1 + 1 = 2 \rightarrow x''(0)$$

$$x'(4) + W^2 x'(6) = 1 + (-j) = 1 - j \rightarrow x''(4)$$

$$x'(0) - W^0 x'(2) = 1 - 1 = 0 \rightarrow x''(2)$$

$$x'(4) - W^2 x'(6) = 1 - (-j) = 1 + j \rightarrow x''(6)$$

$$x'(1) + W^0 x'(3) = 1 + 1 = 2 \rightarrow x''(1)$$

$$x'(5) + W^2 x'(7) = 1 + (-j)(1) = 1 - j \rightarrow x''(5)$$

$$x'(1) - W^0 x'(3) = 1 - 1 = 0 \rightarrow x''(3)$$

$$x'(5) - W^2 x'(7) = 1 - (-j)(1) = 1 + j \rightarrow x''(7)$$

where the intermediate second-stage output sequence x'' becomes the input sequence to the final stage.

Stage 3

$$X(0) = x''(0) + W^0 x''(1) = 4$$

$$X(1) = x''(4) + W^1 x''(5) = 1 - j2.414$$

$$X(2) = x''(2) + W^2 x''(3) = 0$$

$$X(3) = x''(6) + W^3 x''(7) = 1 - j0.414$$

$$X(4) = x''(0) - W^0 x''(1) = 0$$

$$X(5) = x''(4) - W^1 x''(5) = 1 + j0.414$$

$$X(6) = x''(2) - W^2 x''(3) = 0$$

$$X(7) = x''(6) - W^3 x''(7) = 1 + j2.414$$

which is the same output sequence found in Exercise 6.1.

6.5 BIT REVERSAL FOR UNSCRAMBLING

A bit-reversal procedure allows a scrambled sequence to be reordered. To illustrate this bit-swapping process, let $N = 8$, represented by three bits. The first and third bits are swapped. For example, $(100)_b$ is replaced by $(001)_b$. As such, $(100)_b$ specifying the address of $X(4)$ is replaced by or swapped with $(001)_b$ specifying the address of $X(1)$. Similarly, $(110)_b$ is replaced/swapped with $(011)_b$, or the addresses of $X(6)$ and $X(3)$ are swapped. In this fashion, the output sequence in Figure 6.5 with the DIF, or the input sequence in Figure 6.10 with the DIT, can be reordered.

This bit-reversal procedure can be applied for larger values of N. For example, for $N = 64$, represented by six bits, the first and sixth bits, the second and fifth bits, and the third and fourth bits are swapped.

Several examples in this chapter illustrate the FFT algorithm, incorporating algorithms for unscrambling.

6.6 DEVELOPMENT OF THE FFT ALGORITHM WITH RADIX-4

A radix-4 (base 4) algorithm can increase the execution speed of the FFT. FFT programs on higher radices and split radices have been developed. We use a DIF decomposition process to introduce the development of the radix-4 FFT. The last or lowest decomposition of a radix-4 algorithm consists of four inputs and four outputs. The order or length of the FFT is 4^M, where M is the number of stages. For a 16-point FFT, there are only two stages or iterations, compared with four stages with the radix-2 algorithm. The DFT in (6.1) is decomposed into four summations instead of two as follows:

$$X(k) = \sum_{n=0}^{(N/4)-1} x(n)W^{nk} + \sum_{n=N/4}^{(N/2)-1} x(n)W^{nk} + \sum_{n=N/2}^{(3N/4)-1} x(n)W^{nk} + \sum_{n=3N/4}^{N-1} x(n)W^{nk} \quad (6.30)$$

Let $n = n + N/4$, $n = n + N/2$, and $n = n + 3N/4$ in the second, third, and fourth summations, respectively. Then (6.30) can be written

$$X(k) = \sum_{n=0}^{(N/4)-1} x(n)W^{nk} + W^{kN/4} \sum_{n=0}^{(N/4)-1} x(n+N/4)W^{nk}$$
$$+ W^{kN/2} \sum_{n=0}^{(N/4)-1} x(n+N/2)W^{nk} + W^{3kN/4} \sum_{n=0}^{(N/4)-1} x(n+3N/4)W^{nk} \quad (6.31)$$

which represents four $(N/4)$-point DFTs. Using

$$W^{kN/4} = (e^{-j2\pi/N})^{kN/4} = e^{\,jk\pi/2} = (-j)^k$$
$$W^{kN/2} = e^{-jk\pi} = (-1)^k$$
$$W^{3kN/4} = (j)^k$$

(6.31) becomes

$$X(k) = \sum_{n=0}^{(N/4)-1} [x(n)+(-j)^k x(n+N/4)+(-1)^k x(n+N/2)+(j)^k x(n+3N/4)]W^{nk} \quad (6.32)$$

Let $W_N^4 = W_{N/4}$. Equation (6.32) can be written

$$X(4k) = \sum_{n=0}^{(N/4)-1} [x(n)+x(n+N/4)+x(n+N/2)+x(n+3N/4)]W_{N/4}^{nk} \quad (6.33)$$

$$X(4k+1) = \sum_{n=0}^{(N/4)-1} [x(n)-jx(n+N/4)-x(n+N/2)+jx(n+3N/4)]W_N^n W_{N/4}^{nk} \quad (6.34)$$

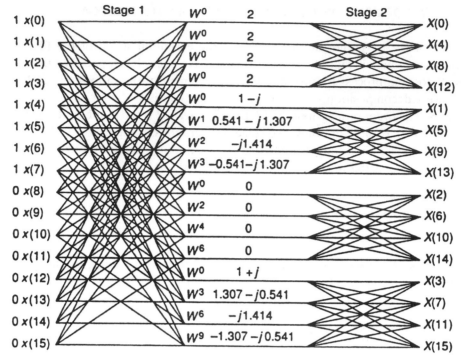

FIGURE 6.11. Sixteen-point radix-4 FFT flow graph using DIF.

$$X(4k+2) = \sum_{n=0}^{(N/4)-1} [x(n) - x(n+N/4) + x(n+N/2) - x(n+3N/4)]W_N^{2n}W_{N/4}^{nk} \quad (6.35)$$

$$X(4k+3) = \sum_{n=0}^{(N/4)-1} [x(n) + jx(n+N/4) - x(n+N/2) - jx(n+3N/4)]W_N^{3n}W_{N/4}^{nk} \quad (6.36)$$

for $k = 0, 1, \ldots, (N/4) - 1$. Equations (6.33) through (6.36) represent a decomposition process yielding four four-point DFTs. The flow graph for a 16-point radix-4 DIF FFT is shown in Figure 6.11. Note the four-point butterfly in the flow graph. The $\pm j$ and -1 are not shown in Figure 6.11. The results shown in the flow graph are for the following exercise.

Exercise 6.4: Sixteen-Point FFT with Radix-4

Given the input sequence $x(n)$ as in Exercise 6.2, representing a rectangular sequence $x(0) = x(1) = \cdots = x(7) = 1$, and $x(8) = x(9) = \cdots = x(15) = 0$, we will find the output sequence for a 16-point FFT with radix-4 using the flow graph in Figure 6.11. The twiddle constants are shown in Table 6.1.

TABLE 6.1   Twiddle Constants for 16-Point FFT with Radix-4

m	W_N^m	$W_{N/4}^m$
0	1	1
1	$0.9238 - j0.3826$	$-j$
2	$0.707 - j0.707$	-1
3	$0.3826 - j0.9238$	$+j$
4	$0 - j$	1
5	$-0.3826 - j0.9238$	$-j$
6	$-0.707 - j0.707$	-1
7	$-0.9238 - j0.3826$	$+j$

The intermediate output sequence after stage 1 is shown in Figure 6.11. For example, after stage 1:

$$[x(0) + x(4) + x(8) + x(12)]W^0 = 1 + 1 + 0 + 0 = 2 \rightarrow x'(0)$$
$$[x(1) + x(5) + x(9) + x(13)]W^0 = 1 + 1 + 0 + 0 = 2 \rightarrow x'(1)$$
$$\vdots \qquad\qquad\qquad \vdots$$
$$[x(0) - jx(4) - x(8) + jx(12)]W^0 = 1 - j - 0 - 0 = 1 - j \rightarrow x'(4)$$
$$\vdots \qquad\qquad\qquad \vdots$$
$$[x(3) - x(7) + x(11) - x(15)]W^6 = 0 \rightarrow x'(11)$$
$$[x(0) + jx(4) - x(8) - jx(12)]W^0 = 1 + j - 0 - 0 = 1 + j \rightarrow x'(12)$$
$$\vdots \qquad\qquad\qquad \vdots$$
$$[x(3) + jx(7) - x(11) - jx(15)]W^9 = [1 + j - 0 - 0](-W^1)$$
$$= -1.307 - j0.541 \rightarrow x'(15)$$

For example, after stage 2:

$$X(3) = (1 + j) + (1.307 - j0.541) + (-j1.414) + (-1.307 - j0.541) = 1 - j1.496$$

and

$$X(15) = (1 + j)(1) + (1.307 - j0.541)(-j) + (-j1.414)(1)$$
$$+ (-1.307 - j0.541)(-j) = 1 + j5.028$$

The output sequence $X(0)$, $X(1)$, ..., $X(15)$ is identical to the output sequence obtained with the 16-point FFT with the radix-2 in Figure 6.6.

The output sequence is scrambled and needs to be resequenced or reordered. This can be done using a digit-reversal procedure, in a similar fashion as a bit

```c
//dft.c N-point DFT of sequence read from lookup table
#include <stdio.h>
#include <math.h>

#define PI 3.14159265358979
#define N 100
#define TESTFREQ 800.0
#define SAMPLING_FREQ 8000.0

typedef struct
{
  float real;
  float imag;
} COMPLEX;

COMPLEX samples[N];

void dft(COMPLEX *x)
{
  COMPLEX result[N];
  int k,n;

  for (k=0 ; k<N ; k++)
  {
    result[k].real=0.0;
    result[k].imag = 0.0;

    for (n=0 ; n<N ; n++)
    {
      result[k].real += x[n].real*cos(2*PI*k*n/N) +
x[n].imag*sin(2*PI*k*n/N);
      result[k].imag += x[n].imag*cos(2*PI*k*n/N) -
x[n].real*sin(2*PI*k*n/N);
    }
  }
  for (k=0 ; k<N ; k++)
  {
    x[k] = result[k];
  }
}

void main()
{
  int n;

  for(n=0 ; n<N ; n++)
  {
  samples[n].real = cos(2*PI*TESTFREQ*n/SAMPLING_FREQ);
  samples[n].imag = 0.0;
  }
  printf("real input data stored in array samples[]\n");
  printf("\n"); // place breakpoint here
  dft(samples);          //call DFT function
  printf("done!\n");
}
```

FIGURE 6.12. Listing of program dft.c.

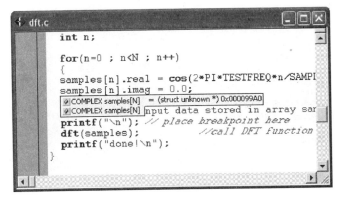

FIGURE 6.16. Pop-up window showing address in memory of array `samples`.

and so on. The real parts of $X(k)$ are displayed by setting the *DSP Data Type* to 32-bit floating point, the *Index Increment* to 2, and the *Start Address* to *samples*, that is, the address of the first value of type `float` in the array `samples`, in the *Graph Property Dialog* window. In order to display the imaginary (rather than the real) parts of the sequence $X(k)$, the *Start Address* must be set to the address of the second value of type `float` in the array `samples`. That address can be found by moving the cursor over an occurrence of the identifier `samples` in the source file `dft.c`. Its hexadecimal address will appear in a pop-up box as shown in Figure 6.16. Entering this value in the *Start Address* field of the *Graph Property Dialog* window in place of the identifier *samples* will result in the same *Graphical Display*. Adding four (the number of bytes used to store one 32-bit floating point value) to the *Start Address* value will result in the imaginary parts of the sequence of complex values being displayed.

Twiddle Factors

Whereas the radix-2 FFT is applicable if N is an integer power of 2, the DFT can be applied to an arbitrary length sequence (e.g., $N = 100$), as illustrated by program `dft.c`. However, the FFT is widely used because of its computational efficiency. Part of that efficiency is due to the use of precalculated twiddle factors, stored in a lookup table, rather than the repeated evaluation of `sin()` and `cos()` functions during computation of the FFT. The use of precalculated twiddle factors can be applied to the function `dft()` to give significant efficiency improvements to program `dft.c`. Calls to the math library functions `sin()` and `cos()` are computationally very expensive and are made a total of $4N^2$ times in function `dft()` (listed in Figure 6.12). In program `dftw.c`, listed in Figure 6.17, these function calls are replaced by reading precalculated twiddle factors from array `twiddle`.

The source file `dftw.c` is stored in folder `dft` and can be substituted for source file `dft.c` in project `dft`. Verify that program `dftw.c` gives similar results. (Change the *Output Filename* to `dftw.out`.)

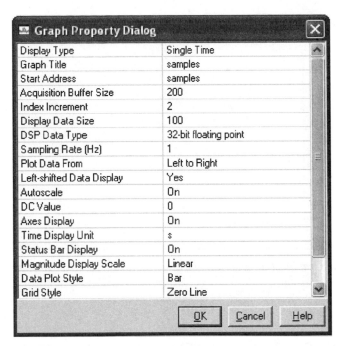

FIGURE 6.13. *Graph Properties* used to display real part of array `samples` in program `dft.c`.

2. Select *View→Graph→Time/Frequency* and set the *Graph Properties* as shown in Figure 6.13. Note that this will display only the real part of the complex values stored in array `samples`. The *Graph Property Data Plot Style* is set to *Bar* in order to emphasize that the DFT operates on discrete data.

3. Select *Debug→ Run*. The program should halt at the breakpoint just before calling function `dft()` and at this point the initial, time-domain contents of array `samples` will be displayed in the *Graphical Display* window.

4. Select *Debug→Run* again. The program should run to completion at which point the contents of array `samples` will be equal to the frequency-domain representation $X(k)$ of the input data $x(n)$. The real part of $X(k)$ will now be displayed in the *Graphical Display* window and you should be able to see two distinct spikes at $k = 10$ and $k = 90$, representing frequency components at $\pm 800\,\mathrm{Hz}$, as shown in Figure 6.14.

Change the frequency of the input waveform to $900\,\mathrm{Hz}$ (`#define TESTFREQ 900.0`) and repeat the procedure listed above. You should see a number of nonzero values in the frequency-domain sequence $X(k)$, as shown in Figure 6.15. This effect is referred to as spectral leakage and is due to the fact that the N sample time-domain sequence stored in array `samples` does not now contain an integer number

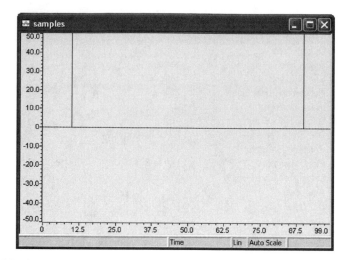

FIGURE 6.14. *Graphical Display* of real part of array `samples` produced by program `dft.c` (TESTFREQ = 800).

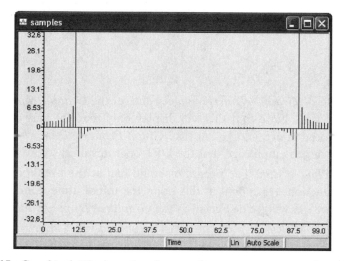

FIGURE 6.15. *Graphical Display* of real part of array `samples` produced by program `dft.c` (TESTFREQ = 900).

of cycles of a sinusoid. Correspondingly, the frequency of that sinusoid is not exactly equal to one of the N discrete frequency components, spaced at intervals of $(8000.0/N)$ Hz in the frequency-domain representation $X(k)$.

The nature of the structured data type COMPLEX is such that array `samples` comprises $2N$ values of type `float` ordered so that the first value is the real part of $X(0)$, the second is the imaginary part of $X(0)$, the third is the real part of $X(1)$,

```c
//dftw.c N-point DFT of sequence read from lookup table
//using pre-computed twiddle factors

#include <stdio.h>
#include <math.h>

#define PI 3.14159265358979
#define N 100
#define TESTFREQ 800.0
#define SAMPLING_FREQ 8000.0

typedef struct
{
    float real;
    float imag;
} COMPLEX;

COMPLEX samples[N];
COMPLEX twiddle[N];

void dftw(COMPLEX *x, COMPLEX *w)
{
    COMPLEX result[N];
    int k,n;

    for (k=0 ; k<N ; k++)
    {
        result[k].real=0.0;
        result[k].imag = 0.0;

        for (n=0 ; n<N ; n++)
        {
            result[k].real += x[n].real*w[(n*k)%N].real -
x[n].imag*w[(n*k)%N].imag;
            result[k].imag += x[n].imag*w[(n*k)%N].real +
x[n].real*w[(n*k)%N].imag;
        }
    }
    for (k=0 ; k<N ; k++)
    {
        x[k] = result[k];
    }
}

void main()
{
    int n;

    for(n=0 ; n<N ; n++)
    {
        twiddle[n].real = cos(2*PI*n/N);
        twiddle[n].imag = -sin(2*PI*n/N);
    }
    for(n=0 ; n<N ; n++)
    {
    samples[n].real = cos(2*PI*TESTFREQ*n/SAMPLING_FREQ);
    samples[n].imag = 0.0;
    }
    printf("real input data stored in array samples[]\n");
    printf("\n"); // place breakpoint here
    dftw(samples,twiddle);                      //call DFT function
    printf("done!\n");
}
```

FIGURE 6.17. Listing of program `dftw.c`.

Example 6.2: Estimating Execution Times for DFT and FFT Functions (fft)

The computational expense of function `dft()` can be illustrated using Code Composer's *Profile Clock* (see Example 1.3). In this example, the functions `dft()` and `dftw()` used in Example 6.1 are compared with a third function, `fft()`, which implements the FFT in C.

Edit the lines in programs `dft.c` and `dftw.c` that read

```
#define N 100
```

to read

```
#define N 128
```

Then

1. Ensure that source file `dft.c` and not `dftw.c` is present in the project.
2. Select *Project→Build Options*. In the *Compiler* tab in the *Basic* category set the *Opt Level* to *Function(–o2)* and in the *Linker* tab set the *Output Filename* to `.\Debug\dft.out`.
3. *Build* the project and load `dft.out`.
4. *Open* source file `dft.c` by double-clicking on its name in the *Project View* window and set breakpoints at the lines `dft(samples);` and `printf("done!\n");`.
5. Select *Profile→Clock→Enable*.
6. Select *Profile→Clock View*.
7. Run the program. It should halt at the first breakpoint.
8. Reset the *Profile Clock* by double-clicking on its icon in the bottom right-hand corner of the CCS window.
9. Run the program. It should stop at the second breakpoint.

The number of instruction cycles counted by the *Profile Clock* (23,828,053) gives an indication of the computational expense of executing function `dft()`. On a 225-MHz C6713, 23,828,053 instruction cycles correspond to an execution time of 105 ms.

Repeat the preceding experiment substituting file `dftw.c` for file `dft.c`. The modified DFT function using twiddle factors, `dftw()`, uses 89,407 instruction cycles, corresponding to 0.397 ms, and representing a decrease in execution time by a factor of 266. At a sampling rate of 8 kHz, 0.397 ms corresponds to just over three sampling periods.

Finally, repeat the experiment using file `fft.c` (also stored in folder `dft`) (see Figure 6.18). This program computes the FFT using a function written in C and

```
//fft.c N-point FFT of sequence read from lookup table

#include <stdio.h>
#include <math.h>
#include "fft.h"

#define PI 3.14159265358979
#define N 128
#define TESTFREQ 800.0
#define SAMPLING_FREQ 8000.0

COMPLEX samples[N];
COMPLEX twiddle[N];

void main()
{
  int n;
  for (n=0 ; n<N ; n++)               //set up DFT twiddle factors
  {
    twiddle[n].real = cos(PI*n/N);
    twiddle[n].imag = -sin(PI*n/N);
  }

  for(n=0 ; n<N ; n++)
  {
  samples[n].real = cos(2*PI*TESTFREQ*n/SAMPLING_FREQ);
  samples[n].imag = 0.0;
  }
  printf("real input data stored in array samples[]\n");
  printf("\n"); // place breakpoint here
  fft(samples,N,twiddle);                          //call DFT function
  printf("done!\n");
}
```

FIGURE 6.18. Listing of program fft.c.

defined in the file fft.h (Figure 6.19). Function fft() takes 24,089 instruction cycles, or 0.107 ms (less than one sampling period at 8 kHz) to execute. The advantage, in terms of execution time, of the FFT over the DFT should increase with the number of points, N, used. Repeat this example using different values of N (e.g., 256 or 512).

6.8.1 Frame-Based Processing

Rather than processing one sample at a time, the DFT and the FFT algorithms process blocks, or frames, of samples. Using the FFT in a real-time program therefore requires a slightly different approach from that used for input and output in previous chapters.

Frame-based processing divides continuous sequences of input and output samples into frames of N samples. Rather than processing one input sample at each

```
//fft.h complex FFT function taken from Rulph's C31 book
//this file contains definition of complex dat structure also

struct cmpx                    //complex data structure used by FFT
{
 float real;
 float imag;
};
typedef struct cmpx COMPLEX;

void fft(COMPLEX *Y, int M, COMPLEX *w)
{
 COMPLEX temp1,temp2;       //temporary storage variables
 int i,j,k;                 //loop counter variables
 int upper_leg, lower_leg;  //index of upper/lower butterfly leg
 int leg_diff;              //difference between upper/lower leg
 int num_stages=0;          //number of FFT stages, or iterations
 int index, step;           //index and step between twiddle factor
 i=1;                       //log(2) of # of points = # of stages
 do
 {
  num_stages+=1;
  i=i*2;
 } while (i!=M);

 leg_diff=M/2;              //difference between upper & lower legs
 step=2;                    //step between values in twiddle.h
 for (i=0;i<num_stages;i++)
 {
  index=0;
  for (j=0;j<leg_diff;j++)
  {
   for (upper_leg=j;upper_leg<M;upper_leg+=(2*leg_diff))
   {
    lower_leg=upper_leg+leg_diff;
    temp1.real=(Y[upper_leg]).real + (Y[lower_leg]).real;
    temp1.imag=(Y[upper_leg]).imag + (Y[lower_leg]).imag;
    temp2.real=(Y[upper_leg]).real - (Y[lower_leg]).real;
    temp2.imag=(Y[upper_leg]).imag - (Y[lower_leg]).imag;
    (Y[lower_leg]).real=temp2.real*(w[index]).real
    -temp2.imag*(w[index]).imag;
    (Y[lower_leg]).imag=temp2.real*(w[index]).imag
    +temp2.imag*(w[index]).real;
    (Y[upper_leg]).real=temp1.real;
    (Y[upper_leg]).imag=temp1.imag;
   }
   index+=step;
  }
  leg_diff=leg_diff/2;
```

FIGURE 6.19. Listing of header file fft.h.

//**dft128c.c**

```c
#include "DSK6713_AIC23.h"              //codec support
Uint32 fs=DSK6713_AIC23_FREQ_8KHZ;    //set sampling rate
#define DSK6713_AIC23_INPUT_MIC 0x0015
#define DSK6713_AIC23_INPUT_LINE 0x0011
Uint16 inputsource=DSK6713_AIC23_INPUT_LINE;

#include <math.h>
#define PI 3.14159265358979
#define TRIGGER 32000
#define N 128
#include "hamm128.h"

typedef struct
{
  float real;
  float imag;
} COMPLEX;

short buffercount = 0;                 //no of samples in iobuffer
short bufferfull = 0;                  //indicates buffer full
COMPLEX A[N], B[N], C[N];
COMPLEX *input_ptr, *output_ptr, *process_ptr, *temp_ptr;
COMPLEX twiddle[N];
short outbuffer[N];

void dft(COMPLEX *x, COMPLEX *w)
{
 COMPLEX result[N];
 int k,n;

  for (k=0 ; k<N ; k++)
  {
    result[k].real=0.0;
    result[k].imag = 0.0;

    for (n=0 ; n<N ; n++)
    {
      result[k].real += x[n].real*w[(n*k)%N].real
                      - x[n].imag*w[(n*k)%N].imag;
      result[k].imag += x[n].imag*w[(n*k)%N].real
                      + x[n].real*w[(n*k)%N].imag;
    }
  }
  for (k=0 ; k<N ; k++)
  {
    x[k] = result[k];
  }
}
```

FIGURE 6.21. DFT program with real-time input (dft128c.c).

```
interrupt void c_int11(void)              //ISR
{
  output_left_sample((short)((output_ptr + buffercount)->real));
  outbuffer[buffercount] =
                      -(short)((output_ptr + buffercount)->real);
  (input_ptr + buffercount)->real = (float)(input_left_sample());
  (input_ptr + buffercount++)->imag = 0.0;
  if (buffercount >= N)
  {
    buffercount = 0;
    bufferfull = 1;
  }
}

main()
{
  int n;

  for (n=0 ; n<N ; n++)                   //set up twiddle factors
  {
    twiddle[n].real = cos(2*PI*n/N);
    twiddlc[n].imag = -sin(2*PI*n/N);
  }
  input_ptr = A;
  output_ptr = B;
  process_ptr = C;
  comm_intr();                            //initialise DSK
  while(1)                                //frame processing loop
  {
    while(bufferfull==0);                 //wait for new frame
    bufferfull = 0;                       //of input samples

    temp_ptr = process_ptr;               //rotate frame pointers
    process_ptr = input_ptr;
    input_ptr = output_ptr;
    output_ptr = temp_ptr;

    dft(process_ptr,twiddle);             //process contents of buffer

    for (n=0 ; n<N ; n++)                 // compute magnitude
    {                                     // and place in real part
      (process_ptr+n)->real =
            -sqrt((process_ptr+n)->real*(process_ptr+n)->real
            + (process_ptr+n)->imag*(process_ptr+n)->imag)/16.0;
    }
    (process_ptr)->real = TRIGGER;        // add oscilloscope trigger
  }                                       //end of while(1)
}                                         //end of main()
```

FIGURE 6.21. (*Continued*)

```
//fft128c.c

#include "DSK6713_AIC23.h"              //codec support
Uint32 fs=DSK6713_AIC23_FREQ_8KHZ;    //set sampling rate
#define DSK6713_AIC23_INPUT_MIC 0x0015
#define DSK6713_AIC23_INPUT_LINE 0x0011
Uint16 inputsource=DSK6713_AIC23_INPUT_LINE;

#include <math.h>
#include "fft.h"
#define PI 3.14159265358979
#define TRIGGER 32000
#define N 128
#include "hamm128.h"

short buffercount = 0;                 //no of samples in iobuffer
short bufferfull = 0;                  //indicates buffer full
COMPLEX A[N], B[N], C[N];
COMPLEX *input_ptr, *output_ptr, *process_ptr, *temp_ptr;
COMPLEX twiddle[N];
short outbuffer[N];

interrupt void c_int11(void)           //ISR
{
  output_left_sample((short)((output_ptr + buffercount)->real));
  outbuffer[buffercount] =
                    -(short)((output_ptr + buffercount)->real);
  (input_ptr + buffercount)->real = (float)(input_left_sample());
  (input_ptr + buffercount++)->imag = 0.0;
  if (buffercount >= N)
  {
    buffercount = 0;
    bufferfull = 1;
  }
}

main()
{
  int n;

  for (n=0 ; n<N ; n++)                //set up twiddle factors
  {
    twiddle[n].real = cos(PI*n/N);
    twiddle[n].imag = -sin(PI*n/N);
  }
  input_ptr = A;
  output_ptr = B;
  process_ptr = C;
  comm_intr();                         //initialise DSK
  while(1)                             //frame processing loop
```

FIGURE 6.28. FFT program with real-time input calling a C-coded FFT function (fft128c.c).

```
{
  while(bufferfull==0);                //wait for new frame
  bufferfull = 0;                      //of input samples

  temp_ptr = process_ptr;              //rotate frame pointers
  process_ptr = input_ptr;
  input_ptr = output_ptr;
  output_ptr = temp_ptr;

  fft(process_ptr,N,twiddle);          //process contents of buffer

  for (n=0 ; n<N ; n++)                // compute magnitude
  {                                    // and place in real part
    (process_ptr+n)->real =
          -sqrt((process_ptr+n)->real*(process_ptr+n)->real
          + (process_ptr+n)->imag*(process_ptr+n)->imag)/16.0;
  }
  (process_ptr)->real = TRIGGER;   // add oscilloscope trigger
  }                                    //end of while(1)
}                                      //end of main()
```

FIGURE 6.28. (*Continued*)

tion `digitrev_index.c`, to produce the index for bit reversal, and `bitrev.sa`, to perform the bit reversal on the twiddle constants, are called before the FFT function is invoked. These two support files for bit reversal are again called to bit-reverse the resulting scrambled output.

N is the number of complex input (note that the input data consist of 2N elements) or output data, so that an N-point FFT is performed. FREQ determines the frequency of the input sine data by selecting the number of points per cycle within the data table. With FREQ set at 8, every eighth point from the table is selected, starting with the first data point. The modulo operator is used as a flag to reinitialize the index. The following four points (scaled) within one period are selected: 0, 1000, 0, and −1000. Example 2.10 (`sine2sliders`) illustrates this indexing scheme to select different numbers of data points within a table.

The magnitude of the resulting FFT is taken. The line of code

```
output_sample (32000);
```

outputs a negative spike. It is used to trigger an oscilloscope. The input data are scaled so that the output magnitude is positive. The sampling rate is achieved through polling.

Build and run this project as **FFTsinetable**. The two support files for bit reversal and the complex FFT function also are included in the project. Figure 6.30 shows a time-domain plot of the resulting output.

Since an output occurs every T_s, the time interval for 32 points corresponds to $32T_s$, or $32(0.125\,\text{ms}) = 4\,\text{ms}$. A negative spike is then repeated every 4 ms. This

```
//FFTsinetable.c FFT{sine}from table. Calls TI FFT function

#include "dsk6713_aic23.h"          //codec support
Uint32 fs=DSK6713_AIC23_FREQ_8KHZ; //set sampling rate
#define DSK6713_AIC23_INPUT_MIC 0x0015
#define DSK6713_AIC23_INPUT_LINE 0x0011
Uint16 inputsource=DSK6713_AIC23_INPUT_LINE; //select line in
#include <math.h>
#define N 32                       //number of FFT points
#define FREQ 8                     //select # of points/cycle
#define RADIX 2                    //radix or base
#define DELTA (2*PI)/N             //argument for sine/cosine
#define TAB_PTS 32                 //# of points in sine_table
#define PI 3.14159265358979
short i = 0;
short iTwid[N/2];                  //index for twiddle constants
short iData[N];                    //index for bitrev X
float Xmag[N];                     //magnitude spectrum of x
typedef struct Complex_tag {float re,im;}Complex;
Complex W[N/RADIX];               //array for twiddle constants
Complex x[N];                      //N complex data values
#pragma DATA_ALIGN(W,sizeof(Complex))    //align W
#pragma DATA_ALIGN(x,sizeof(Complex))    //align x

short sine_table[TAB_PTS] = {0,195,383,556,707,831,924,981,1000,
981,924,831,707,556,383,195,-0,-195,-383,-556,-707,-831,-924,-981,
-1000,-981,-924,-831,-707,-556,-383,-195};

void main()
{
 for( i = 0 ; i < N/RADIX ; i++ )
  {
   W[i].re = cos(DELTA*i);         //real component of W
   W[i].im = sin(DELTA*i);         //neg imag component
  }                                //see cfftr2_dit
 for( i = 0 ; i < N ; i++ )
  {
   x[i].re=sine_table[FREQ*i % TAB_PTS]; //wrap when i=TAB_PTS
   x[i].im = 0 ;                   //zero imaginary part
  }
 digitrev_index(iTwid, N/RADIX, RADIX); //get index for bitrev()
 bitrev(W, iTwid, N/RADIX);        //bit reverse W
 cfftr2_dit(x, W, N );             //TI floating-pt complex FFT

 digitrev_index(iData, N, RADIX);  //get index for bitrev()
 bitrev(x, iData, N);             //freq scrambled->bit-reverse X
 for(i = 0 ; i < N ; i++)
  Xmag[i] = -sqrt(x[i].re*x[i].re+x[i].im*x[i].im ); //mag of X

 comm_poll( ) ;                    //init DSK,codec,McBSP
 while (1)                         //infinite loop
  {
   output_left_sample(32000);      //negative spike as reference
   for (i = 1; i < N; i++)
    output_left_sample((short)Xmag[i]); //output magnitude samples
  }
}
```

FIGURE 6.29. FFT program with input read from a lookup table and using TI's optimized complex FFT function (FFTsinetable.c).

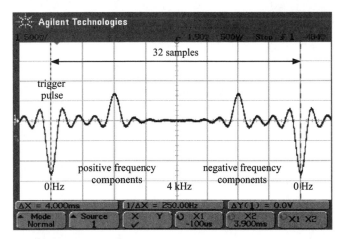

FIGURE 6.30. Time-domain plot representing the magnitude of the FFT of a 2-kHz input signal read from a lookup table and using TI's optimized complex FFT function.

provides a reference, since the time interval between the two negative spikes corresponds to the sampling frequency of 8 kHz. The center of this time interval then corresponds to the Nyquist frequency of 4 kHz (2 ms from the negative spike). The first positive spike occurs at 1 ms from the first negative spike. This corresponds to a frequency of $f = F_s/4 = 2$ kHz. The second positive spike occurs at 3 ms from the first negative spike and corresponds to the folding frequency of $(F_s - f) = 6$ kHz.

Change FREQ to 4 in order to select eight sine data values within the table. Verify that the output is a 1-kHz signal (obtain a plot similar to that in Figure 6.30 from an oscilloscope). A FREQ value of 12 produces an output of 3 kHz. A FREQ value of 15 shows the two positive spikes at the center (between the two negative spikes). Note that aliasing occurs for frequencies larger than 4 kHz. To illustrate that, change FREQ to a value of 20. Verify that the output is an aliased signal at 3 kHz, in lieu of 5 kHz. A FREQ value of 24 shows an aliased signal of 2 kHz in lieu of 6 kHz.

The number of cycles is documented within the function cfftr2_dit.sa (by TI) as

$$\text{Cycles} = ((2N) + 23)\log 2(N) + 6$$

For a 1024-point FFT, the number of cycles would be (2071) (10) + 6 = 20,716. This corresponds to a time of $t = 20{,}716$ cycles/(225 MHz) = 92 μs. That is considerably less time than would be available to process a frame of 1024 samples collected at a sampling rate of 8 kHz (i.e., 1024/8000 = 128 ms).

Example 6.7: FFT of Real-Time Input Using TI's C Callable Optimized Radix-2 FFT Function (FFTr2)

This example extends Example 6.6 for real-time external input in lieu of a sine table as input. Figure 6.31 shows a listing of the C source program FFTr2.c that

```
//FFTr2.c FFT using TI optimized FFT function and real-time input

#include "dsk6713_aic23.h"
Uint32 fs=DSK6713_AIC23_FREQ_8KHZ;  //set sampling rate
#define DSK6713_AIC23_INPUT_MIC 0x0015
#define DSK6713_AIC23_INPUT_LINE 0x0011
Uint16 inputsource=DSK6713_AIC23_INPUT_LINE; //select line in
#include <math.h>
#define N 256                      //number of FFT points
#define RADIX 2                    //radix or base
#define DELTA (2*PI)/N             //argument for sine/cosine
#define PI 3.14159265358979
short i = 0;
short iTwid[N/2];                  //index for twiddle constants
short iData[N];                    //index for bitrev X
float Xmag[N];                     //magnitude spectrum of x
typedef struct Complex_tag {float re,im;}Complex;
Complex W[N/RADIX];               //array for twiddle constants
Complex x[N];                      //N complex data values
#pragma DATA_ALIGN(W,sizeof(Complex)) //align W on boundary
#pragma DATA_ALIGN(x,sizeof(Complex)) //align input x on boundary

void main()
{
 for( i = 0 ; i < N/RADIX ; i++ )
  {
   W[i].re = cos(DELTA*i);         //real component of W
   W[i].im = sin(DELTA*i);         //neg imag component
  }                                //see cfftr2_dit
 digitrev_index(iTwid,N/RADIX,RADIX); //get index for bitrev() W
 bitrev(W, iTwid, N/RADIX);        //bit reverse W

 comm_poll();                      //init DSK,codec,McBSP
 for(i=0; i<N; i++)
  Xmag[i] = 0;                     //init output magnitude
 while (1)                         //infinite loop
 {
  for( i = 0 ; i < N ; i++ )
   {
    x[i].re = (float)((short)input_left_sample()); //get input
    x[i].im = 0.0;                 //zero imaginary part
    if(i==0) output_sample(32000); //negative spike for reference
    else
      output_left_sample((short)Xmag[i]); //output magnitude
   }
  cfftr2_dit(x, W, N );            //TI floating-pt complex FFT
  digitrev_index(iData, N, RADIX); //produces index for bitrev()
  bitrev(x, iData, N);             //bit-reverse x
  for (i =0; i<N; i++)
    Xmag[i] = -sqrt(x[i].re*x[i].re+x[i].im*x[i].im)/32; //mag X
 }
}
```

FIGURE 6.31. FFT program with real-time input using TI's optimized complex FFT function (FFTr2.c).

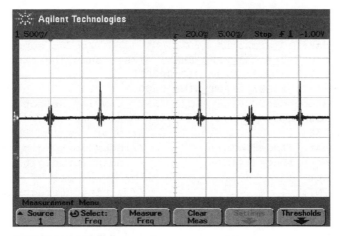

FIGURE 6.32. Output waveform generated by program FFTr2.c.

implements this project. The same FFT support files are used as in Example 6.6, that is, TI's radix-2 optimized FFT function (cfftr2_dit), the function for generating the index for bit reversal (digitrev_index), and the function for the bit-reversal procedure (bitrev). Since the FFT function assumes that the twiddle constants are in reverse order while the input data are in normal order, the index generation and bit reversal associated with the twiddle constants are performed (as in Example 6.6) before the complex FFT function is invoked.

Build this project as **FFTr2**. Input a 2-kHz sinusoidal signal with an amplitude of approximately 2 V p-p and verify the results shown in Figure 6.32. These results are similar to those in Example 6.4 except that in this case $N = 256$.

A project application in Chapter 10 makes use of this example to display a spectrum to a bank of LEDs connected to the DSK through the EMIF 80-pin connector.

Example 6.8: Radix-4 FFT of Real-Time Input Using TI's C Callable Optimized FFT Function (FFTr4)

Figure 6.33 shows the C source program FFTr4.c that calls a radix-4 FFT function to take the FFT of a real-time input signal.

Build this project as **FFTr4**. Input a 2-kHz sinusoidal signal with an amplitude of approximately 2 V p-p. Verify an output similar to that shown in Figure 6.34. These results are similar to those obtained with the radix-2 FFT function in Example 6.7.

6.8.2 Fast Convolution

A major use of frame-based processing and of the FFT is the efficient implementation of FIR filters.

```
//FFTr4.c FFT using TI optimized FFT function and real-time input

#include "dsk6713_aic23.h"                 //codec support
Uint32 fs=DSK6713_AIC23_FREQ_8KHZ;  //set sampling rate
#define DSK6713_AIC23_INPUT_MIC 0x0015
#define DSK6713_AIC23_INPUT_LINE 0x0011
Uint16 inputsource=DSK6713_AIC23_INPUT_LINE; //select input
#include <math.h>
#define N 256                             //no of complex FFT points
unsigned short JIndex[4*N];               //index for digit reversal
unsigned short IIndex[4*N];               //index for digit reversal
int i, count;
float Xmag[N];                            //magnitude spectrum of x
typedef struct Complex_tag {float re,im;}Complex;
Complex W[3*N/2];                         //array for twiddle constants
Complex x[N];                             //N complex data values
double delta = 2*3.14159265359/N;
#pragma DATA_ALIGN(x,sizeof(Complex)); //align x on boundary
#pragma DATA_ALIGN(W,sizeof(Complex)); //align W on boundary

void main()
{
 R4DigitRevIndexTableGen(N,&count,IIndex,JIndex); //for digit rev
 for(i = 0; i < 3*N/4; i++)
  {
   W[i].re = cos(delta*i);              //real component of W
   W[i].im = sin(delta*i);              //Im component of W
  }
 comm_poll();                            //init DSK, codec, McBSP
 for(i=0; i<N; i++)
 Xmag[i] = 0;                            //init output magnitude
 while (1)                               //infinite loop
 {
  output_left_sample(32000);            //-ve spike for reference
  for( i = 0 ; i < N ; i++ )
  {
   x[i].re = (float)((short)input_left_sample()); //get input
   x[i].im = 0.0;                        //zero imaginary part
   if(i>0) output_left_sample((short)Xmag[i]);//output magnitude
  }
  cfftr4_dif(x, W, N);                   //radix-4 FFT function
  digit_reverse((double *)x,IIndex,JIndex,count);//unscramble
  for (i =0; i<N; i++)
   Xmag[i] = -sqrt(x[i].re*x[i].re+x[i].im*x[i].im)/32; //mag X
 }
}
```

FIGURE 6.33. FFT program that calls TI's optimized radix-4 FFT function using real-time input (FFTr4.c).

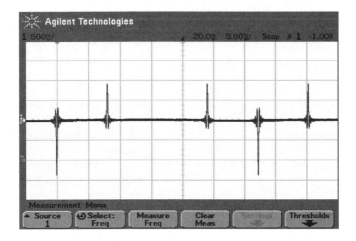

FIGURE 6.34. Output waveform generated by program `FFTr4.c`.

Note that in the following examples, the number of samples in a frame is designated *PTS/2* and the number of coefficients in an FIR filter is designated *N*.

Example 6.9: Frame-Based Implementation of FIR Filters Using Time-Domain Convolution (*timeconvdemo*)

Program `timeconvdemo.c`, listed in Figure 6.35, illustrates frame-based implementation of an FIR filter.

The operation of an FIR filter is described by the convolution sum

$$y(n) = \sum_{i=0}^{N-1} h(i)x(n-i) \tag{6.42}$$

where $h(i)$ is the ith out of N filter coefficients and $x(n)$ is the nth input sample. In a sample-by-sample implementation, the nth output sample $y(n)$ is computed at the nth sampling instant using the convolution sum and a store of past input samples. In a frame-based approach the convolution sum is applied to a frame of *PTS/2* samples every *PTS/2* sampling instants.

The result of convolving a length *PTS/2* sequence of input samples with a length N sequence of filter coefficients is a length ($PTS/2 + N - 1$) sequence of output samples. That is a response (output sequence) that is longer than the frame of *PTS/2* input samples from which it was computed and, as it stands, the basic frame processing mechanism used in programs `frames.c`, `dft128c.c`, and `fft128c.c` cannot be used. That was suited to situations in which each frame of input samples could be processed independently of the frames immediately preceding and succeeding it.

The solution to this problem is to store the section of the response that extends beyond the *PTS/2* samples of the current frame and to add that section of the response to the output computed during processing of the next frame of

```
//timeconvdemo.c overlap-add convolution demonstration program

#include <math.h>
#include <stdio.h>
#include "lp55f.cof"                    //low pass filter coeffs
#define PI 3.14159265358979
#define PTS 128                         //frame size is PTS/2

float coeffs[PTS/2];                    //zero-padded filter coeffs
float A[PTS], B[PTS], C[PTS];           //buffers
float *input_ptr, *output_ptr, *temp_ptr, *process_ptr;
float result[PTS];                      //temporary storage
char in_buffer, proc_buffer, out_buffer, temp_buffer;
short  i;
int wt = 0;

// convolution function - z = conv(x,y)
void conv(float *x, float *y, float *z, int n)
{
  int i, n_lo, n_hi;
  float *xp, *yp;
  for(i=0;i<(2*n-1);i++)
  {
    *z=0.0;
    n_lo=i-(n)+1;
    if(n_lo<0)n_lo=0;
    n_hi=i;
    if(n_hi>(n-1))n_hi=(n-1);
    for(xp=x+n_lo,yp=y+i-n_lo;xp<=x+n_hi;xp++,yp--)
      *z+=*xp * *yp;
    z++;
  }
  *z=0.0;                               //final value in result array
}

main()
{
  input_ptr = A;                        //initialise pointers
  output_ptr = B;
  process_ptr = C;
  in_buffer = 'A';                      //initialise names of buffers
  out_buffer = 'B';                     //for diagnostic messages
  proc_buffer = 'C';

  for (i=0 ; i<PTS/2 ; i++) coeffs[i] = 0.0; //zero pad filter
  for (i=0 ; i<N ; i++) coeffs[i] = h[i];    //in array coeffs

  for (i=0 ; i<PTS ; i++)               //zero all buffer contents
  {
```

FIGURE 6.35. Program illustrating frame-based implementation of FIR filter using time-domain convolution (timeconvdemo.c).

```
        *(output_ptr + i) = 0.0;
        *(process_ptr + i) = 0.0;
        *(input_ptr + i) = 0.0;
    }
    while (1)                           //loop forever
    {
      conv(process_ptr,coeffs,result,PTS/2); //convolve contents
      for (i=0 ; i<PTS ; i++)               //of process buffer
        *(process_ptr+i)=*(result+i);       //with filter coeffs
      for (i=0 ; i<PTS/2 ; i++)             //read new input
      {
        *(input_ptr + i) = (float)(sin(2*PI*wt/50))
                          + 0.25*sin(2*PI*wt/3);
        wt++;
      }
      printf("convolution completed in process buffer (%c)\n"
             ,proc_buffer);
      printf("new input samples read into input buffer (%c)\n"
             ,in_buffer);
      printf("output written from first part");
      printf(" of output buffer (%c)\n",out_buffer);
      printf("\n");                     //insert breakpoint here
      for (i=0 ; i<PTS/2 ; i++)         //add overlapping output
      {                                 //sections in process buffer
        *(process_ptr + i) += *(output_ptr + i + PTS/2);
      }
        printf("second part of output buffer (%c) ", out_buffer);
        printf("has been added to first part");
        printf(" of process buffer (%c)\n",proc_buffer);
      printf("\n");                     //insert breakpoint here
      temp_ptr = process_ptr;           //rotate input, output,
      process_ptr = input_ptr;          //and process buffers
      input_ptr = output_ptr;
      output_ptr = temp_ptr;
      temp_buffer = proc_buffer;        //rotate names of buffer
        proc_buffer = in_buffer;
        in_buffer = out_buffer;
        out_buffer = temp_buffer;
        printf("buffer pointers rotated - ");
        printf("for next section of input\n");
        printf("input buffer is (%c), process buffer is (%c)"
               , in_buffer, proc_buffer);
        printf(", output buffer is (%c)\n", out_buffer);
        printf("\n");
    }                                   // end of while(1)
}                                       //end of main()
```

FIGURE 6.35. (*Continued*)

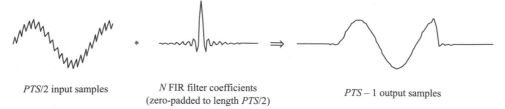

PTS/2 input samples   *N* FIR filter coefficients   *PTS* − 1 output samples
(zero-padded to length *PTS*/2)

FIGURE 6.36. Convolution of one frame of *PTS*/2 input samples and *N* filter coefficients (zero-padded to length *PTS*/2) results in length (*PTS* − 1) section of output samples.

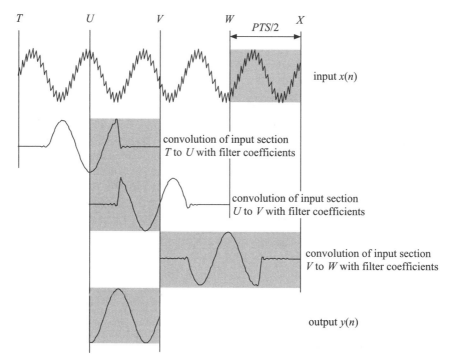

FIGURE 6.37. Overlapping sections of output samples, corresponding to successive frames of input samples are summed to form a continuous output sequence.

input samples. Two alternative forms of this approach are named *overlap-save* and *overlap-add*. Program `timeconvdemo.c` illustrates the overlap-add approach.

The basic mechanism of frame-based, overlap-add, FIR filtering is illustrated in Figures 6.36 and 6.37. Successive *PTS*/2-sample sections of the input sequence $x(n)$ are convolved with the FIR filter coefficients (zero-padded to length *PTS*/2) to produce overlapping sections of the output sequence $y(n)$. Each overlapping convolution result contains (*PTS* − 1) samples. The overlapping sections are summed, point by point, to form the overall output sequence $y(n)$.

The complementary overlap-save method achieves the same result by convolving overlapping sections of the input sequence $x(n)$ with the filter coefficients and discarding parts of the result.

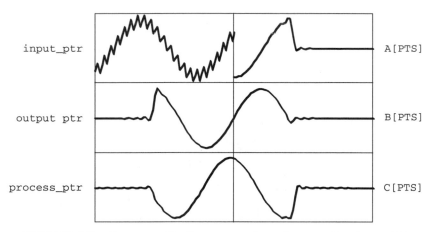

FIGURE 6.38. Contents of buffers A, B, and C at instant X in Figure 6.37.

Using the triple-buffering technique introduced in Example 6.3, it is convenient to use three buffers each of length *PTS* samples, each capable of storing a length $(PTS - 1)$ convolution result. The roles of the three buffers are exchanged after each new frame of *PTS*/2 input samples has been collected.

At the same time as *PTS*/2 input samples are being collected and stored in one buffer (input buffer), and a previously collected set of *PTS*/2 input samples is being convolved with the filter coefficients to give a length $(PTS - 1)$ response (process buffer), the overlapping sum of *PTS*/2 samples from two previously computed convolution operations are output.

Assuming the use of three *PTS*-sample buffers A, B, and C and three pointers `input_ptr`, `process_ptr`, and `output_ptr` (as in program `timeconvdemo.c`), Figure 6.38 shows the contents of those buffers corresponding to Figure 6.37 at instant X, just prior to exchanging pointer values. Over the previous *PTS*/2 sampling instants:

1. *PTS*/2 input samples (section W to X) have been stored in the first *PTS*/2 elements of buffer A (pointed to by `input_ptr`).
2. *PTS*/2 previously collected input samples (section V to W) have been convolved with the filter coefficients and the result stored in the *PTS* elements of buffer C (pointed to by `process_ptr`).
3. *PTS*/2 samples have been output, formed by summing values from the first *PTS*/2 values stored in buffer B (pointed to by `output_ptr`) and the second *PTS*/2 values stored in buffer A.

The buffer contents at instant X are:

1. **buffer A** (`input_ptr`) *PTS*/2 input samples (section W to X) and the last *PTS*/2 samples of the convolution result corresponding to input samples in section T to U.

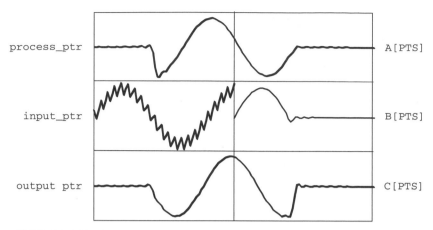

FIGURE 6.39. Contents of buffers A, B, and C at instant Y in Figure 6.37.

2. **buffer B** (`output_ptr`) length $(PTS - 1)$ convolution result corresponding to input samples in section U to V.

3. **buffer C** (`process_ptr`) length $(PTS - 1)$ convolution result corresponding to input samples in section V to W.

At this point the pointer values are exchanged so that the new value of `output_ptr` is equal to the old value of `process_ptr`, the new value of `process_ptr` is equal to the old value of `input_ptr`, and the new value of `input_ptr` is equal to the old value of `output_ptr`.

Figure 6.39 shows the buffer contents corresponding to Figure 6.37 $PTS/2$ sampling instants later at instant Y, just prior to exchanging pointer values. Over the previous $PTS/2$ sampling instants:

1. $PTS/2$ input samples (section X to Y) have been stored in the first $PTS/2$ elements in buffer B (pointed to by `input_ptr`).

2. $PTS/2$ previously collected input samples (section W to X) have been convolved with the filter coefficients and the result stored in the PTS elements of buffer A (pointed to by `process_ptr`).

3. $PTS/2$ samples have been output, formed by summing values from the first $PTS/2$ values stored in buffer C (pointed to by `output_ptr`) and the second $PTS/2$ values stored in buffer B.

The buffer contents at instant Y are:

1. **buffer A** (`process_ptr`) length PTS convolution result corresponding to input samples in section W to X.

2. **buffer B** (`input_ptr`) $PTS/2$ input samples (section X to Y) and the last $PTS/2$ samples of the convolution result corresponding to input samples in section U to V.

3. **buffer C** (`output_ptr`) length PTS convolution result corresponding to input samples in section V to W.

Between instants X and Y (*PTS/2* sampling instants) the contents of buffer C and the second half of buffer B have not changed. Meanwhile, the first *PTS/2* samples in buffer B have been overwritten with new input samples and the *PTS* samples in buffer A have been replaced by a new convolution result.

The process is illustrated by program timeconvdemo.c, which applies a lowpass filter to an internally generated input signal comprising the sum of two sinusoids of different frequencies.

Select *File→Workspace→Load Workspace* and open the file timeconvdemo.wks (in folder timeconvdemo). Note that the saved workspace file will load correctly only if folder timeconvdemo is stored in folder c:\CCStudio_v3.1\MyProjects. Load and Run timeconvdemo.out. Ignore any warning messages warning that identifiers A, B, and C have not been found, and close the *Disassembly* window that appears. A number of breakpoints have been set so that the evolution of the contents of buffers A, B, and C can be observed. Repeatedly clicking on the running man will step through the program from breakpoint to breakpoint. At each breakpoint the *Graphical Display* of the contents of each of the three buffers is updated and explanatory messages are displayed in the *Stdout* window. The Code Composer window should appear as shown in Figure 6.40.

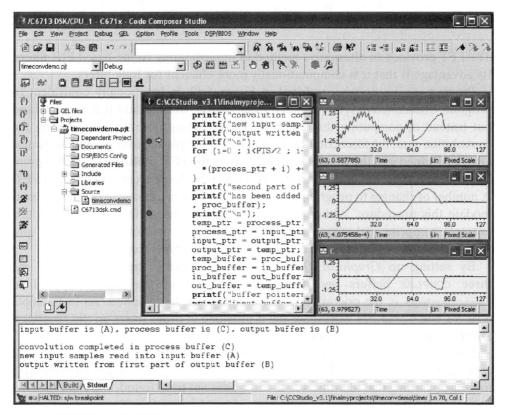

FIGURE 6.40. CCS window during execution of program timeconvdemo.c.

```
void main()
{
  int i;
  for (i=0 ; i<PTS/2 ; i++) coeffs[i] = 0.0; //zero pad filter
  for (i=0 ; i<N ; i++) coeffs[i] = h[i];     //in array coeffs
  input_ptr = A;                    //initialise pointers
  output_ptr = B;
  process_ptr = C;
  comm_intr();
  while(1)                          //frame processing loop
  {
    while (bufferfull == 0);        //wait for buffer full
    bufferfull = 0;
    temp_ptr = process_ptr;
    process_ptr = input_ptr;
    input_ptr = output_ptr;
    output_ptr = temp_ptr;
    conv(process_ptr,coeffs,result,PTS/2); //convolve contents
    for (i=0 ; i<PTS ; i++)               //of process buffer
      *(process_ptr+i)=*(result+i);        //with filter coeffs
    for (i=0 ; i<PTS/2 ; i++)      //add overlapping output
    {                              //sections in process buffer
      *(process_ptr + i) += *(output_ptr + i + PTS/2);
    }
  }                                //end of while
}                                  //end of main()
```

FIGURE 6.41. (*Continued*)

evolution of the contents of buffers A, B, and C can be observed. Repeatedly clicking on the running man will step through the program from breakpoint to breakpoint. At each breakpoint the *Graphical Display* of the contents of each of the three buffers is updated and explanatory messages are displayed in the *Stdout* window.

The basic difference between program fastconvdemo.c and program timeconvdemo.c is the method used to compute the overlapping sections of output samples. The triple-buffering and overlap-add mechanisms used by the two programs are identical. The FFT method of convolution comprises more distinct stages than time-domain convolution and breakpoints have been placed so that the buffer contents after each stage are displayed. Figure 6.43 illustrates the stage at which a sequence of zero-padded input samples in the process buffer (B) have just been transformed into the frequency domain. Although only the real part of the frequency-domain representation is displayed, it is possible to discern two distinct frequency components corresponding to the 160-Hz and 2667-Hz sine waves that make up the input signal.

Example 6.12: Real-Time Frame-Based Fast Convolution (*fastconv*)

This program (Figure 6.44) is the functional equivalent of timeconv.c but uses the FFT method of convolution used in program fastconvdemo.c. Build the project as **fastconv** and verify the implementation of a lowpass filter. The program has

```
//fastconvdemo.c overlap-add convolution demonstration program

#include <math.h>
#include <stdio.h>
#include "lp55f.cof"                //time domain FIR coefficients
#define PI 3.14159265358979

#define PTS 128                     //number of points used in FFT
#define FREQHI 2666.67
#define FREQLO 156.25
#define SAMPLING_FREQ 8000

#include "fft.h"

short buffercount = 0;              //no of new samples in buffer
COMPLEX twiddle[PTS];               //twiddle factors stored in w
COMPLEX coeffs[PTS];                //zero padded freq coeffs
COMPLEX A[PTS], B[PTS], C[PTS];
short i;                            //general purpose index
float a,b;                          //used in complex multiply
COMPLEX *input_ptr, *output_ptr, *temp_ptr, *process_ptr;
char in_buffer, proc_buffer, out_buffer, temp_buffer;
int wt = 0;


main()
{
  input_ptr = A;                    //initialise pointers
  output_ptr = B;
  process_ptr = C;
  in_buffer = 'A';                  //initialise names of buffers
  out_buffer = 'B';                 //for diagnostic messages
  proc_buffer = 'C';

  for (i=0 ; i<PTS ; i++)           //set up twiddle factors
  {
    twiddle[i].real = cos(PI*(i)/PTS);
    twiddle[i].imag = -sin(PI*(i)/PTS);
  }
  for (i=0 ; i<PTS ; i++)           //set up freq domain coeffs
  {
    coeffs[i].real = 0.0;
    coeffs[i].imag = 0.0;
  }
  for (i=0 ; i<N ; i++)
  {
    coeffs[i].real = h[i];
  }
```

FIGURE 6.42. Program illustrating frame-based implementation of FIR filter using fast convolution (fastconvdemo.c).

```
    fft(coeffs,PTS,twiddle);            //transform filter coeffs
                                        //to freq domain
    for (i=0 ; i<PTS ; i++)            //zero all buffer contents
      {
         (output_ptr + i)->real = 0.0;
         (output_ptr + i)->imag = 0.0;
         (process_ptr + i)->real = 0.0;
         (process_ptr + i)->imag = 0.0;
         (input_ptr + i)->real = 0.0;
         (input_ptr + i)->imag = 0.0;
      }
  while (1)
  {
      for(i=0 ; i< PTS ; i++)
         (process_ptr + i)->imag = 0.0;
      for(i=PTS/2 ; i< PTS ; i++)
         (process_ptr + i)->real = 0.0;
      fft(process_ptr,PTS,twiddle);  //transform samples into
                                     //frequency domain
     printf("frequency domain representation of ");
     printf("zero padded input data");
     printf(" in process buffer (%c) \n", proc_buffer);
     printf("\n");                  //insert breakpoint
     for (i=0 ; i<PTS ; i++)        //filter in frequency domain
       {                            //i.e. complex multiply
         a = (process_ptr + i)->real; //samples by coeffs
         b = (process_ptr + i)->imag;
         (process_ptr + i)->real = coeffs[i].real*a
                               - coeffs[i].imag*b;
         (process_ptr + i)->imag = -(coeffs[i].real*b
                               + coeffs[i].imag*a);
       }
     printf("frequency domain result of ");
     printf("multiplying by filter response");
     printf(" in process buffer (%c) \n", proc_buffer);
     printf("\n");                  //insert breakpoint
     fft(process_ptr,PTS,twiddle);
     for (i=0 ; i<PTS ; i++)
       {
         (process_ptr + i)->real /= PTS;
         (process_ptr + i)->imag /= -PTS;
       }
     printf("time domain result of processing now");
     printf(" in process buffer (%c) \n", proc_buffer);
     printf("\n");                  //insert breakpoint
```

FIGURE 6.42. (*Continued*)

```
    for (i=0 ; i<PTS/2 ; i++)          //read new input into buffer
    {
      (input_ptr + i)->real =
        (float)(sin(2*PI*wt*FREQLO/SAMPLING_FREQ))
        + 0.25*sin(2*PI*wt*FREQHI/SAMPLING_FREQ);
      wt++;
    }
    printf("new input samples read into input buffer ");
    printf("(%c)\n",in_buffer);
    printf("output written from first part of output buffer ");
    printf("(%c)\n",out_buffer);
    printf("\n");                        //insert breakpoint here

    for (i=0 ; i<PTS/2 ; i++)          //overlap add (real part only)
    {
      (process_ptr + i)->real += (output_ptr + i + PTS/2)->real;
    }
    printf("second part of output buffer (%c) ", out_buffer);
    printf("has been added to first part of process buffer ");
    printf("(%c)\n", proc_buffer);
    printf("\n");                        //insert breakpoint here


    temp_ptr = process_ptr;            //rotate input, output
    process_ptr = input_ptr;           //and process buffers
    input_ptr = output_ptr;
    output_ptr = temp_ptr;
    temp_buffer = proc_buffer;         //rotate names of buffer
    proc_buffer = in_buffer;
    in_buffer = out_buffer;
    out_buffer = temp_buffer;
    printf("buffer pointers rotated - ");
    printf("for next section of input\n");
    printf("input buffer is (%c)", in_buffer);
    printf(" process buffer is (%c)", proc_buffer);
    printf(", output buffer is (%c)\n", out_buffer);
    printf("\n");
  }                                     // end of while(1)
}                                       //end of main()
```

FIGURE 6.42. (*Continued*)

been written so that different FIR filter coefficient (.cof) files can be used simply by changing the line that reads

```
#include "lp55f.cof"
```

Note that the maximum possible value of N (the number of filter coefficients) is *PTS/2*. For longer FIR filter impulse responses, the value of PTS will have to be increased by changing the line that reads

```
#define PTS 128
```

```
    }
    fft(coeffs,PTS,twiddle);              //transform filter coeffs
                                          //to freq domain
    input_ptr = A;                        //initialise frame pointers
    process_ptr = B;
    output_ptr = C;
    comm_intr();
    while(1)                              //frame processing loop
    {
        while (bufferfull == 0);          //wait for buffer full
        bufferfull = 0;
        temp_ptr = process_ptr;
        process_ptr = input_ptr;
        input_ptr = output_ptr;
        output_ptr = temp_ptr;

        for (i=0 ; i< PTS ; i++) (process_ptr + i)->imag = 0.0;
        for (i=PTS/2 ; i< PTS ; i++) (process_ptr + i)->real = 0.0;
        fft(process_ptr,PTS,twiddle);   //transform samples
                                        //into frequency domain
        for (i=0 ; i<PTS ; i++)         //filter in frequency domain
        {                               //i.e. complex multiply
            a = (process_ptr + i)->real; //samples by coeffs
            b = (process_ptr + i)->imag;
            (process_ptr + i)->real = coeffs[i].real*a
                                - coeffs[i].imag*b;
            (process_ptr + i)->imag = -(coeffs[i].real*b
                                + coeffs[i].imag*a);
        }
        fft(process_ptr,PTS,twiddle);
        for (i=0 ; i<PTS ; i++)
        {
            (process_ptr + i)->real /= PTS;
            (process_ptr + i)->imag /= -PTS;
        }
        for (i=0 ; i<PTS/2 ; i++)       //overlap add (real part only)
        {
            (process_ptr + i)->real += (output_ptr + i + PTS/2)->real;
        }
    }                                   // end of while
}                                       //end of main()
```

FIGURE 6.44. (*Continued*)

Verify that the low and high frequency components are accentuated, while the midrange frequency components are attenuated. This is because the filter coefficients are scaled in the program by bass_gain and treble_gain, initially set to 1, and by mid_gain, initially set to 0. The slider file graphicEQ.gel allows you to control the three frequency bands independently. Figure 6.46 shows the output spectrum obtained with a signal analyzer using noise as input and three different gain settings.

```
//graphicEQ.c Graphic Equalizer using TI floating-point FFT functions

#include "DSK6713_AIC23.h"              //codec-DSK support file
Uint32 fs=DSK6713_AIC23_FREQ_8KHZ;          //set sampling rate
#include <math.h>
#define DSK6713_AIC23_INPUT_MIC 0x0015
#define DSK6713_AIC23_INPUT_LINE 0x0011
Uint16 inputsource=DSK6713_AIC23_INPUT_LINE; // select mic in
#include "GraphicEQcoeff.h"          //time-domain FIR coefficients
#define PI 3.14159265358979
#define PTS 256                  //number of points for FFT
//#define SQRT_PTS 16
#define RADIX 2
#define DELTA (2*PI)/PTS
typedef struct Complex_tag {float real,imag;} COMPLEX;
#pragma DATA_ALIGN(W,sizeof(COMPLEX))
#pragma DATA_ALIGN(samples,sizeof(COMPLEX))
#pragma DATA_ALIGN(h,sizeof(COMPLEX))
COMPLEX W[PTS/RADIX] ;              //twiddle array
COMPLEX samples[PTS];
COMPLEX h[PTS];
COMPLEX bass[PTS], mid[PTS], treble[PTS];
short buffercount = 0;                //buffer count for iobuffer samples
float iobuffer[PTS/2];                //primary input/output buffer
float overlap[PTS/2];                //intermediate result buffer
short i;                      //index variable
short flag = 0;                  //set to indicate iobuffer full
float a, b;                    //variables for complex multiply
short NUMCOEFFS = sizeof(lpcoeff)/sizeof(float);
short iTwid[PTS/2] ;
float bass_gain = 1.0;            //initial gain values
float mid_gain = 0.0;            //change with GraphicEQ.gel
float treble_gain = 1.0;

interrupt void c_int11(void)        //ISR
{
 output_left_sample((short)(iobuffer[buffercount]));
 iobuffer[buffercount++] = (float)((short)input_left_sample());
 if (buffercount >= PTS/2)        //for overlap-add method iobuffer
  {                    //is half size of FFT used
   buffercount = 0;
   flag = 1;
  }
}

main()
{
 digitrev_index(iTwid, PTS/RADIX, RADIX);
 for( i = 0; i < PTS/RADIX; i++ )
  {
   W[i].real = cos(DELTA*i);
   W[i].imag = sin(DELTA*i);
  }
 bitrev(W, iTwid, PTS/RADIX);      //bit reverse W

 for (i=0 ; i<PTS ; i++)
  {
   bass[i].real = 0.0;
```

FIGURE 6.45. Equalizer program using TI's floating-point FFT support functions
(graphicEQ.c).

```
    bass[i].imag = 0.0;
    mid[i].real = 0.0;
    mid[i].imag = 0.0;
    treble[i].real = 0.0;
    treble[i].imag = 0.0;
  }
  for (i=0; i<NUMCOEFFS; i++)        //same # of coeff for each filter
  {
    bass[i].real = lpcoeff[i];       //lowpass coeff
    mid[i].real =  bpcoeff[i];       //bandpass coeff
    treble[i].real = hpcoeff[i];     //highpass coef
  }

  cfftr2_dit(bass,W,PTS);            //transform each band
  cfftr2_dit(mid,W,PTS);                //into frequency domain
  cfftr2_dit(treble,W,PTS);

  comm_intr();                       //initialise DSK, codec, McBSP
  while(1)                           //frame processing infinite loop
  {
    while (flag == 0);               //wait for iobuffer full
          flag = 0;
    for (i=0 ; i<PTS/2 ; i++)        //iobuffer into samples buffer
      {
        samples[i].real = iobuffer[i];
        iobuffer[i] = overlap[i];    //previously processed output
      }                              //to iobuffer
    for (i=0 ; i<PTS/2 ; i++)
      {                              //upper-half samples to overlap
        overlap[i] = samples[i+PTS/2].real;
        samples[i+PTS/2].real = 0.0; //zero-pad input from iobuffer
      }
    for (i=0 ; i<PTS ; i++)
      samples[i].imag = 0.0;         //init samples buffer

    cfftr2_dit(samples,W,PTS);

    for (i=0 ; i<PTS ; i++)          //construct freq domain filter
      {                              //sum of bass,mid,treble coeffs
      h[i].real = bass[i].real*bass_gain + mid[i].real*mid_gain
              + treble[i].real*treble_gain;
      h[i].imag = bass[i].imag*bass_gain + mid[i].imag*mid_gain
              + treble[i].imag*treble_gain;
      }
    for (i=0; i<PTS; i++)            //frequency-domain representation
      {                              //complex multiply samples by h
      a = samples[i].real;
      b = samples[i].imag;
      samples[i].real = h[i].real*a - h[i].imag*b;
      samples[i].imag = h[i].real*b + h[i].imag*a;
      }

    icfftr2_dif(samples,W,PTS);

    for (i=0 ; i<PTS ; i++)
      samples[i].real /= PTS;
    for (i=0 ; i<PTS/2 ; i++)        //add 1st half to overlap
      overlap[i] += samples[i].real;
  }                                  //end of infinite loop
}                                    //end of main()
```

FIGURE 6.45. (*Continued*)

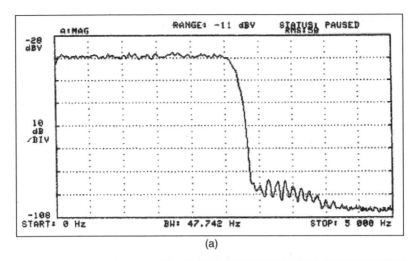

(a)

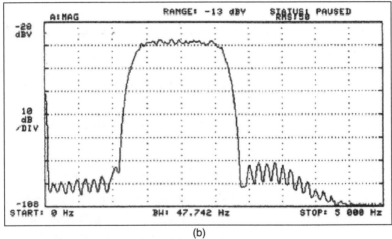

(b)

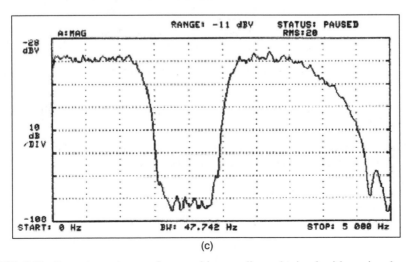

(c)

FIGURE 6.46. Output spectrum of a graphic equalizer obtained with a signal analyzer: (*a*) bass_gain = treble_gain = 1, mid_gain = 0; (*b*) bass_gain = treble_gain = 0, mid_gain = 1; (*c*) bass_gain = mid_gain = 1, treble_gain = 0.

REFERENCES

1. J. W. Cooley and J. W. Tukey, An algorithm for the machine calculation of complex Fourier series, *Mathematics of Computation*, Vol. 19, pp. 297–301, 1965.

2. J. W. Cooley, How the FFT gained acceptance, *IEEE Signal Processing*, pp. 10–13, Jan. 1992.

3. J. W. Cooley, The structure of FFT and convolution algorithms, from a tutorial, *IEEE 1990 International Conference on Acoustics, Speech, and Signal Processing*, Apr. 1990.

4. C. S. Burrus and T. W. Parks, *DFT/FFT and Convolution Algorithms: Theory and Implementation*, Wiley, Hoboken, NJ, 1988.

5. G. D. Bergland, A guided tour of the fast Fourier transform, *IEEE Spectrum*, Vol. 6, pp. 41–51, 1969.

6. E. O. Brigham, *The Fast Fourier Transform*, Prentice Hall, Upper Saddle River, NJ, 1974.

7. S. Winograd, On computing the discrete Fourier transform, *Mathematics of Computation*, Vol. 32, pp. 175–199, 1978.

8. H. F. Silverman, An introduction to programming the Winograd Fourier transform algorithm (WFTA), *IEEE Transactions on Acoustics, Speech, and Signal Processing*, Vol. ASSP-25, pp. 152–165, Apr. 1977.

9. P. E. Papamichalis, Ed., *Digital Signal Processing Applications with the TMS320 Family: Theory, Algorithms, and Implementations, Vol. 3*, Texas Instruments, Dallas, TX, 1990.

10. R. N. Bracewell, Assessing the Hartley transform, *IEEE Transactions on Acoustics, Speech, and Signal Processing*, Vol. ASSP-38, pp. 2174–2176, 1990.

11. R. N. Bracewell, *The Hartley Transform*, Oxford University Press, New York, 1986.

12. H. V. Sorensen, D. L. Jones, M. T. Heidman, and C. S. Burrus, Real-valued fast Fourier transform algorithms, *IEEE Transactions on Acoustics, Speech, and Signal Processing*, Vol. ASSP-35, pp. 849–863, 1987.

13. R. Chassaing, *Digital Signal Processing Laboratory Experiments Using C and the TMS320C31 DSK*, Wiley, Hoboken, NJ, 1999.

14. R. Chassaing, *Digital Signal Processing with C and the TMS320C30*, Wiley, Hoboken, NJ, 1992.

15. A. V. Oppenheim and R. Schafer, *Discrete-Time Signal Processing*, Prentice Hall, Upper Saddle River, NJ, 1989.

16. J. G. Proakis and D. G. Manolakis, *Digital Signal Processing: Principles, Algorithms and Applications*, Prentice Hall, Upper Saddle River, NJ, 2002.

7

Adaptive Filters

- Adaptive structures
- The linear adaptive combiner
- The least mean squares (LMS) algorithm
- Programming examples for noise cancellation and system identification using C code

Adaptive filters are best used in cases where signal conditions or system parameters are slowly changing and the filter is to be adjusted to compensate for this change. A very simple but powerful filter is called the *linear adaptive combiner*, which is nothing more than an adjustable FIR filter. The LMS criterion is a search algorithm that can be used to provide the strategy for adjusting the filter coefficients. Programming examples are included to give a basic intuitive understanding of adaptive filters.

7.1 INTRODUCTION

In conventional FIR and IIR digital filters, it is assumed that the process parameters to determine the filter characteristics are known. They may vary with time, but the nature of the variation is assumed to be known. In many practical problems, there may be a large uncertainty in some parameters because of inadequate prior test data about the process. Some parameters might be expected to change with time, but the exact nature of the change is not predictable. In such cases it is highly

Digital Signal Processing and Applications with the TMS320C6713 and TMS320C6416 DSK,
Second Edition By Rulph Chassaing and Donald Reay
Copyright © 2008 John Wiley & Sons, Inc.

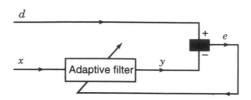

FIGURE 7.1. Basic adaptive filter structure.

desirable to design the filter to be self-learning so that it can adapt itself to the situation at hand.

The coefficients of an adaptive filter are adjusted to compensate for changes in input signal, output signal, or system parameters. Instead of being rigid, an adaptive system can learn the signal characteristics and track slow changes. An adaptive filter can be very useful when there is uncertainty about the characteristics of a signal or when these characteristics change.

Conceptually, the adaptive scheme is fairly simple. Most of the adaptive schemes can be described by the structure shown in Figure 7.1. This is a basic adaptive filter structure in which the adaptive filter's output y is compared with a desired signal d to yield an error signal e, which is fed back to the adaptive filter. The error signal is input to the adaptive algorithm, which adjusts the variable filter to satisfy some predetermined criteria or rules. The desired signal is usually the most difficult one to obtain. One of the first questions that probably comes to mind is: Why are we trying to generate the desired signal at y if we already know it? Surprisingly, in many applications the desired signal does exist somewhere in the system or is known *a priori*. The challenge in applying adaptive techniques is to figure out where to get the desired signal, what to make the output y, and what to make the error e.

The coefficients of the adaptive filter are adjusted, or optimized, using an LMS algorithm based on the error signal. Here we discuss only the LMS searching algorithm with a linear combiner (FIR filter), although there are several strategies for performing adaptive filtering. The output of the adaptive filter in Figure 7.1 is

$$y(n) = \sum_{k=0}^{N-1} w_k(n)x(n-k) \qquad (7.1)$$

where $w_k(n)$ represent N weights or coefficients for a specific time n. The convolution equation (7.1) was implemented in Chapter 4 in conjunction with FIR filtering. It is common practice to use the terminology of weights w for the coefficients associated with topics in adaptive filtering and neural networks.

A performance measure is needed to determine how good the filter is. This measure is based on the error signal,

$$e(n) = d(n) - y(n) \qquad (7.2)$$

which is the difference between the desired signal $d(n)$ and the adaptive filter's output $y(n)$. The weights or coefficients $w_k(n)$ are adjusted such that a mean squared error function is minimized. This mean squared error function is $E[e^2(n)]$, where E represents the expected value. Since there are k weights or coefficients, a gradient of the mean squared error function is required. An estimate can be found instead using the gradient of $e^2(n)$, yielding

$$w_k(n+1) = w_k(n) + 2\beta e(n)x(n-k) \qquad k = 0, 1, \ldots, N-1 \qquad (7.3)$$

which represents the LMS algorithm [1–3]. Equation (7.3) provides a simple but powerful and efficient means of updating the weights, or coefficients, without the need for averaging or differentiating, and will be used for implementing adaptive filters. The input to the adaptive filter is $x(n)$, and the rate of convergence and accuracy of the adaptation process (adaptive step size) is β.

For each specific time n, each coefficient, or weight, $w_k(n)$ is updated or replaced by a new coefficient, based on (7.3), unless the error signal $e(n)$ is zero. After the filter's output $y(n)$, the error signal $e(n)$ and each of the coefficients $w_k(n)$ are updated for a specific time n, a new sample is acquired (from an ADC) and the adaptation process is repeated for a different time. Note that from (7.3), the weights are not updated when $e(n)$ becomes zero.

The linear adaptive combiner is one of the most useful adaptive filter structures and is an adjustable FIR filter. Whereas the coefficients of the frequency-selective FIR filter discussed in Chapter 4 are fixed, the coefficients, or weights, of the adaptive FIR filter can be adjusted based on a changing environment such as an input signal. Adaptive IIR filters (not discussed here) can also be used. A major problem with an adaptive IIR filter is that its poles may be updated during the adaptation process to values outside the unit circle, making the filter unstable.

The programming examples developed later will make use of equations (7.1)–(7.3). In (7.3) we simply use the variable β in lieu of 2β.

7.2 ADAPTIVE STRUCTURES

A number of adaptive structures have been used for different applications in adaptive filtering.

1. *For noise cancellation.* Figure 7.2 shows the adaptive structure in Figure 7.1 modified for a noise cancellation application. The desired signal d is corrupted by uncorrelated additive noise n. The input to the adaptive filter is a noise n' that is correlated with the noise n. The noise n' could come from the same source as n but modified by the environment. The adaptive filter's output y is adapted to the noise n. When this happens, the error signal approaches the desired signal d. The overall output is this error signal and not the adaptive filter's output y. If d is uncorrelated with n, the strategy is to minimize $E(e^2)$,

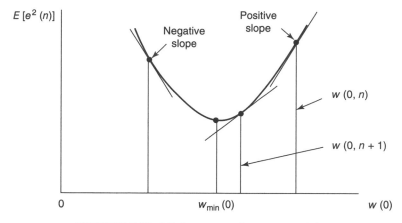

FIGURE 7.10. Minimum search on one weight.

be adjusted in a stepwise fashion until the minimum is reached (Figure 7.10), The size and direction of the step are the two things that must be chosen when making a step. Each step will consist of adding an increment to $w(0, n)$. Note that if the current value of $w(0, n)$ is to the right of the minimum, the step must be negative (but the derivative of the curve is positive); similarly, if the current value is to the left of the minimum, the increment must be positive (but the derivative is negative). This observation leads to the conclusion that the negation of the derivative indicates the proper direction of the increment. Since the derivative vanishes at the minimum, it can also be used to adjust the step size. With these observations we conclude that the step size and direction can be made proportional to the negative of the derivative and the iteration for the weights can be expressed as

$$w(0, n+1) = w(0, n) - \beta \frac{dE[e^2]}{dw(0)} \tag{7.25}$$

where β is an arbitrary positive constant. As shown in Figure 7.10, repeated application of (7.25) will cause $w(0)$ to move by steps from its initial value until it reaches the minimum.

The derivative of the function used in the one-dimensional search can be extended to an N-dimensional surface by replacing it with the gradient of the function. The gradient is a vector of first derivatives with respect to each of the weights:

$$\text{grad}\{E[e^2]\} = \text{grad}\{P\} = \left[\frac{\partial P}{\partial w(0)} \frac{\partial P}{\partial w(1)} \cdots \frac{\partial P}{\partial w(K)} \right]^{\text{T}} \tag{7.26}$$

The gradient points in the direction in which the function, in this case P, increases most rapidly. Therefore, the step size and direction can be made proportional to the gradient of the performance function.

Similarly, the minimum of the N-dimensional performance curve occurs when the gradient vanishes,

$$\text{grad}\{P\} = 0 \tag{7.27}$$

or when the partial derivative with respect to each weight vanishes,

$$\frac{\partial P}{\partial w(0)} = 0, \quad \frac{\partial P}{\partial w(1)} = 0, \quad \cdots, \quad \frac{\partial P}{\partial w(K)} = 0 \tag{7.28}$$

Replacing the single weight with a vector of weights and the derivative with the gradient in (7.25) gives the multiple weight iteration rule,

$$\mathbf{W}(n+1) = \mathbf{W}(n) - \beta \, \text{grad}\{P\} \tag{7.29}$$

The only issue left to resolve is how to find grad$\{P\}$. To get a simple yet practical way to find grad$\{P\}$, we will use an estimate for it rather than the exact gradient. Instead of using the gradient of the expected squared error, we will approximate it with the grad$\{e^2\}$:

$$\text{grad}\{P\} \approx \text{grad}\{e^2\} \tag{7.30}$$

To get a workable expression, let us perform the gradient operation on the squared-error function,

$$\text{grad}\{e^2\} = 2e \, \text{grad}\{e\} \tag{7.31}$$

where

$$e(n) = [d(n) - \mathbf{X}^{\mathrm{T}}(n)\mathbf{W}(n)] \tag{7.32}$$

Substitution yields

$$\text{grad}\{e^2\} = 2e \, \text{grad}[d(n) - \mathbf{X}^{\mathrm{T}}(n)\mathbf{W}(n)] \tag{7.33}$$

Expanding the gradient term gives

$$\text{grad}\{e^2\} = 2e \begin{bmatrix} \dfrac{\partial e}{\partial w(0)} \\[2mm] \dfrac{\partial e}{\partial w(1)} \\[2mm] \vdots \\[2mm] \dfrac{\partial e}{\partial w(K)} \end{bmatrix} = -2e \begin{bmatrix} x(0) \\ x(1) \\ \vdots \\ x(K) \end{bmatrix} \tag{7.34}$$

and

$$\text{grad}\{e^2(n)\} = -2e(n)\mathbf{X}(n) \tag{7.35}$$

Substituting this result for grad{P} in equation (7.29) results in

$$\mathbf{W}(n+1) = \mathbf{W}(n) + 2\beta e(n)\mathbf{X}(n) \tag{7.36}$$

The time index n has been included in the last two equations, implying that e will be updated every sample time. Note that if e goes to zero, then $\mathbf{W}(n + 1) = \mathbf{W}(n)$ and the weights remain constant.

Equation (7.36) forms the single most important result of this chapter, and it is the basis for the LMS algorithm. This equation allows the weights to be updated without squaring, averaging, or differentiating, yet it is powerful and efficient. This equation, as in (7.3), will be used in the following examples.

7.6 PROGRAMMING EXAMPLES FOR NOISE CANCELLATION AND SYSTEM IDENTIFICATION

The following programming examples illustrate adaptive filtering using the LMS algorithm.

Example 7.1: Adaptive Filter Using C Code (adaptc)

This example applies the LMS algorithm using a C-coded program. It illustrates the following steps for the adaptation process using the adaptive structure shown in Figure 7.1:

1. Obtain new samples of the desired signal d and the reference input to the adaptive filter x, which represents a noise signal.
2. Calculate the adaptive FIR filter's output y, applying (7.1) as in Chapter 4 with an FIR filter. In the structure of Figure 7.1, the overall output is the same as the adaptive filter's output y.
3. Calculate the error signal applying (7.2).
4. Update/replace each coefficient or weight applying (7.3).
5. Update the input data samples for the next time n with the data move scheme used in Chapter 4. Such a scheme moves the data instead of a pointer.
6. Repeat the entire adaptive process for the next output sample point.

```
//adaptc.c - non real-time adaptation demonstration
#include <stdio.h>
#include <math.h>
#define beta 0.01                          //convergence rate
#define N   21                             //order of filter
#define NS  60                             //number of samples
#define Fs  8000                           //sampling frequency
#define pi  3.1415926
#define DESIRED 2*cos(2*pi*T*1000/Fs)  //desired signal
#define NOISE sin(2*pi*T*1000/Fs)      //noise signal

float desired[NS], Y_out[NS], error[NS];

void main()
{
  long I, T;
  float D, Y, E;
  float W[N+1] = {0.0};
  float X[N+1] = {0.0};

  for (T = 0; T < NS; T++)              //start adaptive algorithm
  {
    X[0] = NOISE;                       //new noise sample
    D = DESIRED;                        //desired signal
    Y = 0;                             //filter'output set to zero
    for (I = 0; I <= N; I++)
      Y += (W[I] * X[I]);              //calculate filter output
    E = D - Y;                         //calculate error signal
    for (I = N; I >= 0; I--)
    {
      W[I] = W[I] + (beta*E*X[I]);     //update filter coefficients
      if (I != 0) X[I] = X[I-1];       //update data sample
    }
    desired[T] = D;
    Y_out[T] = Y;
    error[T] = E;
  }
  printf("done!\n");
}
```

FIGURE 7.11. Adaptive filter program(adaptc.c).

Figure 7.11 shows a listing of the program adaptc.c, which implements the LMS algorithm for the adaptive filter structure in Figure 7.1. A desired signal is chosen as $2\cos(2n\pi f/Fs)$, and a reference noise input to the adaptive filter is chosen as $\sin(2n\pi f/Fs)$, where f is 1 kHz and $Fs = 8$ kHz. The adaptation rate, filter order, and number of samples are 0.01, 21, and 60, respectively.

The overall output is the adaptive filter's output y, which adapts or converges to the desired cosine signal d.

Build the project as **adaptc**. Because the program does not use any real-time input or output, it is not necessary to add the files c6713dskinit.c or vectors_ intr.asm to the project.

Figure 7.12 shows a plot of the adaptive filter output Y_out, desired output desired, and error error, plotted using CCS. The filter output converges to the desired cosine signal. Change the adaptation or convergence rate beta to 0.02 and verify a faster rate of adaptation.

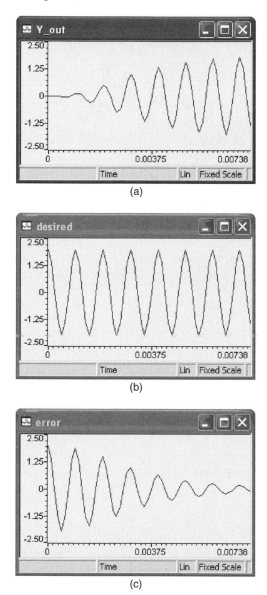

FIGURE 7.12. Plots of adaptive filter output, desired output, and error using program adaptc.c.

Example 7.2: Adaptive Filter for Sinusoidal Noise Cancellation (adaptnoise)

This example illustrates the application of the LMS criterion to cancel an undesirable sinusoidal noise. Figure 7.13 shows a listing of the program adaptnoise.c, which implements an adaptive FIR filter using the structure in Figure 7.2.

A desired sine wave of 1500 Hz with an additive (undesired) sine wave noise of 312 Hz forms one of two inputs to the adaptive filter structure. A reference (template) cosine signal, with a frequency of 312 Hz, is the input to a 30-coefficient adaptive FIR filter. The 312-Hz reference cosine signal is correlated with the 312-Hz additive sine noise but not with the 1500-Hz desired sine signal.

At each sampling instant, the output of the adaptive FIR filter is calculated and the 30 weights or coefficients are updated along with the delay samples. The error signal E is the overall desired output of the adaptive structure. This error signal is the difference between the desired signal and additive noise (dplusn) and the adaptive filter output, yn.

All signals used are from a lookup table generated using MATLAB. No external inputs are used in this example. Figure 7.14 shows the MATLAB m-file adaptnoise.m (a more complete version is on the CD) used to calculate the data values for the desired sine signal of 1500 Hz, the additive noise as a sine of 312 Hz, and the reference signal as a cosine of 312 Hz. The files generated are:

1. dplusn.h: sine(1500 Hz) + sine(312 Hz)
2. refnoise.h: cosine(312 Hz)

Figure 7.15 shows the file dplusn.h with data values that represent the desired 1500-Hz sine wave signal plus additive noise. The constant beta determines the rate of convergence.

Build and run this project as **adaptnoise**. Verify the following output result: The undesired 312-Hz sinusoidal signal is being gradually reduced (canceled), while the desired 1500-Hz signal remains. Note that in this application the output desired is the error signal E, which adapts (converges) to the desired signal. A faster rate of cancellation can be observed with a larger value of beta. However, if beta is too large, the adaptation may become unstable. A GEL slider (adaptnoise.gel) is provided that allows either the error signal or the 1500-Hz sine wave with additive noise signal to be output.

Example 7.3: Adaptive FIR Filter for Noise Cancellation Using External Inputs (adaptnoise_2IN)

This example extends the previous one to cancel undesired sinusoidal noise using external inputs. Figure 7.16 shows the source program adaptnoise_2IN.c that

```
//adaptnoise.c  Adaptive FIR filter for noise cancellation

#include "DSK6713_AIC23.h"
Uint32 fs= DSK6713_AIC23_FREQ_8KHZ;
#define DSK6713_AIC23_INPUT_MIC 0x0015
#define DSK6713_AIC23_INPUT_LINE 0x0011
Uint16 inputsource=DSK6713_AIC23_INPUT_LINE;

#include "refnoise.h"                   //cosine 312 Hz
#include "dplusn.h"                     //sin(1500) + sin(312)
#define beta 1E-9                       //rate of convergence
#define N 30                            //# of weights (coefficients)
#define NS 128                          //# of output sample points
float w[N];                             //buffer weights of adapt filter
float delay[N];                         //input buffer to adapt filter
short output;                           //overall output
short out_type = 1;                     //output type for slider

interrupt void c_int11()                //ISR
{
 short i;
 static short buffercount=0;            //init count of # out samples
 float yn, E;                           //output filter/"error" signal

 delay[0] = refnoise[buffercount];      //cos(312Hz) input to adapt FIR
 yn = 0;                                //init output of adapt filter
 for (i = 0; i < N; i++)                //to calculate out of adapt FIR
     yn += (w[i] * delay[i]);           //output of adaptive filter

 E = dplusn[buffercount] - yn;          //"error" signal=(d+n)-yn

 for (i = N-1; i >= 0; i--)             //to update weights and delays
   {
     w[i] = w[i] + beta*E*delay[i];     //update weights
     delay[i] = delay[i-1];             //update delay samples
   }
 buffercount++;                         //increment buffer count
 if (buffercount >= NS)                 //if buffercount=# out samples
     buffercount = 0;                   //reinit count

 if (out_type == 1)                     //if slider in position 1

     output = ((short)E*10);            //"error" signal overall output
 else if (out_type == 2)                //if slider in position 2
     output=dplusn[buffercount]*10;     //desired(1500)+noise(312)
 output_left_sample(output);            //overall output result
 return;                                //return from ISR
}

void main()
{
 short T=0;
 for (T = 0; T < 30; T++)
   {
     w[T] = 0;                          //init buffer for weights
     delay[T] = 0;                      //init buffer for delay samples
   }
 comm_intr();                           //init DSK, codec, McBSP
 while(1);                              //infinite loop
}
```

FIGURE 7.13. Adaptive FIR filter program for sinusoidal noise cancellation (adaptnoise.c).

```
%adaptnoise.m Generates: dplusn.h (s312+s1500), refnoise.h
cos(312),and sin1500.h

for i=1:128
  desired(i) = round(100*sin(2*pi*(i-1)*1500/8000)); %sin(1500)
  addnoise(i) = round(100*sin(2*pi*(i-1)*312/8000)); %sin(312)
  refnoise(i) = round(100*cos(2*pi*(i-1)*312/8000)); %cos(312)
end
dplusn = addnoise + desired;
%sin(312)+sin(1500)

fid=fopen('dplusn.h','w');
%desired + noise
fprintf(fid,'short dplusn[128]={');
fprintf(fid,'%d, ' ,dplusn(1:127));
fprintf(fid,'%d' ,dplusn(128));
fprintf(fid,'};\n');
fclose(fid);

fid=fopen('refnoise.h','w');
        %reference noise
fprintf(fid,'short refnoise[128]={');
fprintf(fid,'%d, ' ,refnoise(1:127));
fprintf(fid,'%d' ,refnoise(128));
fprintf(fid,'};\n');
fclose(fid);

fid=fopen('sin1500.h','w');
        %desired sin(1500)
fprintf(fid,'short sin1500[128]={');
fprintf(fid,'%d, ' ,desired(1:127));
fprintf(fid,'%d' ,desired(128));
fprintf(fid,'};\n');
fclose(fid);
```

FIGURE 7.14. MATLAB m-file used to generate data values for sine(1500 Hz), sine(1500 Hz) + sine(312 Hz), and cosine(312 Hz) (adaptnoise.m).

```
short dplusn[128]={0, 116, 118, 29, -17, 56, 170, 191, 93, -11,
-7, 81, 120, 34, -100, -143, -70, 7, -24, -138, -198, -129, -7,
32, -39, -108, -62, 71, 155, 111, 17, 5, 100, 189, 160, 37, -43,
-3, 82, 79, -37, -150, -147, -52, 2, -62, -167, -179, -72, 39,
40, -45, -82, 3, 133, 171, 92, 7, 29, 133, 184, 107, -22, -65,
3, 70, 26, -103, -182, -131, -28, -7, -93, -174, -137, -7, 78,
40, -45, -43, 68, 176, 166, 62, -1, 54, 150, 154, 41, -74, -77,
8, 47, -34, -157, -188, -100, -6, -19, -115, -159, -75, 57, 103,
34, -36, 4, 127, 197, 138, 26, -4, 74, 147, 104, -29, -115, -77,
11, 15, -90, -190, -171, -58, 14, -33, -122, -121};
```

FIGURE 7.15. MATLAB header file generated for sine(1500 Hz) + sine(312 Hz) with 128 points (dplusn.h).

//**adaptnoise_2IN.c** Adaptive FIR for sinusoidal noise interference

```
#include "DSK6713_AIC23.h"            //codec support
Uint32 fs=DSK6713_AIC23_FREQ_8KHZ; //set sampling rate
#define DSK6713_AIC23_INPUT_MIC 0x0015
#define DSK6713_AIC23_INPUT_LINE 0x0011
Uint16 inputsource=DSK6713_AIC23_INPUT_LINE;
#define beta 1E-12                   //rate of convergence
#define N 30                         //# of weights (coefficients)
#define LEFT 0                       //left channel
#define RIGHT 1                      //right channel
float w[N];                          //weights for adapt filter
float delay[N];                      //input buffer to adapt filter
short output;                        //overall output
short out_type = 1;                  //output type for slider
volatile union{unsigned int uint; short channel[2];}AIC23_data;

interrupt void c_int11()             //ISR
{
 short i;
 float yn=0, E=0, dplusn=0, desired=0, noise=0;

 AIC23_data.uint = input_sample(); //input from both channels
 desired =(AIC23_data.channel[LEFT]); //input left channel
 noise = (AIC23_data.channel[RIGHT]); //input right channel

 dplusn = desired + noise;           //desired+noise
 delay[0] = noise;                   //noise as input to adapt FIR

 for (i = 0; i < N; i++)             //calculate out of adapt FIR
  yn += (w[i] * delay[i]);           //output of adaptive filter
 E = (desired + noise) - yn;         //"error" signal=(d+n)-yn
 for (i = N-1; i >= 0; i--)          //to update weights and delays
  {
   w[i] = w[i] + beta*E*delay[i];    //update weights
   delay[i] = delay[i-1];            //update delay samples
  }
 if(out_type == 1)                   //if slider in position 1
  output=((short)E);                 //error signal as output
 else if(out_type==2)                //if slider in position 2
  output=((short)dplusn);            //output (desired+noise)
 output_left_sample(output);         //overall output result
 return;
}

 void main()
{
 short T=0;
 for (T = 0; T < 30; T++)
 {
  w[T] = 0;                          //init buffer for weights
  delay[T] = 0;                      //init delay sample buffer
 }
 comm_intr();                        //init DSK, codec, McBSP
 while(1);                           //infinite loop
}
```

FIGURE 7.16. Adaptive filter program for noise cancellation using external inputs (adaptnoise_2IN.c).

allows two external inputs: a desired signal and a sinusoidal interference. The program uses the function `input_sample()` to return both left- and right-hand channels of input as 16-bit signed integer components of a 32-bit structure. The desired signal is input through the left channel and the undesired noise signal through the right channel. A stereo 3.5-mm jack plug to dual RCA jack plug cable and RCA to BNC adapters are useful for implementing this example. The basic adaptive structure shown in Figure 7.2 is applied here along with the LMS algorithm.

Build this project as **adaptnoise_2IN**.

1. Desired: 1.5 kHz; undesired: 2 kHz. Input a desired sinusoidal signal (with a frequency of 1.5 kHz) into the left channel and an undesired sinusoidal noise signal of 2 kHz into the right channel. Run the program. Verify that the 2-kHz noise signal is being canceled gradually. You can adjust the rate of convergence by changing `beta` by a factor of 10 in the program. Load the GEL slider program `adaptnoise_2IN.gel` and change the slider position from 1 to 2.

   Verify the output as the two original sinusoidal signals at 1.5 and at 2 kHz.

2. Desired: wideband random noise; undesired: 2 kHz. Input random noise (from a noise generator, *Goldwave*, etc.) as the desired wideband signal into the left input channel and the undesired 2-kHz sinusoidal noise signal into the right input channel. Restart/run the program. Verify that the 2-kHz sinusoidal noise signal is being canceled gradually, with the wideband random noise signal remaining. With the slider in position 2, observe that both the undesired and desired input signals are as shown in Figure 7.17a. Figure 7.17b shows only the desired wideband random noise signal after the adaptation process.

Example 7.4: Adaptive FIR Filter for System ID of a Fixed FIR as an Unknown System (`adaptIDFIR`)

Figure 7.18 shows a listing of the program `adaptIDFIR.c`, which uses an adaptive FIR filter to identify an unknown system. See also Examples 7.2 and 7.3, which implement an adaptive FIR filter for noise cancellation. A block diagram of the system used in this example is shown in Figure 7.19.

The unknown system to be identified is an FIR bandpass filter with 55 coefficients centered at $F_s/4 = 2$ kHz. The coefficients of this fixed FIR filter are read from the file `bp55f.cof`, previously used in Example 4.5. A 60-coefficient adaptive FIR filter is used to identify the fixed (unknown) FIR bandpass filter.

A pseudorandom binary noise sequence, generated within the program (see also Examples 2.16 and 4.4), is input to both the fixed (unknown) and the adaptive FIR filters and an error signal is formed from their outputs. The adaptation process seeks to minimize the variance of that error signal. It is important to use wideband noise as an input signal in order to identify the characteristics of the

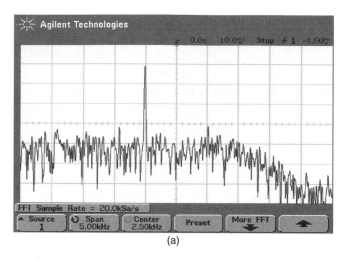

(a)

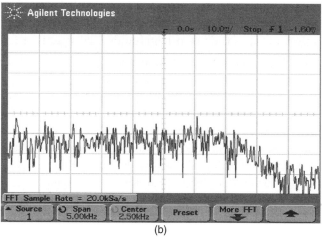

(b)

FIGURE 7.17. Plots illustrating the adaptation process obtained using program `adapt-noise_2IN.c`: (a) 2-kHz undesired sinusoidal interference and desired wideband noise signal before adaptation; and (b) cancellation of 2-kHz interference after adaptation.

unknown system over the entire frequency range from zero to half the sampling frequency.

An extra memory location is used in each of the two delay sample buffers (fixed and adaptive FIR). These are used to update the delay samples.

Build and run this project as **adaptIDFIR**. Load the GEL file `adaptIDFIR.gel` and bring up a GEL slider by selecting *GEL→Output Type*. The slider can be used to select `fir_out` (the output from the fixed (unknown) FIR filter), `adaptfir_out` (the output from the adaptive FIR filter), or `E` (the error) as the

//**adaptIDFIR.c** Adaptive FIR for system ID of an FIR

```c
#include "DSK6713_AIC23.h"          //codec-DSK support file
Uint32 fs=DSK6713_AIC23_FREQ_8KHZ; //set sampling rate
#define DSK6713_AIC23_INPUT_MIC 0x0015
#define DSK6713_AIC23_INPUT_LINE 0x0011
Uint16 inputsource=DSK6713_AIC23_INPUT_LINE;
#include "bp55f.cof"               //fixed FIR filter coefficients
#include "noise_gen.h"             //noise generation support file
#define beta 1E-12                 //rate of convergence
#define WLENGTH 60                 //# of coefffor adaptive FIR
float w[WLENGTH+1];                //buffer coeff for adaptive FIR
int dly_adapt[WLENGTH+1];          //adaptive FIR samples buffer
int dly_fix[N+1];                  //buffer samples of fixed FIR
short out_type = 1;                //output for adaptive/fixed FIR
int fb;                            //feedback variable
shift_reg sreg;                    //shift register

int prand(void)                    //pseudo-random sequence {-1,1}
{
 int prnseq;
 if(sreg.bt.b0)
  prnseq = -8000;                  //scaled negative noise level
 else
  prnseq = 8000;                   //scaled positive noise level
 fb =(sreg.bt.b0)^(sreg.bt.b1);    //XOR bits 0,1
 fb^=(sreg.bt.b11)^(sreg.bt.b13);  //with bits 11,13 -> fb
 sreg.regval<<=1;
 sreg.bt.b0=fb;                    //close feedback path
 return prnseq;                    //return noise sequence
}

interrupt void c_int11()           //ISR
{
 int i;
 int fir_out = 0;                  //init output of fixed FIR
 int adaptfir_out = 0;            //init output of adapt FIR
 float E;                          //error=diff of fixed/adapt out

 dly_fix[0] = prand();             //input noise to fixed FIR
 dly_adapt[0]=dly_fix[0];          //as well as to adaptive FIR
 for (i = N-1; i>= 0; i--)
   {
    fir_out +=(h[i]*dly_fix[i]);   //fixed FIR filter output
    dly_fix[i+1] = dly_fix[i];     //update samples of fixed FIR
   }
 for (i = 0; i < WLENGTH; i++)
   adaptfir_out +=(w[i]*dly_adapt[i]); //adaptive FIR output
```

FIGURE 7.18. Program to implement an adaptive FIR filter that models (identifies) a fixed FIR filter (adaptIDFIR.c).

```
  E = fir_out - adaptfir_out;           //error signal

  for (i = WLENGTH-1; i >= 0; i--)
    {
    w[i]=w[i]+(beta*E*dly_adapt[i]);//update adaptive FIR weights
    dly_adapt[i+1] = dly_adapt[i];  //update adaptive FIR samples
    }
  if (out_type == 1)                     //slider position for adapt FIR
    output_left_sample((short)adaptfir_out); //output adaptive FIR
  else if (out_type == 2)                //slider position for fixed FIR
    output_left_sample((short)fir_out); //output fixed FIR filter
  else if (out_type == 3)                //slider position for fixed FIR
    output_left_sample((short)E);       //output of fixed FIR filter
  return;
}

void main()
{
 int T=0, i=0;
 for (i = 0; i < WLENGTH; i++)
   {
   w[i] = 0.0;                          //init coeff for adaptive FIR
   dly_adapt[i] = 0;                    //init buffer for adaptive FIR
   }
 for (T = 0; T < N; T++)
   dly_fix[T] = 0;                      //init buffer for fixed FIR
 sreg.regval=0xFFFF;                    //initial seed value
 fb = 1;                                //initial feevack value
 comm_intr();                           //init DSK, codec, McBSP
 while (1);                             //infinite loop
}
```

FIGURE 7.18. (*Continued*)

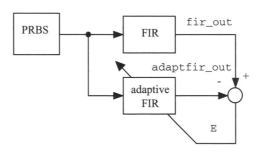

FIGURE 7.19. Block diagram representation of system identification scheme implemented by program adaptIDFIR.c.

signal written to the LINE OUT and HEADPHONE connectors. Verify that the output of the adaptive FIR filter (adaptfir_out) converges to bandlimited noise similar in frequency content to the output of the fixed FIR filter (fir_out) and that the variance of the error signal (E) gradually diminishes. Reload and run the program in order to view the adaptation process again and to observe a different signal (fir_out, adaptfir_out, or E) during the adaptation process.

Edit the program to include the coefficient file bs55f.cof (in place of bp55f.cof), which represents an FIR bandstop filter with 55 coefficients centered at 2 kHz. The FIR bandstop filter represents the unknown system to be identified.

Rebuild and run the program and verify that the output of the adaptive FIR filter (with the slider in position 1) is almost identical to that of the FIR bandstop filter (with the slider in position 2). Increase (decrease) the value of beta by a factor of 10 to observe a faster (slower) rate of convergence. Change the number of weights (coefficients) from 60 to 40 and verify a slight degradation of the identification process.

Example 7.5: Adaptive FIR for System ID of a Fixed FIR as an Unknown System with Weights of an Adaptive Filter Initialized as an FIR Bandpass (adaptIDFIRw)

In this example, program adaptIDFIR.c is modified slightly to create the program adaptIDFIRW.c (Figure 7.20). This new program initializes the weights, w, of the adaptive FIR filter with the coefficients of an FIR bandpass filter centered at 3 kHz, read from the coefficient file bp3000.cof rather than initializing the weights to zero.

Build this project as **adaptIDFIRw**. Initially, the frequency content of the output of the adaptive FIR filter is centered at 3 kHz. Then, gradually, as the adaptive filter identifies the fixed (unknown) FIR bandpass filter (bp55.cof), its output changes to bandlimited noise centered on frequency 2 kHz.

The adaptation process is illustrated in Figures 7.21 and 7.22. Figure 7.21 shows the frequency content of the output of the adaptive filter at different stages in the adaptation process (captured using an oscilloscope) and Figure 7.22 shows the magnitude FFT of the adaptive filter coefficients at corresponding points in time.

Example 7.6: Adaptive FIR for System ID of Fixed IIR as an Unknown System (iirsosadapt)

An adaptive FIR filter can be used to identify the characteristics not only of other FIR filters but of IIR filters (provided that the substantial part of the IIR filter impulse response is shorter than that possible using the adaptive FIR filter). Program iirsosadapt.c (Figure 7.23) combines parts of programs iirsos.c (Example 5.1) and adaptIDFIR.c in order to illustrate this.

```
//adaptIDFIRW.c Adaptive FIR for system ID of an FIR

#include "DSK6713_AIC23.h"          //codec-DSK support file
Uint32 fs=DSK6713_AIC23_FREQ_8KHZ; //set sampling rate
#define DSK6713_AIC23_INPUT_MIC 0x0015
#define DSK6713_AIC23_INPUT_LINE 0x0011
Uint16 inputsource=DSK6713_AIC23_INPUT_LINE;
#include "bp55.cof"                 //fixed FIR filter coefficients
#include "bp3000.cof"
#include "noise_gen.h"              //noise generation support file
#define beta 1E-12                  //rate of convergence
#define WLENGTH 60                  //# of coefffor adaptive FIR
float w[WLENGTH+1];                 //buffer coeff for adaptive FIR
int dly_adapt[WLENGTH+1];           //adaptive FIR samples buffer
int dly_fix[N+1];                   //buffer samples of fixed FIR
short out_type = 1;                 //output for adaptive/fixed FIR
int fb;                             //feedback variable
shift_reg sreg;                     //shift register

int prand(void)                     //pseudo-random sequence {-1,1}
{
 int prnseq;
 if(sreg.bt.b0)
  prnseq = -8000;                   //scaled negative noise level
 else
  prnseq = 8000;                    //scaled positive noise level
 fb =(sreg.bt.b0)^(sreg.bt.b1);     //XOR bits 0,1
 fb^=(sreg.bt.b11)^(sreg.bt.b13);   //with bits 11,13 -> fb
 sreg.regval<<=1;
 sreg.bt.b0=fb;                     //close feedback path
 return prnseq;                     //return noise sequence
}

interrupt void c_int11()            //ISR
{
 int i;
 int fir_out = 0;                   //init output of fixed FIR
 int adaptfir_out = 0;             //init output of adapt FIR
 float E;                           //error=diff of fixed/adapt out

 dly_fix[0] = prand();             //input noise to fixed FIR
 dly_adapt[0]=dly_fix[0];          //as well as to adaptive FIR
 for (i = N-1; i>= 0; i--)
  {
   fir_out +=(h[i]*dly_fix[i]);    //fixed FIR filter output
   dly_fix[i+1] = dly_fix[i];      //update samples of fixed FIR
  }
 for (i = 0; i < WLENGTH; i++)
   adaptfir_out +=(w[i]*dly_adapt[i]); //adaptive FIR output
```

FIGURE 7.20. Program to implement an adaptive FIR filter that models (identifies) a fixed FIR filter with initialised coefficients (adaptIDFIRw.c).

```
 E = fir_out - adaptfir_out;        //error signal

 for (i = WLENGTH-1; i >= 0; i--)
   {
    w[i]=w[i]+(beta*E*dly_adapt[i]);//update adaptive FIR weights
    dly_adapt[i+1] = dly_adapt[i];  //update adaptive FIR samples
   }
 if (out_type == 1)                 //slider position for adapt FIR
    output_left_sample((short)adaptfir_out); //output adaptive FIR
 else if (out_type == 2)            //slider position for fixed FIR
    output_left_sample((short)fir_out); //output fixed FIR filter
 return;
}

void main()
{
 int T=0, i=0;
 for (i = 0; i < WLENGTH; i++)
   {
    w[i] = coeffs[i];               //init coeff for adaptive FIR
    dly_adapt[i] = 0;               //init buffer for adaptive FIR
   }
 for (T = 0; T < N; T++)
   dly_fix[T] = 0;                  //init buffer for fixed FIR
 sreg.regval=0xFFFF;                //initial seed value
 fb = 1;                            //initial feevack value
 comm_intr();                       //init DSK, codec, McBSP
 while (1);                         //infinite loop
}
```

FIGURE 7.20. (*Continued*)

The IIR filter coefficients used are those of a fourth order lowpass elliptic filter (see Example 5.5) and are read from file `elliptic.cof`.

Build and run this project as **iirsosadapt**. Verify that the adaptive filter converges to a state in which the frequency content of its output matches that of the (unknown) IIR filter. Figure 7.24 shows the filtered noise at the output of the adaptive filter (displayed using the FFT function of an *Agilent 54621A* oscilloscope) and the magnitude FFT of the coefficients of the adaptive FIR filter (displayed using CCS).

Example 7.7: Adaptive FIR Filter for System Identification of System External to DSK (sysid)

Program `sysid.c` (Figure 7.25) extends the previous examples to allow the identification of a system external to the DSK, connected between the LINE OUT and

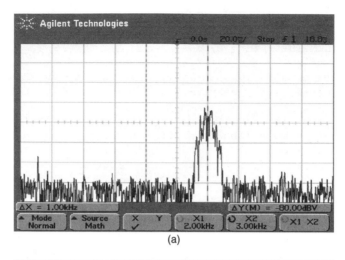

(a)

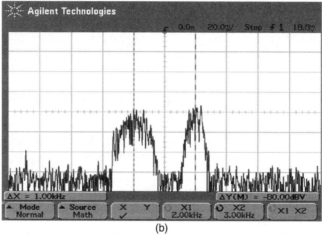

(b)

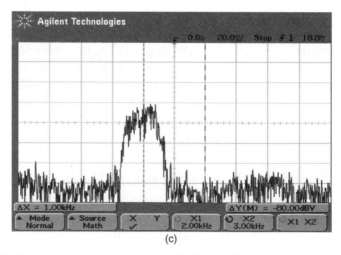

(c)

FIGURE 7.21. Frequency content of output of adaptive filter implemented using program `adaptIDFIRw.c` at three different instants. Captured using FFT function of *Agilent 54621A* oscilloscope.

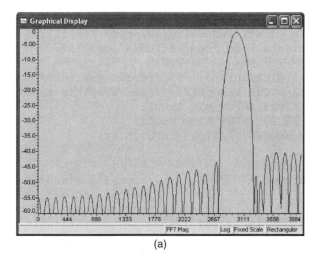

(a)

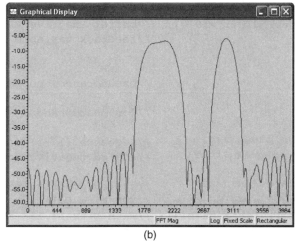

(b)

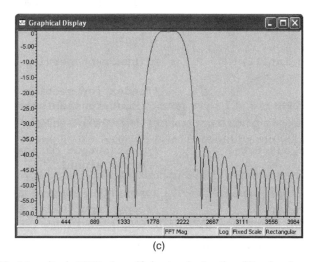

(c)

FIGURE 7.22. Magnitude FFT of coefficients of adaptive filter implemented using program adaptIDFIRw.c at three different instants. Plotted using CCS.

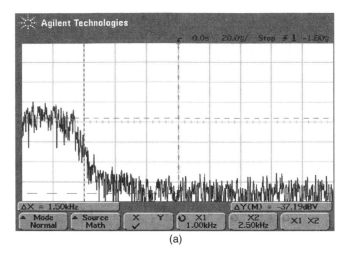

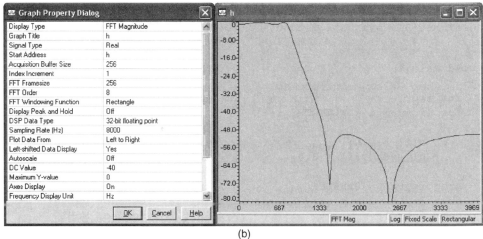

FIGURE 7.24. (a) Frequency content of adaptive filter output and (b) magnitude FFT of adaptive filter coefficients after adaptation in program `iirsosadapt.c`.

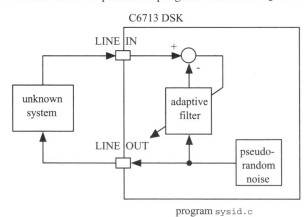

program `sysid.c`

FIGURE 7.25. Use of program `sysid.c` to identify unknown system.

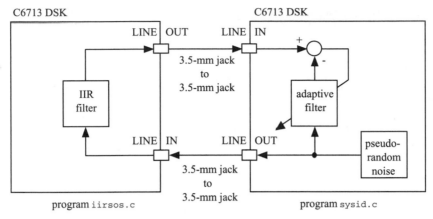

FIGURE 7.26. Connection diagram for Example 7.7.

example, change the number of coefficients to 256 by changing the line that reads

```
#define WLENGTH 128
```

to read

```
#define WLENGTH 256
```

Connect the two DSKs as shown in Figure 7.26. Load and run program `iirsos.c`, including coefficient file `elliptic.cof` on the first DSK. Close CCS and disconnect the USB cable from the DSK. Program `iirsos.c` will continue to run on the DSK as long as the board is powered up. Connect the USB cable to the second DSK, start CCS, and then load program `sysid.c`. (If you have two host computers running CCS there is no need to disconnect the USB on the first DSK.) Run program `sysid.c` on the second DSK. Halt the program after a few seconds and select *View→Graph* in order to examine the coefficients of the adaptive filter. Figure 7.27 shows typical results. A number of features of the plots shown in Figure 7.27 are worthy of note. Compare Figure 7.27b with Figure 7.24b. As noted in Chapter 4, the characteristics of the codec reconstruction and antialiasing filters, the ac coupling between codec and jack sockets, and the potential divider between the LINE IN socket and the codec input are all included in the signal path identified using `sysid.c`.

The magnitude frequency response shown in Figure 7.27b rolls off at low frequencies (due to the ac coupling) and in the passband has a gain of less than unity (due to the potential divider circuits). Less clear, since in this case the gain of the filter at frequencies greater than 3800 Hz is designed to be low, the magnitude frequency response in Figure 7.27b rolls off significantly beyond 3800 Hz (due to the antialias-

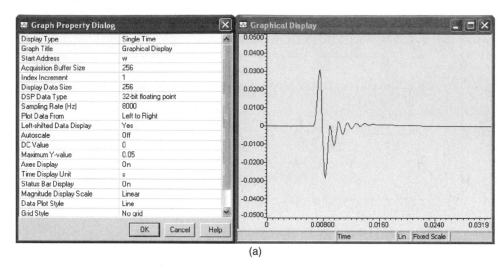

(a)

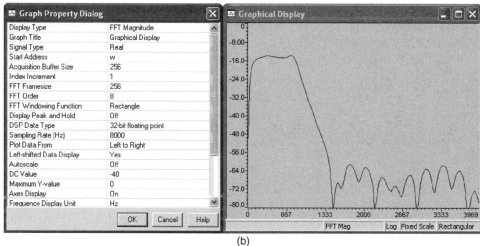

(b)

FIGURE 7.27. Adaptive filter coefficients displayed using CCS (a) time domain and (b) frequency domain.

ing and reconstruction filters in the codecs). Nonetheless, program `sysid.c` has successfully given an indication of the characteristics of the IIR filter implemented on the first DSK.

REFERENCES

1. B. Widrow and S. D. Stearns, *Adaptive Signal Processing*, Prentice Hall, Upper Saddle River, NJ, 1985.

2. B. Widrow and M. E. Hoff, Jr., Adaptive switching circuits, *IRE WESCON*, pp. 96–104, 1960.

3. B. Widrow, J. R. Glover, J. M. McCool, J. Kaunitz, C. S. Williams, R. H. Hearn, J. R. Zeidler, E. Dong, Jr., and R. C. Goodlin, Adaptive noise cancelling: principles and applications, *Proceedings of the IEEE*, Vol. 63, pp. 1692–1716, 1975.

4. R. Chassaing, *Digital Signal Processing with C and the TMS320C30*, Wiley, Hoboken, NJ, 1992.

```
//dotpintrinsic.c Sum of products with C intrinsic functions using C

for (i = 0; i < 100; i++)
    {
            suml = suml + _mpy(a[i], b[i]);
            sumh = sumh + _mpyh(a[i], b[i]);
    }
return (suml + sumh);
```

FIGURE 8.2. Separate sum of products using C intrinsic functions (dotpintrinsic.c).

Example 8.2: Separate Sum of Products with C Intrinsic Functions Using C Code (dotpintrinsic)

Figure 8.2 shows the C code *dotpintrinsic.c* to illustrate the separate sum of products using two C intrinsic functions, *_mpy* and *_mpyh*, which have the equivalent ASM instructions MPY and MPYH, respectively. Whereas the even and odd sums of products are calculated within the loop, the final summation is taken outside the loop and returned to the calling function.

Example 8.3: Sum of Products with Word-Wide Access for Fixed-Point Implementation Using Linear ASM Code (twosumlasmfix.sa)

Figure 8.3 shows the linear ASM code *twosumlasmfix.sa*, which obtains two separate sums of products for a fixed-point implementation. It is not necessary to specify the functional units. Furthermore, symbolic names can be used for registers. The LDW instruction is used to load a 32-bit word-wide data value (which must be word-aligned in memory when using LDW). Lower and upper 16-bit products are calculated separately. The two ADD instructions accumulate separately the even and odd sum of products.

```
;twosumlasmfix.sa Sum of Products. Separate accum of even/odd terms
;With word-wide data for fixed-point implementation using linear ASM

loop:    LDW        *aptr++, ai           ;32-bit word ai
         LDW        *bptr++, bi           ;32-bit word bi
         MPY        ai, bi, prodl         ;lower 16-bit product
         MPYH       ai, bi, prodh         ;higher 16-bit product
         ADD        prodl, suml, suml     ;accum even terms
         ADD        prodh, sumh, sumh     ;accum odd terms
         SUB        count, 1, count       ;decrement count
[count]  B          loop                  ;branch to loop
```

FIGURE 8.3. Separate sum of products using linear ASM code for fixed-point implementation (twosumlasmfix.sa).

```
;twosumlasmfloat.sa Sum of products.Separate accum of even/odd terms
;Using double-word load LDDW for floating-point implementation

loop:      LDDW      *aptr++, ai1:ai0       ;64-bit word ai0 and ai1
           LDDW      *bptr++, bi1:bi0       ;64-bit word  bi0 and bi1
           MPYSP     ai0, bi0, prodl        ;lower 32-bit product
           MPYSP     ai1, bi1, prodh        ;higher 32-bit product
           ADDSP     prodl, suml, suml      ;accum 32-bit even terms
           ADDSP     prodh, sumh, sumh      ;accum 32-bit odd terms
           SUB       count, 1, count        ;decrement count
  [count]  B         loop                   ;branch to loop
```

FIGURE 8.4. Separate sum of products with LDDW using ASM code for floating-point implementation (twosumlasmfloat.sa).

Example 8.4: Sum of Products with Double-Word Load for Floating-Point Implementation Using Linear ASM Code (twosumlasmfloat)

Figure 8.4 shows the linear ASM code twosumlasmfloat.sa used to obtain two separate sums of products for a floating-point implementation. The double-word load instruction LDDW loads a 64-bit data value and stores it in a pair of registers. Each single-precision multiply instruction MPYSP performs a 32 × 32 multiplication. The sums of products of the lower and upper 32 bits are performed to yield a sum of both even and odd terms as 32 bits.

Example 8.5: Dot Product with No Parallel Instructions for Fixed-Point Implementation Using ASM Code (dotpnp)

Figure 8.5 shows the ASM code dotpnp.asm for the dot product with no instructions in parallel for a fixed-point implementation. A fixed-point implementation can be

```
;dotpnp.asm ASM Code, no parallel instructions, fixed-point

           MVK    .S1    200, A1        ;count into A1
           ZERO   .L1    A7             ;init A7 for accum
  LOOP     LDH    .D1    *A4++,A2       ;A2=16-bit data pointed by A4
           LDH    .D1    *A8++,A3       ;A3=16-bit data pointed by A8
           NOP           4              ;4 delay slots for LDH
           MPY    .M1    A2,A3,A6       ;product in A6
           NOP                          ;1 delay slot for MPY
           ADD    .L1    A6,A7,A7       ;accum in A7
           SUB    .S1    A1,1,A1        ;decrement count
    [A1]   B      .S2    LOOP           ;branch to LOOP
           NOP           5              ;5 delay slots for B
```

FIGURE 8.5. ASM code with no parallel instructions for fixed-point implementation (dotpnp.asm).

performed with all C6x devices, whereas a floating-point implementation requires a C67x platform such as the C6713 DSK.

The loop iterates 200 times. With a fixed-point implementation, each pointer register A4 and A8 increments to point at the next half-word (16 bits) in each buffer, whereas with a floating-point implementation, a pointer register increments the pointer to the next 32-bit word. The load, multiply, and branch instructions must use the .D, .M, and .S units, respectively; the add and subtract instructions can use any unit (except .M). The instructions within the loop consume 16 cycles per iteration. This yields $16 \times 200 = 3200$ cycles. Table 8.4 shows a summary of several optimization schemes for both fixed- and floating-point implementations.

Example 8.6: Dot Product with Parallel Instructions for Fixed-Point Implementation Using ASM Code (dotpp)

Figure 8.6 shows the ASM code *dotpp.asm* for the dot product with a fixed-point implementation with instructions in parallel. With code in lieu of NOPs, the number of NOPs is reduced.

The MPY instruction uses a cross-path (with .M1x) since the two operands are from different register files or different paths. The instructions SUB and B are moved up to fill some of the delay slots required by LDH. The branch instruction occurs after the ADD instruction. Using parallel instructions, the instructions within the loop now consume eight cycles per iteration, to yield $8 \times 200 = 1600$ cycles.

Example 8.7: Two Sums of Products with Word-Wide (32-bit) Data for Fixed-Point Implementation Using ASM Code (twosumfix)

Figure 8.7 shows the ASM code *twosumfix.asm*, which calculates two separate sums of products using word-wide access of data for a fixed-point implementation. The loop count is initialized to 100 (not 200) since two sums of products are obtained per iteration. The instruction LDW loads a word or 32-bit data. The

```
;dotpp.asm ASM Code with parallel instructions, fixed-point

        MVK   .S1    200, A1       ;count into A1
     || ZERO  .L1    A7            ;init A7 for accum
LOOP    LDH   .D1    *A4++,A2       ;A2=16-bit data pointed by A4
     || LDH   .D2    *B4++,B2       ;B2=16-bit data pointed by B4
        SUB   .S1    A1,1,A1        ;decrement count
  [A1]  B     .S1    LOOP           ;branch to LOOP (after ADD)
        NOP          2              ;delay slots for LDH and B
        MPY   .M1x   A2,B2,A6       ;product in A6
        NOP                         ;1 delay slot for MPY
        ADD   .L1    A6,A7,A7       ;accum in A7,then branch
;branch occurs here
```

FIGURE 8.6. ASM code with parallel instructions for fixed-point implementaition.

```
;twosumfix.asm ASM code for two sums of products with word-wide data
;for fixed-point implementation

            MVK     .S1   100,A1       ;count/2 into A1
     ||     ZERO    .L1   A7           ;init A7 for accum of even terms
     ||     ZERO    .L2   B7           ;init B7 for accum of odd terms
LOOP        LDW     .D1   *A4++,A2     ;A2=32-bit data pointed by A4
     ||     LDW     .D2   *B4++,B2     ;A3=32-bit data pointed by B4
            SUB     .S1   A1,1,A1      ;decrement count
     [A1]   B       .S1   LOOP         ;branch to LOOP (after ADD)
            NOP           2            ;delay slots for both LDW and B
            MPY     .M1x  A2,B2,A6     ;lower 16-bit product in A6
     ||     MPYH    .M2x  A2,B2,B6     ;upper 16-bit product in B6
            NOP                        ;1 delay slot for MPY/MPYH
            ADD     .L1   A6,A7,A7     ;accum even terms in A7
     ||     ADD     .L2   B6,B7,B7     ;accum odd terms in B7
;branch occurs here
```

FIGURE 8.7. ASM code for two sums of products with 32-bit data for fixed-point implementation (twosumfix.asm).

multiply instruction MPY finds the product of the lower 16 × 16 data, and MPYH finds the product of the upper 16 × 16 data. The two ADD instructions accumulate separately the even and odd sums of products. Note that an additional ADD instruction is needed outside the loop to accumulate A7 and B7. The instructions within the loop consume eight cycles, now using 100 iterations (not 200), to yield 8 × 100 = 800 cycles.

Example 8.8: Dot Product with No Parallel Instructions for Floating-Point Implementation Using ASM Code (dotpnpfloat)

Figure 8.8 shows the ASM code dotpnpfloat.asm for the dot product with a floating-point implementation using no instructions in parallel. The loop iterates 200

```
;dotpnpfloat.asm ASM Code with no parallel instructions for floating-pt

            MVK     .S1   200,A1       ;count into A1
            ZERO    .L1   A7           ;init A7 for accum
LOOP        LDW     .D1   *A4++,A2     ;A2=32-bit data pointed by A4
            LDW     .D1   *A8++,A3     ;A3=32-bit data pointed by A8
            NOP           4            ;4 delay slots for LDW
            MPYSP   .M1   A2,A3,A6     ;product in A6
            NOP           3            ;3 delay slots for MPYSP
            ADDSP   .L1   A6,A7,A7     ;accum in A7
            SUB     .S1   A1,1,A1      ;decrement count
     [A1]   B       .S2   LOOP         ;branch to LOOP
            NOP           5            ;5 delay slots for B
```

FIGURE 8.8. ASM code with no parallel instructions for floating-point implementation (dotpnpfloat.asm).

```
;dotppfloat.asm   ASM Code with parallel instructions for floating-point

             MVK    .S1    200, A1      ;count into A1
       ||    ZERO   .L1    A7           ;init A7 for accum
LOOP         LDW    .D1    *A4++,A2     ;A2=32-bit data pointed by A4
       ||    LDW    .D2    *B4++,B2     ;B2=32-bit data pointed by B4
             SUB    .S1    A1,1,A1      ;decrement count
             NOP           2            ;delay slots for both LDW and B
      [A1]   B      .S2    LOOP         ;branch to LOOP (after ADDSP)
             MPYSP  .M1x   A2,B2,A6     ;product in A6
             NOP           3            ;3 delay slots for MPYSP
             ADDSP  .L1    A6,A7,A7     ;accum in A7,then branch
;branch occurs here
```

FIGURE 8.9. ASM code with parallel instructions for floating-point implementation (dotppfloat.asm).

times. The single-precision floating-point instruction MPYSP performs a 32×32 multiply. Each MPYSP and ADDSP requires three delay slots. The instructions within the loop consume a total of 18 cycles per iteration (without including three NOPs associated with ADDSP). This yields a total of $18 \times 200 = 3600$ cycles. (See Table 8.4 for a summary of several optimization schemes for both fixed- and floating-point implementations.)

Example 8.9: Dot Product with Parallel Instructions for Floating-Point Implementation Using ASM Code (dotppfloat)

Figure 8.9 shows the ASM code *dotppfloat.asm* for the dot product with a floating-point implementation using instructions in parallel. The loop iterates 200 times. By moving the SUB and B instructions up to take the place of some NOPs, the number of instructions within the loop is reduced to 10. Note that three additional NOPs would be needed outside the loop to retrieve the result from ADDSP. The instructions within the loop consume a total of 10 cycles per iteration. This yields a total of $10 \times 200 = 2000$ cycles.

Example 8.10: Two Sums of Products with Double-Word-Wide (64-bit) Data for Floating-Point Implementation Using ASM Code (twosumfloat)

Figure 8.10 shows the ASM code *twosumfloat.asm*, which calculates two separate sums of products using double-word-wide access of 64-bit data for a floating-point implementation. The loop count is initialized to 100 since two sums of products are obtained per iteration. The instruction LDDW loads a 64-bit double-word data value into a register pair. The multiply instruction MPYSP performs a 32×32 multiply. The two ADDSP instructions accumulate separately the even and odd sums of products. The additional ADDSP instruction is needed outside the loop to accumulate A7 and

```
;twosumfloat.asm ASM Code with two sums of products for floating-pt

              MVK    .S1    100, A1       ;count/2 into A1
        ||    ZERO   .L1    A7            ;init A7 for accum of even terms
        ||    ZERO   .L2    B7            ;init B7 for accum of odd terms
LOOP          LDDW   .D1    *A4++,A3:A2   ;64-bit-> register pair A2,A3
        ||    LDDW   .D2    *B4++,B3:B2   ;64-bit-> register pair B2,B3
              SUB    .S1    A1,1,A1       ;decrement count
              NOP           2             ;delay slots for LDW
        [A1]  B      .S2    LOOP          ;branch to LOOP
              MPYSP  .M1x   A2,B2,A6      ;lower 32-bit product in A6
        ||    MPYSP  .M2x   A3,B3,B6      ;upper 32-bit product in B6
              NOP           3             ;3 delay slot for MPYSP
              ADDSP  .L1    A6,A7,A7      ;accum even terms in A7
        ||    ADDSP  .L2    B6,B7,B7      ;accum odd terms in B7
;branch occurs here
              NOP           3             ;delay slots for last ADDSP
              ADDSP  .L1x   A7,B7,A4      ;final sum of even and odd terms
              NOP           3             ;delay slots for ADDSP
```

FIGURE 8.10. ASM code with two sums of products for floating-point implementation (twosumfloat.asm).

B7. The instructions within the loop consume a total of 10 cycles, using 100 iterations (not 200), to yield a total of $10 \times 100 = 1000$ cycles.

8.5 SOFTWARE PIPELINING FOR CODE OPTIMIZATION

Software pipelining is a scheme to write efficient code in ASM so that all the functional units are utilized within one cycle. Optimization levels -o2 and -o3 enable code generation to generate (or attempt to generate) software-pipelined code.

There are three stages associated with software pipelining:

1. *Prolog (warm-up)*. This stage contains instructions needed to build up the loop kernel (cycle).
2. *Loop kernel (cycle)*. Within this loop, all instructions are executed in parallel. The entire loop kernel can be executed in *one* cycle, since all the instructions within the loop kernel stage are in parallel.
3. *Epilog (cool-off)*. This stage contains the instructions necessary to complete all iterations.

8.5.1 Procedure for Hand-Coded Software Pipelining

1. Draw a dependency graph.
2. Set up a scheduling table.
3. Obtain code from the scheduling table.

8.5.2 Dependency Graph

Figure 8.11 shows a dependency graph. A procedure for drawing a dependency graph follows.

1. Draw the nodes and paths.
2. Write the number of cycles to complete an instruction.
3. Assign functional units associated with each node.
4. Separate the data path so that the maximum number of units are utilized.

A node has one or more data paths going into and/or out of the node. The numbers next to each node represent the number of cycles required to complete the associated instruction. A parent node contains an instruction that writes to a variable, whereas a child node contains an instruction that reads a variable written by the parent.

The LDH instructions are considered to be the parents of the MPY instruction since the results of the two load instructions are used to perform the MPY instruction. Similarly, the MPY is the parent of the ADD instruction. The ADD instruction is fed back as input for the next iteration; similarly with the SUB instruction.

Figure 8.12 shows another dependency graph associated with two sums of products for a fixed-point implementation. The length of the prolog section is the longest path from the dependency graph in Figure 8.12. Since the longest path is 8, the length of the prolog is 7 before entering the loop kernel (cycle) at cycle 8.

A similar dependency graph for a floating-point implementation can be obtained using LDDW, MPYSP, and ADDSP in lieu of LDW, MPY/MPYH, and ADD, respectively, in Figure 8.12. Note that the single-precision instructions ADDSP and MPYSP both take four cycles to complete (three delay slots each).

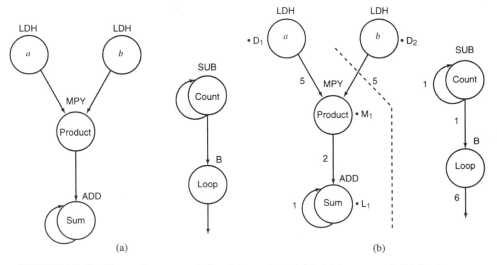

FIGURE 8.11. Dependency graph for dot product: (a) initial stage and (b) final stage.

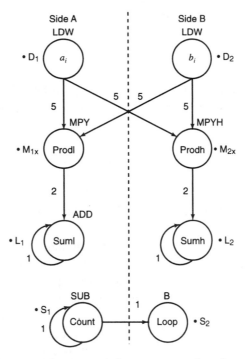

FIGURE 8.12. Dependency graph for two sums of products per iteration.

TABLE 8.1   Schedule Table of Dot Product Before Software Pipelining for Fixed-Point Implementation

Cycles Units	1, 9, ...	2, 10, ...	3, 11, ...	4, 12, ...	5, 13, ...	6, 14, ...	7, 15, ...	8, 16, ...
.D1	LDW							
.D2	LDW							
.M1						MPY		
.M2						MPYH		
.L1								ADD
.L2								ADD
.S1		SUB						
.S2			B					

8.5.3 Scheduling Table

Table 8.1 shows a scheduling table drawn from the dependency graph.

1. LDW starts in cycle 1.
2. MPY and MPYH must start five cycles after the LDWs due to the four delay slots. Therefore, MPY and MPYH start in cycle 6.

3. ADD must start two cycles after MPY/MPYH due to the one delay slot of MPY/MPYH. Therefore, ADD starts in cycle 8.

4. B has five delay slots and starts in cycle 3, since branching occurs in cycle 9, after the ADD instruction.

5. SUB instruction must start one cycle before the branch instruction, since the loop count is decremented before branching occurs. Therefore, SUB starts in cycle 2.

From Table 8.1, the two LDW instructions are in parallel and are issued in cycles 1, 9, 17, ... The SUB instruction is issued in cycles 2, 10, 18, ... This is followed by the branch (B) instruction issued in cycles 3, 11, 19, ... The two parallel instructions MPY and MPYH are issued in cycles 6, 14, 22, ... The ADD instructions are issued in cycles 8, 16, 24, ...

Table 8.1 is extended to illustrate the different stages: prolog (cycles 1 through 7), loop kernel (cycle 8), and epilog (cycles 9, 10, . . . not shown), as shown in Table 8.2. The instructions within the prolog stage are repeated until and including the loop kernel (cycle) stage. Instructions in the epilog stage (cycles 9, 10, ...) complete the functionality of the code.

From Table 8.2, an efficient optimized code can be obtained. Note that it is possible to start processing a new iteration before previous iterations are finished. Software pipelining allows us to determine when to start a new loop iteration.

Loop Kernel (Cycle)

Within the loop kernel, in cycle 8, each functional unit is used only once. The minimum iteration interval is the minimum number of cycles required to wait before the initiation of a successive iteration. This interval is 1. As a result, a new iteration can be initiated every cycle.

Within loop cycle 8, multiple iterations of the loop execute in parallel. In cycle 8, different iterations are processed at the same time. For example, the ADDs add

TABLE 8.2   Schedule Table of Dot Product After Software Pipelining for Fixed-Point Implementation

Cycles Units	Prolog							Loop Kernel
	1	2	3	4	5	6	7	8
.D1	LDW	LDW	LDW	LDW	LDW	LDW	LDW	LDW
.D2	LDW	LDW	LDW	LDW	LDW	LDW	LDW	LDW
.M1						MPY	MPY	MPY
.M2						MPYH	MPYH	MPYH
.L1								ADD
.L2								ADD
.S1		SUB	SUB	SUB	SUB	SUB	SUB	SUB
.S2			B	B	B	B	B	B

data for iteration 1, while MPY and MPYH multiply data for iteration 3, LDWs load data for iteration 8, SUB decrements the counter for iteration 7, and B branches for iteration 6. Note that the values being multiplied are loaded into registers five cycles prior to the cycle when the values are multiplied. Before the first multiplication occurs, the fifth load has just completed. This software pipeline is eight iterations deep.

Example 8.11: Dot Product Using Software Pipelining for a Fixed-Point Implementation

This example implements the dot product using software pipelining for a fixed-point implementation. From Table 8.2, one can readily obtain the ASM code *dotpipedfix. asm* shown in Figure 8.13. The loop count is 100 since two multiplies and two accumulates are calculated per iteration. The following instructions start in the following cycles:

> *Cycle 1*: LDW, LDW (also initialization of count and accumulators A7 and B7)
> *Cycle 2*: LDW, LDW, SUB
> *Cycles 3–5*: LDW, LDW, SUB, B
> *Cycles 6–7*: LDW, LDW, MPY, MPYH, SUB, B
> *Cycles 8–107*: LDW, LDW, MPY, MPYH, ADD, ADD, SUB, B
> *Cycle 108*: LDW, LDW, MPY, MPYH, ADD, ADD, SUB, B

The prolog section is within cycles 1 through 7; the loop kernel is in cycle 8, where all the instructions are in parallel; and the epilog section is in cycle 108. Note that SUB is made conditional to ensure that Al is no longer decremented once it reaches zero.

Example 8.12: Dot Product Using Software Pipelining for a Floating-Point Implementation

This example implements the dot product using software pipelining for a floating-point implementation. Table 8.3 shows a floating-point version of Table 8.2. LDW becomes LDDW, MPY/MPYH become MPYSP, and ADD becomes ADDSP. Both MPYSP and ADDSP have three delays slots. As a result, the loop kernel starts in cycle 10 in lieu of cycle 8. The SUB and B instructions start in cycles 4 and 5, respectively, in lieu of cycles 2 and 3. ADDSP starts in cycle 10 in lieu of cycle 8. The software pipeline for a floating-point implementation is 10 deep.

Figure 8.14 shows the ASM code *dotpipedfloat.asm*, which implements the floating-point version of the dot product. Since ADDSP has three delay slots, the accumulation is staggered by four. The accumulation associated with one of the ADDSP instructions at each loop cycle follows:

```
;dotpipedfix.asm   ASM code for dot product with software pipelining
;For fixed-point implementation
;cycle 1
                MVK         .S1    100,A1              ;loop count
          ||    ZERO.       L1     A7              ;init accum A7
          ||    ZERO        .L2    B7              ;init accum B7
          ||    LDW         .D1    *A4++,A2        ;32-bit data in A2
          ||    LDW         .D2    *B4++,B2        ;32-bit data in B2
;cycle 2
          ||    LDW         .D1    *A4++,A2        ;32-bit data in A2
          ||    LDW         .D2    *B4++,B2        ;32-bit data in B2
          ||    [A1]   SUB       .S1    A1,1,A1       ;decrement count
;cycle 3
          ||    LDW         .D1    *A4++,A2        ;32-bit data in A2
          ||    LDW         .D2    *B4++,B2        ;32-bit data in B2
          ||    [A1]   SUB       .S1    A1,1,A1       ;decrement count
          ||    [A1]   B         .S2    LOOP          ;branch to LOOP
;cycle 4
          ||    LDW         .D1    *A4++,A2        ;32-bit data in A2
          ||    LDW         .D2    *B4++,B2        ;32-bit data in B2
          ||    [A1]   SUB       .S1    A1,1,A1       ;decrement count
          ||    [A1]   B         .S2    LOOP          ;branch to LOOP
;cycle 5
          ||    LDW         .D1    *A4++,A2        ;32-bit data in A2
          ||    LDW         .D2    *B4++,B2        ;32-bit data in B2
          ||    [A1]   SUB       .S1    A1,1,A1       ;decrement count
          ||    [A1]   B         .S2    LOOP          ;branch to LOOP
;cycle 6
          ||    LDW         .D1    *A4++,A2        ;32-bit data in A2
          ||    LDW         .D2    *B4++,B2        ;32-bit data in B2
          ||    [A1]   SUB       .S1    A1,1,A1       ;decrement count
          ||    [A1]   B         .S2    LOOP          ;branch to LOOP
          ||    MPY         .M1x   A2,B2,A6        ;lower 16-bit product into A6
          ||    MPYH        .M2x   A2,B2,B6        ;upper 16-bit product into B6
;cycle 7
          ||    LDW         .D1    *A4++,A2        ;32-bit data in A2
          ||    LDW         .D2    *B4++,B2        ;32-bit data in B2
          ||    [A1]   SUB       .S1    A1,1,A1         ;decrement count
          ||    [A1]   B         .S2    LOOP          ;branch to LOOP
          ||    MPY         .M1x   A2,B2,A6        ;lower 16-bit product into A6
          ||    MPYH        .M2x   A2,B2,B6        ;upper 16-bit product into B6
;cycles 8-107 (loop cycle)
          ||    LDW         .D1    *A4++,A2        ;32-bit data in A2
          ||    LDW         .D2    *B4++,B2        ;32-bit data in B2
          ||    [A1]   SUB       .S1    A1,1,A1       ;decrement count
          ||    [A1]   B         .S2    LOOP          ;branch to LOOP
          ||    MPY         .M1x   A2,B2,A6        ;lower 16-bit product into A6
          ||    MPYH        .M2x   A2,B2,B6        ;upper 16-bit product into B6
          ||    ADD         .L1    A6,A7,A7        ;accum in A7
          ||    ADD         .L2    B6,B7,B7        ;accum in B7
;branch occurs here
;cycle 108 (epilog)
                ADD         .L1x   A7,B7,A4        ;final accum of odd/even
```

FIGURE 8.13. ASM code using software pipelining for fixed-point implementation (dotpipedfix.asm).

Loop
Cycle   Accumulator (one ADDSP)

Cycle	Accumulator	Comment
1	0	
2	0	
3	0	
4	0	
5	p0	;first product
6	p1	;second product
7	p3	
8	p4	
9	p0 + p4	;sum of first and fifth products
10	p1 + p5	;sum of second and sixth products
11	p2 + p6	
12	p3 + p7	
13	p0 + p4 + p8	;sum of first, fifth, and ninth products
14	p1 + p5 + p9	
15	p2 + p6 + p10	
16	p3 + p7 + p11	
17	p0 + p4 + p8 + p12	
.	.	
.	.	
.	.	
99	p2 + p6 + p10 + ... + p94	
100	p3 + p7 + p11 + ... + p95	

This accumulation is shown associated with the loop cycle. The actual cycle is shifted by 9 (by the cycles in the prolog section). Note that the first product, p0, is obtained (available) in loop cycle 5 since the first ADDSP starts in loop cycle 1 and has three delay slots. The first product, p0, is associated with the lower 32-bit term. The second ADDSP (not shown) accumulates the upper 32-bit sum of products.

TABLE 8.3   Schedule Table of Dot Product After Software Pipelining for Floating-Point Implementation

Cycle Units	Prolog									Loop Kernel
	1	2	3	4	5	6	7	8	9	10
.D1	LDDW	LDDW	LDDW	LDDW	LDDW	LDDW	LDDW	LDDW	LDDW	LDDW
.D2	LDDW	LDDW	LDDW	LDDW	LDDW	LDDW	LDDW	LDDW	LDDW	LDDW
.M1						MPYSP	MPYSP	MPYSP	MPYSP	MPYSP
.M2						MPYSP	MPYSP	MPYSP	MPYSP	MPYSP
.L1										ADDSP
.L2										ADDSP
.S1				SUB	SUB	SUB	SUB	SUB	SUB	SUB
.S2					B	B	B	B	B	B

```
;dotpipedfloat.asm  ASM code for dot product with software pipelining
;For floating-point implementation
;cycle 1
            MVK         .S1   100,A1                ;loop count
      ||    ZERO        .L1   A7              ;init accum A7
      ||    ZERO        .L2   B7              ;init accum B7
      ||    LDDW        .D1   *A4++,A3:A2   ;64-bit data in A2 and A3
      ||    LDDW        .D2   *B4++,B3:B2   ;64-bit data in B2 and B3
;cycle 2
      ||    LDDW        .D1   *A4++,A3:A2   ;64-bit data in A2 and A3
      ||    LDDW        .D2   *B4++,B3:B2   ;64-bit data in B2 and B3
;cycle 3
      ||    LDDW        .D1   *A4++,A3:A2   ;64-bit data in A2 and A3
      ||    LDDW        .D2   *B4++,B3:B2   ;64-bit data in B2 and B3
;cycle 4
      ||    LDDW        .D1   *A4++,A3:A2   ;64-bit data in A2 and A3
      ||    LDDW        .D2   *B4++,B3:B2   ;64-bit data in B2 and B3
      || [A1]  SUB      .S1   A1,1,A1               ;decrement count
;cycle 5
      ||    LDDW        .D1   *A4++,A3:A2   ;64-bit data in A2 and A3
      ||    LDDW        .D2   *B4++,B3:B2   ;64-bit data in B2 and B3
      || [A1]  SUB      .S1   A1,1,A1               ;decrement count
      || [A1]  B        .S2   LOOP                  ;branch to LOOP
;cycle 6
      ||    LDDW        .D1   *A4++,A3:A2   ;64-bit data in A2 and A3
      ||    LDDW        .D2   *B4++,B3:B2   ;64-bit data in B2 and B3
      || [A1]  SUB      .S1   A1,1,A1               ;decrement count
      || [A1]  B        .S2   LOOP                  ;branch to LOOP
      ||    MPYSP .M1x  A2,B2,A6        ;lower 32-bit product into A6
      ||    MPYSP .M2x  A3,B3,B6        ;upper 32-bit product into B6
;cycle 7
      ||    LDDW        .D1   *A4++,A3:A2   ;32-bit data in A2 and A3
      ||    LDDW        .D2   *B4++,B3:B2   ;32-bit data in B2 and B3
      || [A1]  SUB      .S1   A1,1,A1               ;decrement count
      || [A1]  B        .S2   LOOP                  ;branch to LOOP
      ||    MPYSP .M1x  A2,B2,A6        ;lower 32-bit product into A6
      ||    MPYSP .M2x  A3,B3,B6        ;upper 32-bit product into B6
;cycle 8
      ||    LDDW        .D1   *A4++,A3:A2   ;32-bit data in A2 and A3
      ||    LDDW        .D2   *B4++,B3:B2   ;32-bit data in B2 and B3
      || [A1]  SUB      .S1   A1,1,A1               ;decrement count
      || [A1]  B        .S2   LOOP                  ;branch to LOOP
      ||    MPYSP .M1x  A2,B2,A6        ;lower 32-bit product into A6
      ||    MPYSP .M2x  A3,B3,B6        ;upper 32-bit product into B6
;cycle 9
      ||    LDDW        .D1   *A4++,A3:A2   ;32-bit data in A2 and A3
      ||    LDDW        .D2   *B4++,B3:B2   ;32-bit data in B2 and B3
      || [A1]  SUB      .S1   A1,1,A1               ;decrement count
      || [A1]  B        .S2   LOOP                  ;branch to LOOP
      ||    MPYSP .M1x  A2,B2,A6        ;lower 32-bit product into A6
      ||    MPYSP .M2x  A3,B3,B6        ;upper 32-bit product into B6
```

FIGURE 8.14. ASM code using software pipelining for floating-point implementation (dotpipedfloat.asm).

```
;cycles 10-109 (loop kernel)
       ||      LDDW        .D1    *A4++,A3:A2   ;32-bit data in A2 and A3
       ||      LDDW        .D2    *B4++,B3:B2   ;32-bit data in B2 and B3
       || [A1]      SUB    .S1    A1,1,A1         ;decrement count
       || [A1]      B      .S2    LOOP            ;branch to LOOP
       ||      MPYSP .M1x  A2,B2,A6      ;lower 32-bit product into A6
       ||      MPYSP .M2x  A3,B3,B6      ;upper 32-bit product into B6
       ||      ADDSP .L1   A6,A7,A7      ;accum in A7
       ||      ADDSP .L2   B6,B7,B7      ;accum in B7
;branch occurs here
;cycles 110-124 (epilog)
            ADDSP .L1x  A7,B7,A0     ;lower/upper sum of products
            ADDSP .L2x  A7,B7,B0     ;
            ADDSP .L1x  A7,B7,A0     ;
            ADDSP .L2x  A7,B7,B0     ;
            NOP                          ;wait for 1ˢᵗ B0
            ADDSP .L1x  A0,B0,A5     ;1st two sum of products
            NOP                          ;wait for 2ⁿᵈ B0
            ADDSP .L2x  A0,B0,B5     ;last two sum of products
            NOP               3          ;3 delay slots for ADDSP
            ADDSP .L1x  A5,B5,A4     ;final sum
            NOP               3          ;3 delay slots for final sum
```

FIGURE 8.14. (*Continued*)

A6 contains the lower 32-bit products and B6 contains the upper 32-bit products. The sums of the lower and upper 32-bit products are accumulated in A7 and B7, respectively.

The epilog section contains the following instructions associated with the actual cycle (not loop cycles), as shown in Figure 8.14.

Cycle	Instruction	
110	ADDSP	
111	ADDSP	
112	ADDSP	
113	ADDSP	
114	NOP	
115	ADDSP	
116	NOP	
117	ADDSP	
118–120	NOP	3
121	ADDSP	
122–124	NOP	3

In cycles 113 through 116, A7 contains the lower 32-bit sum of products and B7 contains the upper 32-bit sum of products, or:

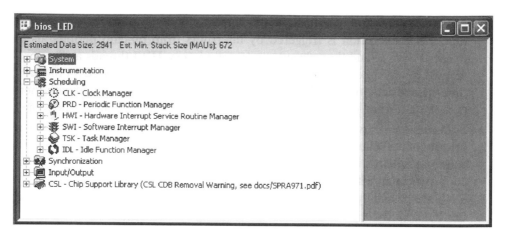

FIGURE 9.1. DSP/BIOS *Configuration Tool* window.

3. Call the function `main()`, defined in a user-supplied source file included in the project. At this point, interrupts are still disabled and application-specific initialization functions (defined in user-supplied source files) can be called. Function `main()` runs to completion.

4. Start DSP/BIOS. At this point hardware and software interrupts are enabled, the clock that is used by PRD threads is started, and TSKs are enabled. Execution of the highest priority TSK will start and all TSKs will eventually be run to completion.

5. When all TSKs have been completed or are blocked, and when no HWI, SWI, or PRD is running, the DSP/BIOS idle loop is entered.

More detailed information about DSP/BIOS can be found in the *TMS320C6000 DSP/BIOS User's Guide* [1].

The following examples illustrate the use of different types of DSP/BIOS threads.

9.1.1 Periodic Functions

Example 9.1: Blinking of LEDs at Different Rates Using DSP/BIOS PRDs (`bios_LED`)

This example illustrates the steps involved in the creation of a simple DSP/BIOS application. The source file `bios_LED.c`, listed in Figure 9.2, is stored in folder `bios_LED` but no project file is supplied. We will create the project from scratch.

1. Create a new project by selecting *Project → New* and typing the *Project Name* `bios_LED` and set the *Target* as TM320C67XX. A new project file `bios_LED.pjt` will be created in the existing folder `bios_LED`.

```
//bios_LED.c DSP/BIOS application to flash LEDs

void blink_LED0()

{
        DSK6713_LED_toggle(0);
}
void blink_LED1()
{
        DSK6713_LED_toggle(1);
}
void blink_LED2()

{
        DSK6713_LED_toggle(2);
}
void blink_LED3()

{
        DSK6713_LED_toggle(3);
}

void main()
{
  DSK6713_LED_init();
  return;
}
```

FIGURE 9.2. Listing of program `bios_LED.c`.

2. Add the source file `bios_LED.c` to the project using *Project → Add Files to Project*.
3. Add a configuration file to the project . Select *File → New → DSP/BIOS Configuration* and select `dsk6713.cdb` as the configuration template.
4. Expand on *Scheduling* in the configuration file window and right-click on *PRD—Periodic Function Manager → Insert PRD*. This adds a periodic function, PRD0, to the application. Rename the periodic function PRDblink_LED0 by right-clicking on its icon in the configuration file window and choosing *Rename*.
5. Right-click on PRDblink_LED0 and select *Properties* to set the *period (ticks)* to 250 and the *function* to _blink_LED0. Note the underscore prefixing the function name. By convention, this identifies it as a C function. Function blink_LED0() is defined in file `bios_LED.c`. Click on *OK* to accept the default settings for all other properties. The properties you have set for periodic function PRDblink_LED0 should now appear at the right-hand side of the configuration file window.

6. Repeat steps 4 and 5 three times, substituting first PRDblink_LED1 for PRDblink_LED0, _blink_LED1 for _blink_LED0, and 500 for 250, and then PRDblink_LED2 for PRDblink_LED0, _blink_LED2 for _blink_LED0, and 1000 for 250, and finally PRDblink_LED3 for PRDblink_LED0, _blink_LED3 for _blink_LED0, and 2000 for 250.

7. Save the configuration file in folder bios_LED as bios_LED.cdb.

8. Add the configuration file to the project (selecting *Project → Add Files to Project*). Note that it is a (.cdb) type file. Verify that the file has been added to the project by expanding *DSP/BIOS Config* in the *Project View* window.

9. Expand on *Generated Files* in the *Project View* window and you will find that when the configuration file was added to the project, three more files, bios_LEDcfg.cmd, bios_LEDcfg.s62, and bios_LEDcfg_c.c, were generated and added to the project automatically.

10. Select *Project → Build Options* and in the *Basic* category in the *Compiler* tab set the *Target Version* to *C671x*. In the *Preprocessor* category, set the *Pre-Define Symbol* option to *CHIP_6713* and the *Include Search Path* option to *c:\CCStudio_v3.1\C6000\dsk6713\include*. In the *Linker* tab, set the *Include Libraries* option to *DSK6713bsl.lib* and the *Library Search Path* to *c:\CCStudio_v3.1\C6000\dsk6713\lib*.

Build the project as **bios_LED**. Load and run bios_led.out and verify that the four LEDs on the DSK flash at rates of 2, 1, 0.5, and 0.25 Hz.

Figure 9.3 shows the configuration settings for the DSP/BIOS application bios_LED. The application comprises four PRD objects scheduled to execute at intervals of 250, 500, 1000, and 2000 PRD clock ticks (by default, one PRD clock tick is equal to 1 ms). The functions called at these instants, blink_LED0(), blink_LED1(), blink_LED2(), and blink_LED3(), are defined in the source file bios_led.c. Each function toggles the state of one of the LEDs on the DSK. Also defined in that source file is the function main(). This function is called at the start of execution of the DSP/BIOS application following DSP/BIOS initialization. In this example, function main() does very little, simply initializing the LEDs on the DSK by calling a function from the Board Support Library DSK6713bsl.lib. When function main() finishes execution, the application falls into the idle loop, which is then preempted, periodically, by the four PRDs. There is nothing in the source file bios_LED.c to indicate the real-time operation of the application. All of the scheduling is handled by DSP/BIOS, as configured using the configuration tool. Source file bios_LED.c simply defines the functions executed by DSP/BIOS objects.

9.1.2 Hardware Interrupts

Many of the example programs in previous chapters made use of hardware interrupts generated by the AIC23 codec in order to perform in real-time. The following example illustrates the use of hardware interrupts in a DSP/BIOS application.

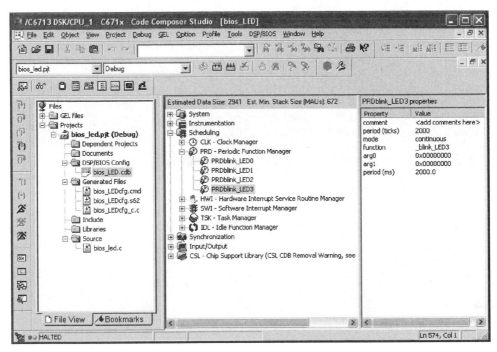

FIGURE 9.3. DSP/BIOS configuration settings for `bios_LED` application.

Example 9.2: Sine Wave Generation Using DSP/BIOS Hardware Interrupts (HWIs) (`bios_sine8_intr`)

This example modifies program `sine8_intr.c`, introduced in Chapter 2, to run as a DSP/BIOS application. That program, listed in Figure 2.14, is quite simple. After calling the initialization function `comm_intr()`, function `main()` enters an endless idle loop (`while(1)`). The interrupt service routine `c_int11()` is assigned to the codec ADC (McBSP_0 receive) interrupt INT11 by means of the interrupt service table in file `vectors_intr.asm`. (See Figure 9.4.)

The following modifications to program `sine8_intr.c` and settings in the DSP/BIOS configuration file `bios_sine8_intr.cdb` make it suitable for use in a DSP/BIOS application.

- Delete the program statement `while(1);` in function `main()`. There is no need to supply an explicit idle loop since in a DSP/BIOS application, after function `main()` has completed execution, the application will enter the DSP/BIOS idle loop (IDL).
- HWI objects corresponding to all hardware interrupt sources are present by default in the DSP/BIOS configuration template *dsk6713.cdb*. Also by default, most have their *function* property set to *HWI_unused*.
- Set the *function* property of HWI object HWI_INT11 to `_c_int11`.

```
//bios_sine8_intr.c DSP/BIOS application to generate sine wave
#include "DSK6713_AIC23.h"              // codec support
Uint32 fs=DSK6713_AIC23_FREQ_8KHZ;      //set sampling rate
#define DSK6713_AIC23_INPUT_MIC 0x0015
#define DSK6713_AIC23_INPUT_LINE 0x0011
Uint16 inputsource=DSK6713_AIC23_INPUT_MIC; // select mic in

#define LOOPLENGTH 8         // size of look up table
short sine_table[LOOPLENGTH]={0,7071,10000,7071,0,-7071,-10000,-
7071};
short loopindex = 0;         // look up table index

void c_int11(void)
{
  output_left_sample(sine_table[loopindex++]);
  if (loopindex >= LOOPLENGTH ) loopindex = 0;
  return;
}

void main()
{
  comm_intr();
}
```

FIGURE 9.4. Listing of Program `bios_sine8_intr.c`.

- Delete the word `interrupt` preceding `void c_int11()` and set the HWI_INT11 object property *Use Dispatcher* to *True*.

The file `vectors_intr.asm` is not required in this example since the mapping of interrupts to interrupt service routines is defined in the configuration file and the files it generates. However, routines `comm_intr()` and `output_left_sample()` defined in file `c6713dskinit.c` are used and that source file must be added to the project. As in the previous example, source file `bios_sine8_intr.c` is supplied in folder `bios_sine8_intr` but no project or configuration file has been provided. In order to create a DSP/BIOS application:

1. Create a new project by selecting *Project → New* and typing the *Project Name* `bios_sine8_intr` and set the *Target* as TM320C67XX. A new project file `bios_sine8_intr.pjt` will be created in the existing folder `bios_sine8_intr`.

2. Add the source file `bios_sine8_intr.c` and the initialization and communication file `c6713dskinit.c` (from folder `Support`) to the project using *Project → Add Files to Project*.

3. Create and add a configuration file to the project . Select *File → New → DSP/BIOS Configuration*. Select `dsk6713.cdb` as the configuration template. By default, this configuration file contains HWI objects corresponding to all hardware interrupt sources.

4. Expand *Scheduling* and HWI—*Hardware Interrupt Service Routine Manager* in the configuration tool window and click on HWI _INT11. Verify that, among the HWI_INT11 properties, by default the *interrupt source is MCBSP_0_ Receive* and the *function* is *HWI_unused*.

5. Right-click on HWI_INT11 and select *Properties* to set the *function* to *_c_ int11*, the interrupt service routine defined in `bios_sine8_intr.c`. Under the *Dispatcher* tab, check *Use Dispatcher*. Click on *OK* to accept the default settings for all other properties.

6. Save the configuration file in folder `bios_sine8_intr` as `bios_sine8_intr.cdb`.

7. Add the configuration file to the project (selecting *Project → Add Files to Project*).

8. Select *Project → Build Options* and in the *Basic* category in the *Compiler* tab set the *Target Version* to *C671x*. In the *Preprocessor* category, set the *Pre-Define Symbol* option to *CHIP_6713* and the *Include Search Path* option to *c:\CCStudio_v3.1\C6000\dsk6713\include*. In the *Linker* tab, set the *Include Libraries* option to *DSK6713bsl.lib* and the *Library Search Path* to *c:\ CCStudio_v3.1\C6000\dsk6713\lib*. In the *Advanced* category of compiler options set the *Memory Model* option to *Far*. (See Figure 9.5.)

Build the project as **bios_sine8_intr**. Load and run `bios_sine8_intr.out` and verify that a 1-kHz tone is generated.

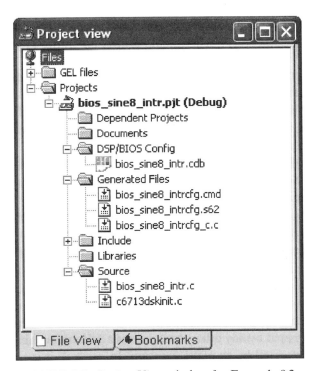

FIGURE 9.5. *Project View* window for Example 9.2.

```
        temp_ptr = process_ptr;
        process_ptr = input_ptr;
        input_ptr = output_ptr;
        output_ptr = temp_ptr;

        for (i=0 ; i< PTS ; i++) (process_ptr + i)->imag = 0.0;
        for (i=PTS/2 ; i< PTS ; i++) (process_ptr + i)->real = 0.0;

        fft(process_ptr,PTS,twiddle);  //transform into freq domain

        for (i=0 ; i<PTS ; i++)         //filter in frequency domain
        {                               //i.e. complex multiply
          a = (process_ptr + i)->real; //samples by coeffs
          b = (process_ptr + i)->imag;
          (process_ptr+i)->real=coeffs[i].real*a-coeffs[i].imag*b;
          (process_ptr+i)->imag=-(coeffs[i].real*b+coeffs[i].imag*a);
        }
        fft(process_ptr,PTS,twiddle);
        for (i=0 ; i<PTS ; i++)
        {
          (process_ptr + i)->real /= PTS;
          (process_ptr + i)->imag /= -PTS;
        }
        for (i=0 ; i<PTS/2 ; i++)       //overlap add (real part only)
        {
          (process_ptr + i)->real += (output_ptr + i + PTS/2)->real;
        }
      }                                 // end of while
}

// attach to HWI
void c_int11(void)                      //ISR
{
  output_left_sample((short)((output_ptr + buffercount)->real));
  (input_ptr + buffercount)->real = (float)(input_left_sample());
  (input_ptr + buffercount++)->imag = 0.0;
  if (buffercount >= PTS/2)
  {
    bufferfull = 1;
    buffercount = 0;
  }
}

void main()
{
  return;
}                                       //end of main()
```

FIGURE 9.10. (*Continued*)

```
//RTDX_MATLAB_sim.c MATLAB-DSK interface using RTDX between PC & DSK

#include <rtdx.h>                    //RTDX support file
#include "target.h"                  //for init interrupt
short buffer[10] = {0};              //init data from PC
RTDX_CreateInputChannel(ichan);      //data transfer PC-->DSK
RTDX_CreateOutputChannel(ochan);     //data transfer DSK-->PC

void main(void)
{
 int i;

 TARGET_INITIALIZE();                         //init for interrupt
 while(!RTDX_isInputEnabled(&ichan))  //for MATLAB to enable RTDX
      puts("\n\n Waiting to read ");  //while waiting
 RTDX_read(&ichan,buffer,sizeof(buffer));//read data by DSK
 puts("\n\n Read Completed");
 for (i = 0; I < 10; i++)
      buffer[i]++;                            //increment by 1 data from PC
 while(!RTDX_isOutputEnabled(&ochan))  //for MATLAB to enable RTDX
      puts("\n\n Waiting to write ");  //while waiting
 RTDX_write(&ochan,buffer,sizeof(buffer));//send data from DSK to PC
 puts("\n\n Write Completed");
 while(1)    {}                         // infinite loop
}
```

FIGURE 9.11. C program that runs on the DSK to illustrate RTDX with MATLAB. The buffer of data is incremented by one on the DSK and sent back to MATLAB (rtdx_matlab_sim.c).

transfer data from the target DSK to the PC host. When the input channel is enabled, data are *read* (received as input to the DSK) from MATLAB. After each data value in the buffer is incremented by 1, an output channel is enabled to *write* the data (sent as output from the DSK) to MATLAB. Note that the input (read) and output (write) designations are from the target DSK.

Figure 9.12 shows the MATLAB-based program *rtdx_matlab_sim.m*. This program creates a buffer of data values 1, 2, ..., 10. It requests board information, opens CCS, and enables RTDX. It also loads the executable file *rtdx_matlab_sim.out* within CCS and runs the program on the DSK. Two channels are opened through RTDX: an input channel to write/send the data from MATLAB (PC) to the DSK and an output channel to read/receive the data from the DSK.

Build this project as **rtdx_matlab_sim** within CCS. The appropriate support files are included in the folder rtdx_matlab_sim. Add the necessary support files: the C source file *rtdx_matlab_sim.c*, the vector file *intvecs.asm* (from TI), *c6713dsk.cmd* (from TI), *rtdx.lib* (located in *CCStudio_v3.1\c6000\rtdx\lib*), and the interrupt support header file *target.h* (from MATLAB). This process creates the executable file *rtdx_matlab_sim.out*.

```
%RTDX_MATLAB_sim.m MATLAB-DSK interface using RTDX. Calls CCS
%loads .out file.Data transfer from MATLAB->DSK,then DSK->MATLAB

indata(1:10) = [1:10];                      %data to send to DSK
ccsboardinfo                                %board info
cc = ccsdsp('boardnum',0);                  %set up CCS object
reset(cc)                                   %reset board
visible(cc,1);                              %for CCS window
enable(cc.rtdx);                            %enable RTDX
if ~isenabled(cc.rtdx)
    error('RTDX is not enabled')
end
cc.rtdx.set('timeout', 20);                 %set 20sec time out for RTDX
open(cc,'rtdx_matlab_sim.pjt');             %open project
load(cc,'./debug/rtdx_matlab_sim.out');     %load executable file
run(cc);                                    %run
configure(cc.rtdx,1024,4);                  %configure two RTDX channels
open(cc.rtdx,'ichan','w');                  %open input channel
open(cc.rtdx,'ochan','r');                  %open output channel
pause(3)                                    %wait for RTDX channel to
open
enable(cc.rtdx,'ichan');                    %enable channel TO DSK
if isenabled(cc.rtdx,'ichan')
    writemsg(cc.rtdx,'ichan', int16(indata)) %send 16-bit data to DSK
    pause(3)
else
    error('Channel ''ichan'' is not enabled')
end
enable(cc.rtdx,'ochan');                    %enable channel FROM DSK
if isenabled(cc.rtdx,'ochan')
    outdata=readmsg(cc.rtdx,'ochan','int16') %read 16-bit data from DSK
    pause(3)
else
    error('Channel ''ochan'' is not enabled')
end
if isrunning(cc), halt(cc);                 %if DSP running halt
processor
end
disable(cc.rtdx);                           %disable RTDX
close(cc.rtdx,'ichan');                     %close input channel
close(cc.rtdx,'ochan');                     %close output channel
```

FIGURE 9.12. MATLAB program that runs on the host PC to illustrate RTDX with MATLAB. Buffer of data sent from MATLAB to the DSK (rtdx_matlab_sim.m).

Access MATLAB and make the following directory (path) active:

CCStudio_v3.1\myprojects\rtdx_matlab_sim

Within MATLAB, run the (*.m*) file, typing *rtdx_matlab_sim*. Verify that the executable file is being loaded (through the CCS window) and run. Within the CCS window, the following messages should be printed:Waiting to read, Read completed, Waiting to write, and Write completed.Then, within MATLAB, the following should be printed: *outdata* = 2 3 4 . . . 11, indicating that the values (1, 2, . . . , 10) in the buffer *indata* sent initially to the DSK were each incremented by 1 due to the C source program line of code: `buffer[i]`++; executed on the C6x (DSK).

Example 9.8 further illustrates RTDX through MATLAB, acquiring external real-time input data (from the DSK) and sending them to MATLAB for further processing (FFT, plotting).

Example 9.8: MATLAB–DSK Interface Using RTDX, with MATLAB for FFT and Plotting (`rtdx_matlabFFT`)

This example illustrates the interface between MATLAB and the DSK using RTDX. An external input signal is acquired from the DSK, and the input samples are stored in a buffer on the C6x processor. Using RTDX, data from the stored buffer are transferred from the DSK to the PC host running MATLAB. MATLAB takes the FFT of the received data from the DSK and plots it, displaying the FFT magnitude on the PC monitor. The same support tools as in Example 9.7 are required, including The Embedded Target for TI C6000 DSP (2.0) and MATLAB Link for CCS, available from MathWorks. The following support files are also used for this example and provided by TI: (1) the linker command file `c6713dsk.cmd`; (2) the vector file `intvecs.asm`; and (3) the library support file `rtdx.lib`. In the init/comm file `c6713dskinit.c`, the line of code to point at the IRQ vector table is bypassed since the support file `intvecs.asm` handles that.

Figure 9.13 shows the program `rtdx_matlabFFT.c` to illustrate the interface. It is a loop program as well as a data acquisition program, storing 256 input samples. Even though the program is polling-based, interrupt is used for RTDX. An *output* channel is created to provide the real-time data transfer from the C6x on the DSK to the PC host.

Figure 9.14 shows the MATLAB-based program `rtdx_matlabFFT.m`. This program provides board information, opens CCS, and enables RTDX. It also loads the executable file (`rtdx_matlabFFT.out`) within CCS and runs the program on the DSK. Note that the output channel for RTDX is opened and data are *read* (from MATLAB running on the PC).A 256-point FFT of the acquired input data is taken, sampling at 16 kHz.The program obtains a total of 2048 buffers, and execution stops afterwards.

Build this project as **rtdx_matlabFFT** within CCS. The necessary support files are included in the folder rtdx_matlabFFT. Add the necessary support files, including `rtdx_matlabFFT.c`, `c6713dskinit.c`, `intvecs.asm` (from TI), `c6713dsk.cmd` (from TI), and `rtdx.lib` (located in `c6713\c6000\rtdx\lib`). Use the following compiler options: `-g -ml3`. The option `-ml3` (from the Advanced Category) allows for Memory Models: Far Calls and Data. This process yields the executable `.out` file.

```
//RTDX_MATLABFFT.c RTDX-MATLAB for data transfer PC->DSK(with loop)

#include "dsk6713_aic23.h"             //codec-DSK support file
#include <rtdx.h>                      //RTDX support file
Uint32 fs=DSK6713_AIC23_FREQ_16KHZ;   //set sampling rate
RTDX_CreateOutputChannel(ochan);      //create out channel C6x-->PC

void main()
{
 short i, input_data[256]={0};        //input array size 256
 comm_poll();                         //init DSK, codec, McBSP
 IRQ_globalEnable();                  //enable global intr for RTDX
 IRQ_nmiEnable();                     //enable NMI interrupt
 while(!RTDX_isOutputEnabled(&ochan)) //wait for PC to enable RTDX
     puts("\n\n Waiting... ");        //while waiting
 while(1)                             // infinite loop
 {
  i=0;
  while (i<256)                       //for 256 samples
   {
     input_data[i] = input_sample(); //defaults to left channel
     output_sample(input_data[i++]); //defaults to left channel
   }
  RTDX_write(&ochan,input_data,sizeof(input_data));//send 256 samples
 }
}
```

FIGURE 9.13. C program that runs on the DSK to illustrate RTDX with MATLAB. Input from the DSK is sent to MATLAB (`rtdx_matlabFFT.c`).

Access MATLAB and make the following directory (path) active:

CCStudio_v3.1\myprojects\rtdx_matlabFFT

This folder contains the necessary files associated with this project. Within MATLAB, run the (.m) file *rtdx_matlabFFT*. Verify that the executable (.out) file is being loaded and run within CCS. Input a sinusoidal signal with a frequency of 2 kHz and verify that the output is the delayed (attenuated) input signal (a loop program). Within MATLAB the plot shown in Figure 9.15 is displayed on the PC monitor, which is the FFT magnitude of the input sinusoidal signal. Vary the frequency of the input signal to 3 kHz and verify the FFT magnitude displaying a spike at 3 kHz.

The FFT is executed on the PC host. As a result, on an older/slower PC, changing the input signal frequency will not yield a corresponding FFT magnitude plot immediately. *Note*: If it is desired to transfer data from the PC to the DSK, an input channel would be created using

```
%RTDX_MATLABFFT.m MATLAB-DSK interface with loop. Calls CCS,
%loads .out file. Data from DSK→MATLAB for FFT and plotting

ccsboardinfo                              %board info
cc=ccsdsp('boardnum',0);                  %setup CCS object
reset(cc);                                %reset board
visible(cc,1);                            %for CCS window
enable(cc.rtdx);                          %enable RTDX
if ~isenabled(cc.rtdx);
    error('RTDX is not enabled')
end
cc.rtdx.set('timeout', 20);               %set 20sec timeout for RTDX
open(cc,'rtdx_matlabFFT.pjt');            %open project
load(cc,'./debug/rtdx_matlabFFT.out');    %load executable file
run(cc);                                  %run program
configure(cc.rtdx,1024,1);                %configure one RTDX channel
open(cc.rtdx,'ochan','r');                %open output channel
pause(3)                                  %wait for RTDX channel to open
fs=16e3;                                  %set sample rate in MATLAB
fftlen=256;                               %FFT length
fp=[0:fs/fftlen:fs/2-1/fftlen];           %for plotting within MATLAB
enable(cc.rtdx,'ochan');                  %enable channel from DSK
isenabled(cc.rtdx,'ochan');
for i=1:2048                              %obtain 2048 buffers then stop
  outdata=readmsg(cc.rtdx,'ochan','int16'); %read 16-bit data from DSK
  outdata=double(outdata);                %32-bit data for FFT
  FFTMag=abs(fftshift(fft(outdata)));     %FFT using MATLAB
  plot(fp,FFTMag(129:256))
  title('FFT Magnitude of data from DSK');
  xlabel('Frequency');
  ylabel('Amplitude');
  drawnow;
end
halt(cc);                                 %halt processor
close(cc.rtdx,'ochan');                   %close channel
clear cc                                  %clear object
```

FIGURE 9.14. MATLAB program that runs on the host PC to illustrate RTDX with MATLAB. MATLAB's FFT and plotting functions are used (rtdx_matlabFFT.m).

```
RTDX_CreateInputChannel(ichan);
While(!RTDX_isInputEnabled(&ichan));
RTDX_read(&ichan, . . .)
```

This creates an input channel, waits for the input channel to be enabled, and reads the data (input to the C6x on the DSK). In the MATLAB program, the following lines of code

```
open(cc.rtdx,'ichan', 'w' );
enable(cc.rtdx,'ichan' );
writemsg(. . .);
```

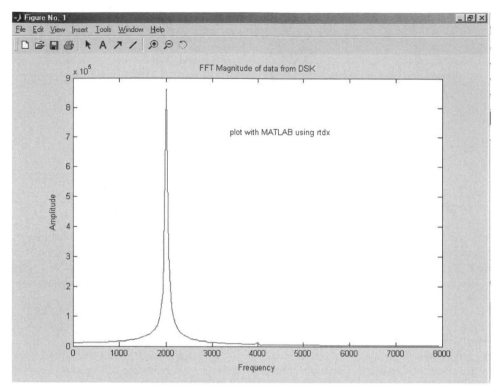

FIGURE 9.15. MATLAB's plot of the FFT magnitude of data received from the DSK.

open and enable an input channel and then write (send) the data from MATLAB running on the host PC to the C6x on the DSK. See Example 9.7.

Example 9.9: MATLAB–DSK Interface Using RTDX for FIR Filter Implementation (rtdx_matlabFIR)

This example further illustrates RTDX with MATLAB with the implementation of FIR filters. Figure 9.16 shows the C source program FIR3LP_RTDX.c that generates an input signal and implements an FIR filter on the DSK. The input signal consists of the product of random noise and a sine wave from a lookup table. This generated signal is the input to an FIR filter (see Example 4.1). The output of the filter is stored in a buffer, the address of which is transferred to MATLAB through the output RTDX channel. Initially, the implemented filter is a lowpass FIR filter with a cutoff frequency at 600 Hz. The coefficients of this filter are in the file LP600.cof. Two other FIR lowpass filter coefficients can also be selected in this example: LP1500.cof and LP3000.cof. These three sets of coefficients were used in Example 4.2 (FIR3LP). The address of the specific filter to be implemented is read through the RTDX input channel. All the appropriate support

```
//FIR3LP_RTDX.c FIR-3 Lowpass with different BWs using RTDX-MATLAB
#include "lp600.cof"                       //coeff file LP @ 600 Hz
#include <rtdx.h>
#include <stdio.h>
#include "target.h"
int yn = 0;                                //initialize filter's output
short dly[N];                              //delay samples
short h[N];                                //filter characteristics 1xN
short loop = 0;
short sine_table[32]={0,195,383,556,707,831,924,981,1000,981,924,831,
                707,556,383,195,0,-195,-383,-556,-707,-831,-924,-981,
                -1000,-981,-924,-831,-707,-556,-383,-195};//sine values
short amplitude = 10;
#define BUFFER_SIZE 256
int buffer[BUFFER_SIZE];
int inputsample, outputsample;
short j = 0;
RTDX_CreateInputChannel(ichan);           //create input channel
RTDX_CreateOutputChannel(ochan);          //create output channel

void main()
{
 short i;
 TARGET_INITIALIZE();
 RTDX_enableInput(&ichan);                //enable RTDX channel
 RTDX_enableOutput(&ochan);               //enable RTDX channel
 for (i=0; i<N; i++)
  {
     dly[i] = 0;                          //init buffer
     h[i] = hlp600[i];                    //start addr of LP600 coeff
  }
 while(1)                                 //infinite loop
 {
   inputsample=rand()+amplitude*(sine_table[loop]);//generate  input
   if (loop < 31) ++loop;
   else loop = 0;
   dly[0]=inputsample;                    //FIR filter section
   yn = 0;                                //initialize filter output
   if (!RTDX_channelBusy(&ichan))   {
       RTDX_readNB(&ichan,&h[0],N*sizeof(short));} //input coeff
   for (i = 0; i< N; i++)
       yn +=(h[i]*dly[i]);                //y(n) += h(LP#,i)*x(n-i)
   for (i = N-1; i > 0; i--)              //starting @ bottom of buffer
       dly[i] = dly[i-1];                 //update delays
   outputsample = (yn >> 15);             //filter output
   buffer[j] = outputsample;              //store output -> buffer
   j++;
   if (j==BUFFER_SIZE) {
      j = 0;
      while (RTDX_writing != NULL) {}     //wait rtdx write to complete
      RTDX_write( &ochan, &buffer[0], BUFFER_SIZE*sizeof(int) );
   }
 }
}
```

FIGURE 9.16. C program that implements FIR filters and runs on the DSK. It illustrates RTDX with MATLAB.

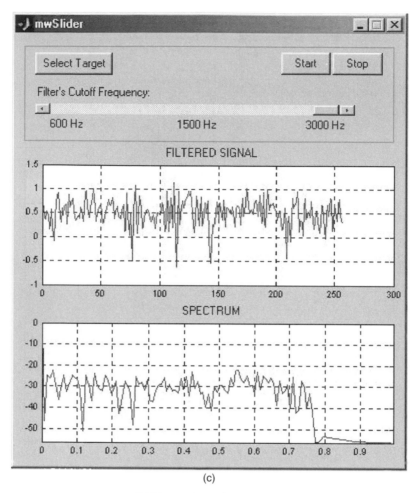

(c)

FIGURE 9.17. (*Continued*)

9.3 RTDX USING VISUAL C++ TO INTERFACE WITH DSK

Two examples are provided to illustrate the use of RTDX with Microsoft's Visual C++, one of which makes use of MATLAB's functions for finding and plotting the FFT magnitude (not for the RTDX interface). Three projects in Chapter 10 (DTMF, FIR, and Radix-4 FFT) make use of RTDX with Visual C++ to obtain a PC-DSK interface.

Example 9.10: Visual C++–DSK Interface Using RTDX for Amplitude Control of the Sine Wave (rtdx_vc_sine)

This example illustrates the use of RTDX with Microsoft Visual C++. The application running on the target DSK generates a sine wave. A procedure follows to illustrate the development of the host application with RTDX support—in particular, the development of a Visual C++ application with a slider control for adjusting the amplitude of the generated sine wave running on the C6x DSK. All the Visual C++ application files are on the CD in the folder *rtdx_vc_sine*.

CCS Component

Figure 9.18 shows the C source program *rtdx_vc_sine.c* that implements the sine generation with amplitude control. This is the same C source program used to illustrate RTDX with Visual Basic in Example 9.12 as well as with LabVIEW in Example 9.16. An RTDX input channel is created and enabled in order to read the slider data from the PC host.

Create, save, and add the configuration file *rtdx_vc_sine.cdb* to the project. Select INT11, *MCSP_1_Transmit* as the interrupt source and *_c_int11* as the function. See Example 9.2. Add the autogenerated linker command file and the BSL library support file. The run-time and the CSL library support files are included

```
//RTDX_vc_sine.c Sine generation.RTDX using Visual C++(or VB/LABVIEW)

#include "rtdx_vc_sinecfg.h"              //generated by .cdb file
#include "dsk6713_aic23.h"               //codec-dsk support file
#include <rtdx.h>                        // for rtdx support
Uint32 fs=DSK6713_AIC23_FREQ_16KHZ;      //set sampling rate
short loop = 0;
short sin_table[8] = {0,707,1000,707,0,-707,-1000,-707};
int gain = 1;
RTDX_CreateInputChannel(control_channel); //create input channel

interrupt void c_int11()                 //ISR set in .cdb
{
 output_sample(sin_table[loop]*gain);
 if (++loop > 7) loop = 0;
}

void main()
{
 comm_intr();                            //init codec,dsk,MCBSP
 RTDX_enableInput(&control_channel);     //enable input channel
 while(1)                                //infinite loop
 {
  if(!RTDX_channelBusy(&control_channel)) //if channel not busy
     RTDX_read(&control_channel,&gain,sizeof(gain));//read from PC
 }
}
```

FIGURE 9.18. C program that runs on the DSK to illustrate RTDX with Visual C++. It generates a sine wave (rtdx_vc_sine.C).

class is responsible for displaying an *About* message dialog, as in most window-based applications.

7. From the main menu, select *View* and select the *ClassWizard* menu item. This pops up the *MFCClassWizard* dialog window. (Make sure to select *CTestpro-jectDlg* from the class name.) Select the *Member Variables* tab, and then select *IDC_SLIDER1* from the list of control IDs.

8. Click the *Add Variable* button to display the *Add Member Variable* dialog. Choose an appropriate member variable name, such as `m_slider`, and make sure that the *Category* field is *Value* and the *Variable type* field is *int*. Click *OK* to return to the *ClassWizard* dialog window.

9. Create a new class for RTDX. Click on the *Add Class* button and select *from a type library*. Browse in the folder `c:\CCStudio_v3.1\cc\bin` and select (or type) the file `Rtdxint.dll`. This pops up the *Confirm classes* dialog. Click *OK* to return to the *ClassWizard* dialog. Click *OK* again to dismiss the *ClassWizard* dialog. The new class *IRtdxExp* has been added for the functionality of RTDX.

10. From the *ClassView* pane (lower-left window):

   (a) Select the class *CTestprojectDlg*. Right-click on the class and select *Add member variable*. For variable type, use *IRtdxExp** (note the pointer notation), and for variable name use *pRTDX* (or another name). Click *OK* to dismiss the dialog. This creates a pointer that represents and manipulates the class *IRtdxExp* created in the previous step.

   (b) Right-click on the class *CTestprojectDlg* and select *Add Windows Message Handler*. This will bring up the *New Windows Message* dialog. From the list, find and select the message *WM_DESTROY*. Click on the *Add and Edit* button to insert the new windows message. Add the following lines of code just after the function

```
CDialog::OnDestroy( ).
if(pRTDX->Close( ))
    MessageBox("Could not close the channel!", "Error");
```

   (c) Right-click on the class *CTestprojectDlg* and choose the *Add Windows Message Handler* to bring up again the *New Windows Message* dialog. Select the `WM_HSCROLL` message and click on the *Add and Edit* button. Add the following lines of code just above the function `CDialog::OnHScroll(nSBCode, nPos, pScrollBar)`. This is shown in Figure 9.21.

```
long buffer;
UpdateData(TRUE);
pRTDX->WriteI4((long)m_slider, &buffer);
UpdateData(FALSE);
```

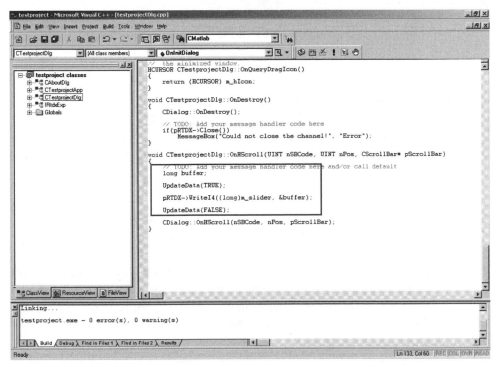

FIGURE 9.21. Visual C++ windows handler for the message *WM_HSCROLL*.

(d) Sclect the class *CTestprojectDlg* and expand it. Locate the function
`OnInitDialog( )` and double-click on it. Add the following lines of code
just above the *return* instruction:

```
CSliderCrtl* pSliderCrtl = (CSliderCrtl*)
   GetDlgItem(IDC_SLIDER1);
pSliderCrtl->SetRange(1,10);
pRTDX = new IRtdxExp;
pRTDX->CreateDispatch(_T("RTDX"));
if(pRTDX->SetProcessor(_T("C6713DSK"),_T("CPU_1")))
   MessageBox("Could not set the processor!",
"Error");if(pRTDX->Open("control_channel", "W"))
   MessageBox("Could not open the channel!", "Error");
```

(e) Double-click on the class *CTestprojectDlg* and add the following line of
code just before the class definition statement:

```
#include "rtdxint.h"
```

(f) Select the class *CTestprojectApp* and expand it. Double-click on the function `InitInstance( )` and add the following line of code:

```
AfxOleInit( );
```

just above the line *CTestAppDlg dlg;*.

The added lines of code can be verified from the file `rtdx_vc_sineDlg.cpp` (on the CD). Select *Build* (menu item from the main project window) → *Rebuild All* to create the application (executable) file.

Example 9.11: Visual C++–DSK Interface Using RTDX with MATLAB Functions for FFT and Plotting (`rtdx_vc_FFTmatlab`)

This example illustrates real-time data communication using RTDX with Microsoft Visual C++, invoking MATLAB's FFT and plotting functions. MATLAB is not used in this example to provide the RTDX communication link between the PC and the DSK, as in Examples 9.7–9.9. Instead, only MATLAB's functions for FFT and plotting are invoked.

The folder **rtdx_vc_FFTmatlab** contains the Visual C++ support files, including the application/executable file `rtdx_vc_FFTmatlab.exe` (already built). See also Example 9.10.

Running Executable from CCS

The folder *rtdx_MatlabFFT* for Example 9.8 includes the main C source program (Figure 9.13) `rtdx_matlabFFT.c`, which implements a loop program. It also creates and enables an output channel to write/send data acquired from the DSK to the PC. It illustrated RTDX with MATLAB in Example 9.8, and it can be used in this example to illustrate this Microsoft Visual C++ application. The (`.m`) MATLAB file that provides the RTDX communication link between the DSK and the PC in Example 9.8 is *not* used in this example. Only MATLAB's FFT and plotting functions are used.

Input into the DSK a 2-kHz sine wave with an approximate amplitude of 1 V p-p. Within the CCS window, select *Tools* → *RTDX* → *Enable RTDX* (check it). Load and run `rtdx_matlabFFT.out`. The RTDX communication link is not yet produced, and "waiting" is printed continuously within the CCS window.

Running Visual C++ Application

Run the Visual C++ application `rtdx_vc_FFTMatlab.exe` located in the folder *rtdx_vc_FFTMatlab\debug* (double-click on it).

Verify a loop program with the DSK output to a scope, and an FFT plot of the 2-kHz sine wave as shown in Figure 9.22, obtained using MATLAB's FFT and

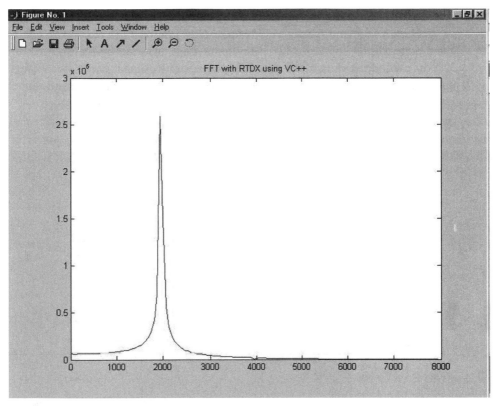

FIGURE 9.22. Plot of FFT magnitude (with MATLAB) to illustrate RTDX using Visual C++ for the project rtdx_vc_matlabFFT.

plotting functions (see also Example 9.8). Change the input sine wave frequency to 3 kHz and verify that the MATLAB plots 3-kHz sine wave.

You can readily add the labels for the x and y axes in Figure 9.22 by modifying the file rtdx_vc_FFTMatlabDlg.cpp. Find the section of code where the MATLAB functions are invoked for FFT and plotting. After the line of code for the figure's title, insert the appropriate xlabel and ylabel functions. Launch Microsoft Visual C++. Select *File* and open the workspace (.dsw) file located in the folder **rtdx_vc_FFTmatlab**. Select *Build* → *Rebuild All* to recreate a new application (.exe) file. Verify that the FFT plot now contains the x and y axis labels.

Creation of Visual C++ Application and Support Files

1. Repeat steps 1–3 in Example 9.10. The Resource Dialog editor should be opened. Resize the main dialog window. Right-click on the *TODO:Place dialog control here* and select the *Properties* menu item. From the resulting property dialog in the *Caption* field, enter any messages that you want displayed in the dialog window (such as RTDX with Visual C++ to . . .), and then close the property dialog window.

(d) Click on the *File View* pane (next to the *ClassView* pane), and expand *Rtdx_vc_fftMatlab* to expose three folders. Expand on *Source Files*, and double-click on *MatlabClass.cpp*. Add this section of code at the end of this file (after the pair of brackets):

```
void CMatlabClass::OpenMatlab(LPCTSTR lpCommand)
{
pEngine = engOpen(lpCommand);
. . .
return engOutputBuffer(pEngine, pOutputBuffer, nLength);
}
```

(e) Right-click on the class *CRtdx_vc_fftMatlabDlg* and select *Add Windows Message Handler*. Find and select the message *WM_DESTROY*, and click on *Add and Edit* to insert the new windows message. Add the following lines of code beneath the function *CDialog::OnDestroy( )*:

```
nFlat = 0;
WaitForSingleObject(pRTDXThread->m_hThread,INFINITE);
```

(f) Right-click on the class *CRtdx_vc_fftMatlabDlg* and click on *Add member function*. For the type, use UINT and for the declaration, type *static RTD XThreadFunction(LPVOID lpVoid)* and then click *OK*.

(g) Expand the class *CRtdx_vc_fftMatlabDlg*, double-click on the member function *RTDXThreadFunction(LPVOID lpVoid)*, and add the following lines of code in the function body (between the pair of brackets):

```
CMatlabClass* pMatlab;
IRtdxExp *pRtdx;
. . .
pMatlab->ExecuteLine(_T("fs = 16e3;"));
. . .
pMatlab->ExecuteLine (_T("plot(fp, fftMag(129: 256))"));
. . .
return 0;
```

Scroll to the top of the file and add the following two `include` files and the global variable `nflag`:

```
#includee "MatlabClass.h"
#include "Rtdxint.h"
int nFlag = 1;
```

(h) With the class *CRtdx_vc_fftMatlabDlg* expanded, double-click on the member function *OnInitDialog( )* and add the following line of code just before the return instruction:

```
pRTDXThread = (CRTDXThread*)AfxBeginThread
             (RTDXThreadFunction,m_hWnd);
```

7. The path of MATLAB libraries and `include` files need to be added before building the project. Select *Tools* → *Options* to display the *Options* dialog, and click on the *Directories* tab. Select the *Include File*s item from *Show directories for*. Click twice on the rectangle below the list of *Directories*, then click on the "...". displayed on the right. Browse in your MATLAB installation directory for the `include` path *c:\Matlab_folder\extern\include* (e.g., *matlabR13* as the *Matlab_folder*). From the *Show directories for* list, select the *library file* item. Click twice on the rectangle below the list of *Directories* and select the "..." (as before). Browse in your MATLAB folder for the path *c:\Matlab_folder\extern\lib\win32\microsoft\msvc60*, and click on OK to save the changes.

Build the Visual C++ application project. Select *Build* → *Rebuild All* to create `rtdx_vc_FFTMatlab.exe`.

9.4 RTDX USING VISUAL BASIC TO PROVIDE INTERFACE BETWEEN PC AND DSK

Two examples are provided to illustrate the interface between the PC host and the DSK with RTDX using Visual Basic.

Example 9.12: Visual Basic–DSK Interface Using RTDX for Amplitude Control of a Sine Wave (rtdx_vbsine)

This example generates a sine wave outputted through the codec on the DSK. It illustrates RTDX using Visual Basic (VB) to create a slider and control the amplitude of the generated sine wave.

CCS Component
Figure 9.24 shows the C source program `rtdx_vbsine.c` that implements the sine generation with amplitude control. This is the same C source program used to illustrate RTDX with Visual C++ in Example 9.7 and LabVIEW in Example 9.16. An RTDX input channel is created and enabled in order to read the slider data from the PC host. This example is not meant to teach the reader VB, but rather to use it.

Create, save, and add a configuration file `rtdx_vbsine.cdb` to the project. Select INT11, *MCSP_1_Transmit* as the interrupt source, and `_c_int11` as the function (see Example 9.2). Add the autogenerated linker command file and the

```
//rtdx_vbsine.c Sine generation.RTDX with Visual Basic(VC++/LABVIEW)

#include "rtdx_vbsinecfg.h"                //generated by .cdb file
#include "dsk6713_aic23.h"                 //codec-dsk support file
#include <rtdx.h>                          // for rtdx support
Uint32 fs=DSK6713_AIC23_FREQ_16KHZ;        //set sampling rate
short loop = 0;
short sin_table[8] = {0,707,1000,707,0,-707,-1000,-707};
int gain = 1;
RTDX_CreateInputChannel(control_channel); //create input channel

interrupt void c_int11()                   //ISR set in .cdb
{
 output_sample(sin_table[loop]*gain);
 if (++loop > 7) loop = 0;
}

void main()
{
 comm_intr();                              //init codec,dsk,MCBSP
 RTDX_enableInput(&control_channel);       //enable input channel
 while(1)                                  //infinite loop
 {
  if(!RTDX_channelBusy(&control_channel)) //if channel not busy
     RTDX_read(&control_channel,&gain,sizeof(gain));//read from PC
 }
}
```

FIGURE 9.24. C program that generates a sine wave. It illustrates RTDX using VB to control the amplitude of the generated sine wave (rtdx_vbsine.c).

BSL library support file. The run-time and the CSL library support files are included in the auto-generated linker command file. Add also the init and communication file, but not the vector file. The necessary files are included in the folder rtdx_vbsine.

Build this project as **rtdx_vbsine**. Within CCS, load and run the executable file *rtdx_vbsine.out*. Verify that a 2-kHz sine wave is generated and outputted through the codec on the DSK.

Enable RTDX within CCS. Select *Tools → RTDX → Configuration Control → Enable RTDX* (activate/check it).

VB Component

The folder *rtdx_vbsine* contains a subfolder PC that contains the support files associated with VB. Click on the (.vbp) VB project file to open VB. The project consists of the file *slider.frm* that describes the slider and the file *boardproc_frm.frm* that describes the board information. These two files are included with

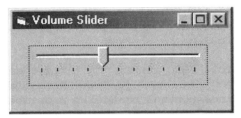

FIGURE 9.25. Volume slider to control the amplitude of the DSK output signal. Object created with VB for the project `rtdx_vbsine`.

CCS. The slider is the same as that used in an example *(hostio1)* included with CCS. Within VB, select *Run → Start*. Press OK for the board information and the slider box shown in Figure 9.25 should pop up. Connect the DSK output to a scope. Vary the slider position and verify the change in the amplitude of the generated output sine wave (keep the mouse cursor on the slider button to change the slider value). Note that the Application (`.exe`) file, included on the CD, also can be used to run the VB project directly. This application file can be recreated within VB after loading the project file and selecting *File → Make* `rtdx_vbsine.exe`.

The next example implements a loop using RTDX with VB, where the amplitude of the output signal is changed using a gain value sent by the PC host to the C6x processor.

Example 9.13: Visual Basic–DSK Interface Using RTDX for Amplitude Control of Output in a Loop Program (`rtdx_vbloop`)

This example extends the previous example with a loop program using VB and RTDX to control the amplitude of an output signal. A window where the user can enter a gain value is built in VB. That gain value is sent from the PC host to the C6x processor. Figure 9.26 shows the C source program `rtdx_vbloop.c` that implements this project example. See also the previous example.

An RTDX input channel is created and enabled. When the RTDX channel is not busy, the C6x processor reads the data from the PC. Create and add a configuration file to set the interrupt service function, and add similar support files to the project, as in the previous example.

Build this project as **rtdx_vbloop**. Input a sine wave with an approximate amplitude and frequency of 0.5 V p-p and 2 kHz, respectively. Verify that the DSK output exhibits the characteristics of a loop program, as in Examples 2.1 and 2.2. Enable RTDX within CCS as in the previous example.

The subfolder *PC* within the folder *rtdx_vbloop* contains the support files associated with VB. The VB project includes the board information file, as in the previous example, and *gain.frm*, a block where the user can enter a gain value to control the amplitude of the output sine wave. The object *gain.frm* was created with VB.

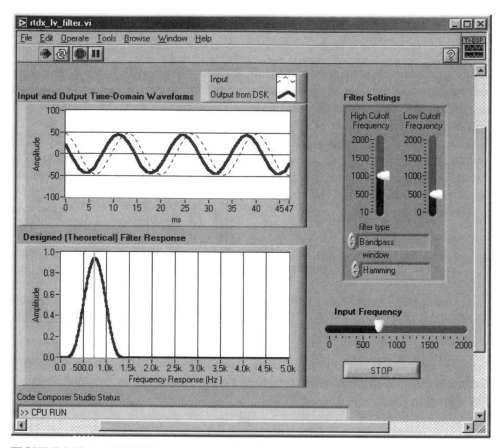

FIGURE 9.28. LabVIEW Instrument window for FIR filter design and plotting to illustrate RTDX for the project `rtdx_lv_filter`.

are the same as those of the input signal for frequencies between 0 and 1300 Hz. The output signal's amplitude decreases toward zero for input frequencies beyond 1300 Hz.

Various windows for the filter design are available, such as Hamming, Hanning, Blackman, and so on. Experiment with different filter characteristics.

2. From Figure 9.28, select *Window → Show Block Diagram*. The LabVIEW tools are required to view the block diagram (the source). Figure 9.29 shows a section of the block diagram that contains various components (smaller blocks). A full description and the function of different blocks can readily be obtained by highlighting each block.

CCS is invoked from LabVIEW to build the project and to load and run the (`.out`) file (from the current directory) on the DSK. (See the CPU status within CCS in Figure 9.28.) Input and output arrays of data, specified as 32

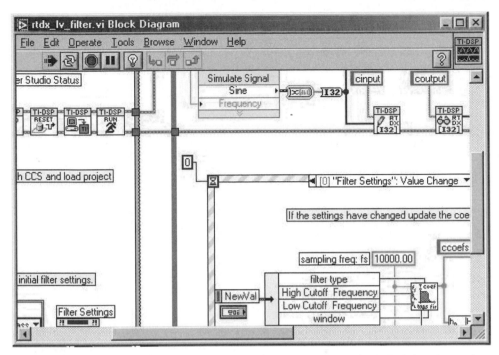

FIGURE 9.29. LabVIEW diagram for FIR filter design through RTDX for the project `rtdx_lv_filter`.

bit integers (`cinput`,`coutput`), are transferred to the DSK through RTDX (Figure 9.29).

3. Figure 9.30 shows the C source program `rtdx_lv_filter.c` that runs on the DSK. It creates two input channels (for the sine wave data and the filter coefficients generated by LabVIEW) and one output channel for the filtered output data (`coutput`). Inputs to the DSK are obtained using `RTDX_read()` or `RTDX_readNB()` to read/input the sine data (`cinput`) and the coefficients (`ccoefs`). The filter is implemented on the DSK by the function `FIR Filter`, and the filtered output (`coutput`) is sent to LabVIEW for plotting using `RTDX_write()`. If the filter characteristics are changed, a new set of coefficients (`ccoefs`) is calculated within LabVIEW and sent to the DSK through RTDX.

Example 9.15: LabVIEW–DSK Interface Using RTDX for Controlling the Gain of a Generated Sinusoid (`rtdx_lv_gain`)

In this example, LabVIEW is used to control the amplitude of a generated sine wave and to plot the scaled output sine wave. An array of data representing the generated sine wave and a gain value are sent from LabVIEW to the DSK. Through RTDX,

```
//rtdx_lv_filter.c RTDX with LABVIEW->filter design/plot DSK output
#include <rtdx.h>                         //RTDX support
#include "target.h"                       //init target
#define kBUFFER_SIZE 48                   //RTDX read/write buffers
#define kTAPS 51
double gFIRHistory [kTAPS+1];
double gFIRCoefficients [kTAPS];
int input[kBUFFER_SIZE],output[kBUFFER_SIZE];
int gain;
double FIRFilter(double val,int nTaps,double* history,double* coefs);
int  ProcessData (int* output, int* input, int gain);
RTDX_CreateInputChannel(cinput);         //create RTDX input data channel
RTDX_CreateInputChannel(ccoefs);         //input channel for coefficients
RTDX_CreateOutputChannel(coutput);       //output channel DSK->PC(Labview)
void main()
{
int i;
TARGET_INITIALIZE();                     //init target for RTDX
RTDX_enableInput(&cinput);               //enable RTDX channels
RTDX_enableInput(&ccoefs);               //for input, coefficients, output
RTDX_enableOutput(&coutput);
gFIRCoefficients[0] = 1.0;
for (i = 1; i<kTAPS; i++)
   gFIRCoefficients[i] = 0.0;
for (;;)                                 //infinite loop
 {
  while(!RTDX_read(&cinput,input,sizeof(input)));//wait for new buffer
  if (!RTDX_channelBusy(&ccoefs))        //if new set of coefficients
     RTDX_readNB(&ccoefs,&gFIRCoefficients,sizeof(gFIRCoefficients));
  ProcessData (output, input, 1);             //filtering on DSK
  RTDX_write(&coutput,&output,szeof(output));//output from DSK->LABVIEW
 }
}
int ProcessData (int *output,int *input,int gain) //calls FIR filter
{
int i;
double filtered;
for(i=0; i<kBUFFER_SIZE; i++) {
  filtered=FIRFilter(input[i]*gain,kTAPS,gFIRHistory,gFIRCoefficients);
  output[i] = (int)(filtered + 0.5);}              //scale output
return 0;
}

double FIRFilter (double val,int nTaps,double* history,double* coefs)
{                                        //FIR Filter
 double temp, filtered_val, hist_elt;
 int  i;
 hist_elt = val;
 filtered_val = 0.0;
 for (i = 0; i <  nTaps; i++)
  {
   temp = history[i];
   filtered_val += hist_elt * coefs[i];
   history[i] = hist_elt;
   hist_elt = temp;
  }
 return filtered_val;
}
```

FIGURE 9.30. C program running on the DSK that implements an FIR filter and illustrates RTDX with LabVIEW (rtdx_lv_filter.c).

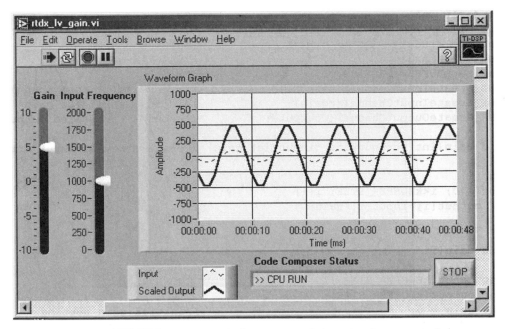

FIGURE 9.31. LabVIEW Instrument window to control the gain of a generated sine wave through RTDX for the project rtdx_lv_gain.

the C6x on the DSK scales the received sine wave input data and sends the resulting scaled output waveform to LabVIEW for plotting. The necessary files for this example are in the folder **rtdx_lv_gain**.

1. Click on the LabVIEW Instrument (.vi) file rtdx_lv_gain to obtain Figure 9.31. Run it as in Example 9.14. The project rtdx_lv_gain.pjt is opened within CCS, and loaded and run on the DSK. See the Code Composer Status in Figure 9.31. Verify that the amplitude of the output sine wave is five times that of the input. You can vary the input signal frequency as well as the gain settings to control the scaled output amplitude waveform. The output frequency is the same as the input frequency. You can readily change the input signal type to a square wave, a triangle, or a sawtooth.

   From the block diagram, one can verify that the input and output data are transferred through RTDX as two arrays (using [I32]), whereas the gain is transferred as a single value (using I32).The brackets represent the array notation (using 32-bit integer format).

2. Figure 9.32 shows the C source program rtdx_lv_gain.c that runs on the DSK. Through RTDX, the input and output channels are enabled and opened for the C6x on the DSK to read the generated sine wave data and the user set gain value and to write the scaled sine wave data to LabVIEW for plotting.

10.1.1 Using a Correlation Scheme and Onboard LEDs for Verifying Detection

The correlation scheme is as follows. Let the input signal be $u(t) = A(\sin(2\pi 697t + \varphi_1) + \sin(2\pi 1209t + \varphi_2))$. Since the input signal includes $\sin(2\pi 697t + \varphi_1)$, the correlation of the input signal with $\sin(2\pi 697t + \varphi_1)$ must be higher than the correlations with $\sin(2\pi 770t + \varphi_1)$, $\sin(2\pi 852t + \varphi_1)$, and $\sin(2\pi 941t + \varphi_1)$. The Fourier transform $\int u(t)e^{-j\omega t}dt$ has a peak at 697 Hz. Using Euler's formula for the exponential function, it becomes a correlation of $u(t)$ with sine and cosine functions. As a result, the input frequency can be determined by correlating the input signal with the sine and cosine for each possible frequency. The algorithm is as follows:

1. For each frequency, find the following correlations:

$$W_{\sin 697} = \sum_{n=1}^{N} u(t_n)\sin(2\pi 697 t_n), \quad W_{\cos 697} = \sum_{n=1}^{N} u(t_n)\cos(2\pi 697 t_n)$$

...

$$W_{\sin 1477} = \sum_{n=1}^{N} u(t_n)\sin(2\pi 1477 t_n), \quad W_{\cos 1477} = \sum_{n=1}^{N} u(t_n)\cos(2\pi 1477 t_n)$$

2. For each frequency, find the maximum between sine weight and cosine weight:

$$W_{697} = \max(|W_{\sin 697}|, |W_{\cos 697}|)$$
$$\cdots$$
$$W_{1477} = \max(|W_{\sin 1477}|, |W_{\cos 1477}|)$$

3. Among the first four weights, choose the largest one; and among the last three weights, choose the largest one:

$$W_1 = \max(|W_{697}|, |W_{770}|, |W_{852}|, |W_{941}|)$$
$$W_2 = \max(|W_{1209}|, |W_{1336}|, |W_{1477}|)$$

4. The frequencies present in the input signal can then be obtained. If both W_1 and W_2, are larger than a threshold, turn on the appropriate LEDs corresponding to each character, as shown in Table 10.2.

Figure 10.1 shows the C source program `partial_dtmf.c` that can be completed readily. Build this project as DTMF. You can test this project first since the complete executable file `DTMF.out` is included on the CD in the folder DTMF. It can be tested using one of the following:

1. A phone to create the DTMF signals and a microphone to capture these signals as input to the DSK's mic input. Alternatively, a microphone with the

TABLE 10.2   Characters and Corresponding LEDs

1	0001
2	0010
3	0011
4	0100
5	0101
6	0110
7	0111
8	1000
9	1001
*	1010
0	1011
#	1100

```
//DTMF.c Core program to decode DTMF signals and turn on LEDs
#define N    100
#define thresh 40000
short i;short buffer[N]; short sin697[N],cos697[N],sin770[N],cos770[N];
...
long weight697,weight697_sin,weight697_cos; long ...weight1477_cos;
long weight1,weight2,choice1,choice2;
interrupt void c_int11()
{
 for (i = N-1; i > 0; i--)
     buffer[i]=buffer[i-1];                  // initialize buffer
 buffer[0] = input_sample();                 //input into buffer
 output_sample(buffer[0]*10);                //output from buffer
 weight697_sin=0;  weight697_cos=0;          //weight @ each freq
 ...
 weight1477_sin = 0;  weight1477_cos =  0;
 for (i = 0; i < N; i++)
 {
  weight697_sin = weight697_sin + buffer[i]*sin697[i];
  weight697_cos = weight697_cos + buffer[i]*cos697[i];
  ...
  weight1477_cos= weight1477_cos + buffer[i]*cos1477[i];
 }
 //for each freq compare sine and cosine weights and choose largest
 if(abs(weight697_sin)>abs(weight697_cos))    weight697=abs(weight697_sin);
 else weight697 = abs(weight697_cos);
 ...
 if(abs(weight1477_sin)>abs(weight1477_cos)) weight1477 = abs(weight1477_sin);
 else weight1477 = abs(weight1477_cos);
 weight1=weight697; choice1=1;//among weight697,..weight941->largest
 if(weight770 > weight1) {weight1 = weight770; choice1=2;} //...
 if(weight941 > weight1) {weight1 = weight941; choice1=4;}
 weight2=weight1209; choice2=1;//among weight1209,..weight1477->largest
 if(weight1336> weight2) {weight2 = weight1336; choice2=2;}
```

FIGURE 10.1. Core C program using correlation to detect DTMF tones (partial_dtmf.c).

```
//DTMF_BIOS_RTDX.c Addtl. code to DTMF.c for RTDX version using VC++

#include <rtdx.h>                            //RTDX support file
RTDX_CreateOutputChannel(ochan);            //output channel for DSK->PC
#define thresh 80000                         //defines a threshold
short value = 0; short w = 0;                //used for RTDX version
.... see DTMF.c
if((weight1>thresh)&&(weight2>thresh))       //set threshold
 if((choice1 == 1)&&(choice2 == 1)) {        //button "1" -> 0001
   DSK6713_LED_off(0);DSK6713_LED_off(1);DSK6713_LED_off(2);DSK6713_LED_on(3);
   value = 1;
 }
 . . . //for button "2", "3",..., "*", "0"
 if((choice1 == 4)&&(choice2 == 3)) { //button "#" -> 1100
   DSK6713_LED_on(0);DSK6713_LED_on(1);DSK6713_LED_off(2);DSK6713_LED_off(3);
    value = 12;
 }
}                          //end of if > than the threshold value (see DTM
else { //weights below threshold, turn LEDs off
 DSK6713_LED_off(0);DSK6713_LED_off(1);DSK6713_LED_off(2);DSK6713_LED_off(3);
 value = 0;
}
w = w + 1;
if w > 50;
{
 w = 0;
 RTDX_write(&ochan,&value,sizeof(value));//send value to PC
}
return;
}                                           //end of interrupt service routine
void main()
{
. . . as in DTMF.c
comm_intr();
while(!RTDX_isOutputEnabled(&ochan))
     puts("\n\n Waiting . . . ");          //wait for output channel->enabled
while(1);                                    //infinite loop
}
```

(a)

(b)

FIGURE 10.3. (a) Core C program to detect DTMF signals with RTDX for PC–DSK interface (DTMF_BIOS_RTDX.c); (b) PC screen displaying detected DTMF signals with RTDX for PC–DSK interface.

Implementation Issues

1. A number is sent to the PC (through RTDX) every 50th time and can be changed.
2. The threshold value can be adjusted.
3. A "length" of 15 is set in the file `numbersDlg.cpp`. This is used to analyze the last 15 numbers and determine if a button was pressed. A smaller value can cause false detection due to noise, whereas it can be more difficult to recognize a short DTMF signal with a larger value of `length`.

If the number 1 is pressed using a Dialpad, dozens of 1's are transmitted through RTDX and appear in the data stream. With no button pressed, a stream of 0's is transmitted. The algorithm distinguishes the actual buttons that are pressed. An array of size `length` stores the last `length` numbers. The number of 1's in the array goes into `Weight1`, the number of 2's in the array goes into `Weight2`, and so on. If any of the weights is greater than 70% of `length`, then it is decided that the number corresponding to that weight was pressed. The character corresponding to this number is then added to the string shown in Figure 10.3b. Note that each weight should be followed by `Weight0` (except `Weight0`).

10.1.3 Using FFT and Onboard LEDs for Verifying Detection

Figure 10.4 shows the core of the C source program that implements this miniproject using an FFT scheme to detect the DTMF signals. Example 6.8 and Section 10.4 illustrates the radix-4 FFT. The FFT is used to estimate the weights associated with the seven frequencies. For example, the 697-Hz signal corresponds to a weight of $697(256/8000) \simeq 22$, and we would use the 22nd value of the FFT array. A 256-point FFT is used with a sampling frequency of 8000 Hz. Similarly, the 770-Hz signal corresponds to a weight of $770(256/8000) \simeq 25$, and we would use the 25th value of the FFT array, and so on for the other weights (28, 31, 39, 43, and 47). We then find the largest weights associated with the first four frequencies to determine the row frequency signal and the largest weights associated with the last three frequencies to determine the column frequency signal. For the largest weights, the corresponding LEDs are turned on (as in Section 10.1.1). As with the previous schemes, the same input (MATLAB, Dialpad, or microphone) can be used. Verify similar results.

10.1.4 Using Goertzel Algorithm

The Goertzel algorithm described in Appendix F may be used for DTMF detection.

10.2 BEAT DETECTION USING ONBOARD LEDs

This miniproject implements a beat detection scheme using the onboard LEDs [32]. Music visualization is a continuously progressing area in audio processing, not only

```
//DTMF_Bios_FFT.c Core program using radix-4 FFT and onboard LEDs
. . .                           //see radix-4 example in Chapter 6
short input_buffer[N] = {0};  //to store input samples...same as x
float output_buffer[7] = {0}; //to store magnitude of FFT
short buffer_count, i, J;
short nFlag;                    //indicator to begin FFT
short nRow, nColumn;
double delta;
float tempvalue;
interrupt void c_int11()
{
 input_buffer[buffer_count] = input_sample();
 output_sample((short)input_buffer[buffer_count++]);
 if(buffer_count >= N)          //if accum more than N points->begin FFT
  {
   buffer_count = 0;            //reset buffer_count
   nFlag = 0;                   //flag to signal completion
   for(i = 0; i < N; i++)
      {
        x[2*i] = (float)input_buffer[i];  //real part of input
        x[2*i+1] = 0;                      //imaginary part of input
      }
  }
}
void main(void)
{
 nFlag = 1;
 buffer_count = 0;
 . . . //generate twiddle constants, then index for digit reversal
 comm_intr();
 while(1)                                   //infinite loop
 {
  while(nFlag);    //wait for ISR to finish buffer accum samples
  nFlag = 1;
  //call radix-4 FFT, then digit reverse function
  output_buffer[0]=(float) sqrt(x[2*22]*x[2*22]+x[2*22+1]*x[2*22+1]);
  . . . //for weigths 25,28,31,39,43
  output_buffer[6]=(float) sqrt(x[2*47]*x[2*47]+x[2*47+1]*x[2*47+1]);
  tempvalue = 0;                           //choose largest row frequency
  nRow = 0;
  for(j = 0; j < 4; j++)
   {
    if(tempvalue < output_buffer[j])
      {
       if(output_buffer[j] > 0.5e4)
        {
            nRow = j + 1;
            tempvalue = output_buffer[j];
        }
      }
   }
 }                                         //end of for loop
```

FIGURE 10.4. Core C program using FFT to detect DTMF tones (partial_dtmf_bios_FFT.c).

```
tempvalue = 0;                          //choose largest column frequency
nColumn = 0;
for(j = 4; j < 7; j++)
  . . .                                 //as with the rows
          nColumn = j - 3;
  . . .                                 //as with the rows
  } //end of for loop
if((nRow != 0) && (nColumn != 0))
{
  if((nRow==1)&&(nColumn==1))          //for button 0001 ("1")
  {DSK6713_LED_off(0);DSK6713_LED_off(1);DSK6713_LED_off(2);DSK6713_LED_on(3);}
  if((nRow==1)&&(nColumn==2))          //for button 0010
  {DSK6713_LED_off(0);DSK6713_LED_off(1);DSK6713_LED_on(2);DSK6713_LED_off(3);}
  //for button "3", "4", ..., "#"
}
else
{DSK6713_LED_off(0);DSK6713_LED_off(1);DSK6713_LED_off(2);DSK6713_LED_off(3);}
};                                      //end of while (1) infinite loop
}                                       //end of main
```

FIGURE 10.4. (*Continued*)

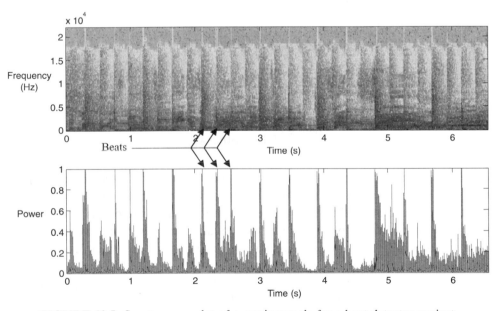

FIGURE 10.5. Spectrogram plot of a music sample for a beat detector project.

for analysis of music but also for entertainment visualization purposes. The scheme is based on the idea that the drum is the most energy-rich component of the music. In this project, the beat of the music is the drum pattern or bass line of the piece of music. Figure 10.5, obtained with MATLAB's capability for plotting the spectrogram of an input .wav file, shows a representative sample section of a piece of music featuring a live drum, a voice, and other instruments. The beat pattern is visible in

flashed whenever a beat is detected. To expand on this project, the beat information can be fed back (from the DSK output) as data or as an audio signal to control, for example, external light effects. Alternatively, it can be fed back to the host PC for further processing, such as calculating beats per minute. RTDX can then be used to provide an interface between the PC host and the DSK (see Chapter 9).

Build this project as **beatdetector** and verify that this detection scheme (with several different types of music) recognizes the drum in most cases, with very few false positives.

10.3 FIR WITH RTDX USING VISUAL C++ FOR TRANSFER OF FILTER COEFFICIENTS

This project implements an FIR filter using VC++ with RTDX to transfer the coefficients. Chapters 4 and 9 discuss FIR filters and RTDX with VC++, respectively. All the appropriate files for this project are on the CD in the folder rtdx_vc_FIR. Figure 10.7 shows the C source program rtdx_vc_FIR.c that runs on the DSK. It implements the FIR filter and creates and enables an input channel through RTDX to read a new set of coefficients. These coefficients are transferred through RTDX from the PC host to the C6x running on the DSK.

1. Build this project as **rtdx_vc_FIR**. A configuration (.cdb) file is created to set INT11. Note that the project includes several autogenerated support files including the linker command file. The init/comm. file is included in the project for real-time input and output. The vector file is not included since INT11 is set within the configuration file. See Example 9.2.

   Within CCS, load and run the executable file. Select *Tools → RTDX → Configuration Control* and *Enable RTDX* (check it).

2. Run the Visual C++ application file included in the folder rtdx_vc_FIR\VC_FIR_RTDX\Debug. A message for the user to load a coefficient file pops up, as shown in Figure 10.8. Load the coefficient file LP600.cof, looking in the folder rtdx_vc_FIR. This coefficient file was designed with MATLAB and used in Example 4.7 to implement a lowpass FIR filter with a cutoff frequency at 600 Hz. Verify this result.

   Load LP1500.cof and LP3000.cof, which represent FIR lowpass filters with 81 coefficients and with cutoff frequencies at 1500 and 3000 Hz, respectively. Verify that these FIR filters can be implemented readily.

The coefficient files are transferred in real time to the C program running on the DSK, using the function RTDX_read( ) in Figure 10.7. The coefficients are stored in the buffer RtdxBuffer, along with N that represents the number of coefficients (81) as the first value in the coefficient file (the lowpass coefficient files in the example FIR3LP have been modified for this project). Experiment with different sets of coefficients.

```
//rtdx_vc_FIR.c FIR with RTDX using VC++ to transfer coefficients file
#include "dsk6713_aic23.h"
#include <rtdx.h>
#define RTDX_BUFFER_SIZE 256                      //change for higher order
Uint32 fs = DSK6713_AIC23_FREQ_8KHZ;
RTDX_CreateInputChannel(control_channel); //create input channel
short* pFir;                                      //->filter's Impulse response
short RtdxBuffer[RTDX_BUFFER_SIZE]={0};    //buffer for RTDX
short dly[RTDX_BUFFER_SIZE] = {0};         //buffer for input samples
short i;
short N;                                   //order of filter
int yn;
interrupt void c_int11()
{
  dly[0] = input_sample();
  yn = 0;
  for(i = 0; i < N; i++)
      yn += pFir[i]*dly[i];
  for(i = N - 1; i > 0; i--)
      dly[i] = dly[i-1];
  output_sample(yn >> 15);
}
void main()
{
  N = 0;                                   //initial filter order
  pFir = &RtdxBuffer[1];                   //-> 2nd element in buffer
  comm_intr();
  RTDX_enableInput(&control_channel);      //enable RTDX input channel
  while(1)                                 //infinite loop
  {
   if(!RTDX_channelBusy(&control_channel)) //if free, read->buffer
   {                                       //read N and coefficients
    RTDX_read(&control_channel,&RtdxBuffer,sizeof(RtdxBuffer));
    N = RtdxBuffer[0];                     //extract filter order
   }
  }
}
```

FIGURE 10.7. C source program that runs on the DSK to implement an FIR filter using RTDX with Visual C++ to transfer the coefficients from the PC to the DSK (rtdx_vc_FIR.c).

10.4 RADIX-4 FFT WITH RTDX USING VISUAL C++ AND MATLAB FOR PLOTTING

This project implements a radix-4 FFT using TI's optimized functions. The resulting FFT magnitude of a real-time input is sent to MATLAB for plotting. In real time, the output data are sent to the PC host using RTDX with Visual C++. Chapter 9 includes two examples using RTDX with Visual C++, and Chapter 6 includes two

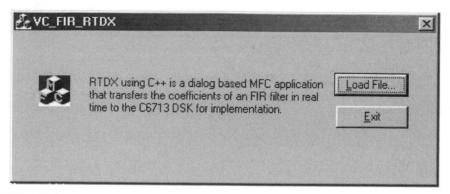

FIGURE 10.8. Visual C++ message to load a file with the FIR coefficients to be transferred through RTDX from the PC to the DSK.

examples (one in real time) to implement a radix-4 FFT. The necessary files are in the folder `rtdx_vc_FFTr4`. This includes the Visual C++ support and executable files in the folder `rtdx_vc_FFTr4\rtdxFFT`.

CCS Component

The C source program `rtdx_vc_FFTr4.c` runs on the DSK and is shown in Figure 10.9a. An output RTDX channel is created and enabled to write (send) the resulting FFT magnitude data in the buffer `output_buffer` to MATLAB running on the PC host for plotting (only). RTDX is achieved using Visual C++. The radix-4 FFT support functions for generating the index for digit reversal, and for digit reversal, were used in Chapter 6. The complex radix-4 FFT function `cfftr4_dif.c` is also on the CD (the ASM version was used in Chapter 6). Note that the real and imaginary components of the input are consecutively arranged in memory (as required by the FFT function). Digit reversal is performed on the resulting FFT since it is scrambled and needs to be resequenced. After the FFT magnitude is calculated and stored in `output_buffer`, it is sent to MATLAB through an output RTDX channel.

The project uses DSP/BIOS only to set interrupt INT 11 using the (`.cdb`) configuration file (see Example 9.2). As a result, a vector file is not required. The BSL file needs to be added (the support files for RTDX and CSL are included in the autogenerated linker command file, which must be added to the project by the user).

Build this project within CCS as **rtdx_vc_FFTr4**. Within CCS, select *Tools →RTDX* and configure the buffer size to 2048 (not 1024), and then enable RTDX (check it). From the configuration (`.cdb`) file, select *Input/Output → RTDX*. Right-click for properties to increase the buffer size from 1024 to 2056. Load and run the (`.out`) file. Input a 2-kHz sine wave with an approximate amplitude of $\frac{1}{2}$ V p-p. The output from the DSK is like a loop program.

```
//rtdx_vc_FFTr4.c Core r4-FFT using RTDX with VC++(MATLAB for plotting)
... N=256,16kHz rate,align x&w,... see Examples in Chapter 6
#include <rtdx.h>
short input_buffer[N] = {0};            //store input samples(same as x)
float output_buffer[N] = {0};           //store magnitude FFT
short buffer_count=0;
short nFlag=1;                          //when to begin the FFT
short i, j;
RTDX_CreateOutputChannel(ochan);        //output channel C6x->PC transfer
interrupt void c_int11()                //ISR
{
 input_buffer[buffer_count] = input_sample(); //input -->buffer
 output_sample(input_buffer[buffer_count++]); //loop
 if(buffer_count >= N)
  {                                     //if more than N pts, begin FFT
   buffer_count = 0;                    //reset buffer_count
   nFlag = 0;                           //flag to signal completion
   for(i = 0; i < N; i++)
    {
     x[2*i]=(float)input_buffer[i]; //real component of input
     x[2*i+1] = 0;                      //imaginary component of input
    }
  }
}
void main(void)
{
 ... //generate twiddle constants and digit reversal index
 comm_intr();                           //init DSK
 while(!RTDX_isOutputEnabled(&ochan));//wait for PC to enable RTDX
 while(1)                               //infinite loop
  {
   while(nFlag);                        //wait to finish accum samples
   nFlag = 1;
   cfftr4_dif(x, w, N);                 //call radix-4 FFT function
   digit_reverse((double *)x, IIndex, JIndex, count);
   for(j = 0; j < N; j++)
     output_buffer[j]=(float)sqrt(x[2*j]*x[2*j]+x[2*j+1]*x[2*j+1]);
   RTDX_write(&ochan,output_buffer,sizeof(output_buffer));//Send DSK>PC
  };
}
```

(a)

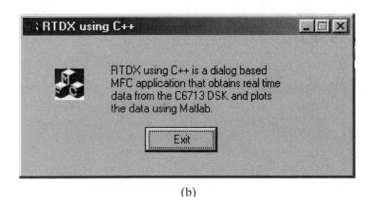

(b)

FIGURE 10.9. (a) C program to implement radix-4 FFT and illustrate RTDX with Visual C++, using MATLAB for FFT and plotting (rtdx_vc_FFTr4.c); (b) message when the VC++ application file is executed.

Visual C++ Component

Execute/run the application file rtdxFFT.exe located in the VC++ folder rtdx_vc_
FFTr4\rtdxFFT (within debug). Figure 10.9b will pop up, followed by the FFT
magnitude plot from MATLAB. Verify that the FFT of the 2-kHz sine wave output
is plotted within MATLAB, as in Example 9.8.

The Visual C++ file rtdxFFTDlg.cpp includes the code section for MATLAB to
set the sampling rate and plot the received data. It is located in the dialog class
within the thread

```
UINT CRtdxFFTDlg::RTDXThreadFunction(LPVOID lpvoid)
```

Recreate the executable (application) file. Launch Microsoft Visual C++ and select
File → Open Workspace to open rtdxFFT.dsw. Build and Rebuild All.

10.5 SPECTRUM DISPLAY THROUGH EMIF USING A BANK OF 32 LEDs

This miniproject takes the FFT of an input analog audio signal and displays the
spectrum of the input signal through a bank of 32 LEDs. The specific LED that
turns on depends on the frequency content of the input signal. The bank of LEDs
is controlled through the external memory interface (EMIF) bus on the DSK. This
EMIF bus is a 32-bit data bus available through the 80-pin connector J4 onboard
the DSK.

The FFT program in Chapter 6 using TI's optimized ASM-coded FFT function is
extended for this project. Figure 10.10 shows the core of the program that implements this project—using a 64-point radix-2 FFT, sampling at 32 kHz—and does not
output the negative spike (32,000) for reference. The executable (.out) file is on the
CD in the folder **graphic_FFT**. and can be used first to test this project. See also the
project used to display the spectrum through EMIF using LCDs in Section 10.6.

EMIF Consideration

To determine whether the data is being outputted through the EMIF bus, the following program is used:

```
# define OUTPUT 0xA0000000      //output address (EMIF)
int *output = (int*) OUTPUT;    //map memory location to variable
void main( )
{
*output = 0x00000001;           //output 0x1 to the bus
}
```

This program defines the output EMIF address and gives the capability to read and
write to the EMIF bus. Test the EMIF by writing different values lighting different
LEDs. The final version of the program includes a header file to define the output
EMIF address.

```
//graphic_FFT.c Core program.Displays spectrum to LEDs through EMIF

#include "output.h"          //contains EMIF address
int *output = (int *)OUTPUT; //EMIF address in header file
. . .
while (1)                    //infinite loop
 {
     .                       //same as in FFTr2.c
     .
     .
  for(i = 0; i < N/2; i++)
   {
     if (Xmag[i] > 20000.0)  //if mag FFT >20000
      {
        out = out + 1 << i;  //shifts one to appropriate bit location
      }
   }
  *output = out;            //output to EMIF bus
  out = 0;                  //reset out variable for next iteration
 }
```

FIGURE 10.10. Core C program to implement radix-2 FFT using TI's optimized FFT support functions. It displays the spectrum to 32 LEDs through EMIF (graphic_FFT.c).

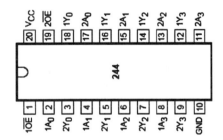

FIGURE 10.11. Line driver used with external LEDs to display the spectrum in project graphic_FFT.

EMIF-LEDs

A total of 32 LEDs connect through four line drivers (74LS244). Current-limiting resistors of 300 ohms are connected between each LED and ground. The line drivers allow for the needed current to light up the LEDs. The current drawn by the LED is limited to 10 mA so that the line drivers are not overloaded. Figure 10.11 shows one of the line drivers. Pin 20 is connected to +5 V and pin 10 to ground. Pins 1 and 19 are also connected to ground to enable the output of the line driver. Each line driver supports eight inputs and eight outputs. The pins labeled with "Y" are output pins. Each of the output pins (on a line driver) is connected to pins 33–40, which correspond to data pins 31–24 on the EMIF bus. The arrangement is the same with the other three line drivers connecting to pins 43–50 (data pins 23–16), pins 53–60 (data pins 15–8), and pins 63–70 (data pins 7–0), respectively. Pin 79 on the EMIF bus is used for universal ground. See also the schematics of connectors J3 and J4 shown in the file *c6713_dsk_schem.pdf*, included with CCS. Table 10.3 shows the EMIF signals.

Note: Pin 75 on J3 (not J4), the 80-pin connector for the external peripheral interface, is to be connected to ground since it is an enable pin for the EMIF interface and enables the output voltages on these pins.

Implementation

The real-time radix-2 FFT program example in Chapter 6 is slightly modified to check the amplitude of a specific frequency and determine whether or not it is above a set threshold value of 20,000. If so, the value of that specific frequency is sent to the EMIF output port to light the appropriate LED(s). From Figure 10.10, when a value of the FFT magnitude is larger than the set threshold, the variable *out* is output. This output corresponds to a bit that is shifted by the value of the index i that is the corresponding frequency location in the FFT array. This bit shift moves a binary 1 to the appropriate bit location corresponding to the specific LED to be lit. This process is repeated for every value in the magnitude FFT array. If multiple values in the FFT array are larger than the set threshold of 20,000, then the appropriate bit-shifted value is accumulated. This process lights up all the LEDs that have frequencies with corresponding amplitudes above the set threshold value. Setting the threshold value at 20,000 creates a range of frequencies from about 150 Hz to 15 kHz.

Build this project as **graphic_FFT** and verify that the lights adapt to the input audio signal in real time. You can also test this program with a signal generator as input to the DSK. Increase the frequency of the input signal and verify the sequence associated with the LEDs that turn on.

10.6 SPECTRUM DISPLAY THROUGH EMIF USING LCDs

This project implements a graphical frequency display through the use of a 2×16 character liquid-crystal display (LCD) (LCM-S01602DTR/M from Lumex). Each LCD character is decomposed into two separate states to form a bar graph displaying the spectrum of an input signal. See also the previous project, which displays a spectrum through EMIF using a bank of 32 LEDs. Figure 10.12 shows the core of the program, *EMIF_LCD.c*, that implements this project. It uses the C-coded FFT function called from *FFT128c.c* in Chapter 6 to obtain the spectrum (for the section of code that is excluded without outputting the negative spike for reference).

FFT Component

One component of the program is based on the FFT program example in Chapter 6 that calls a C-coded FFT function (see *FFT128c.c*). The FFT component uses 256

TABLE 10.3   EMIF Signals

Pin	Signal	I/O	Description	Pin	Signal	I/O	Description
1	5V	Vcc	5 V voltage supply pin	2	5V	Vcc	5 V voltage supply pin
3	EA21	O	EMIF address pin 21	4	EA20	O	EMIF address pin 20
5	EA19	O	EMIF address pin 19	6	EA18	O	EMIF address pin 18
7	EA17	O	EMIF address pin 17	8	EA16	O	EMIF address pin 16
9	EA15	O	EMIF address pin 15	10	EA14	O	EMIF address pin 14
11	GND	Vss	System ground	12	GND	Vss	System ground
13	EA13	O	EMIF address pin 13	14	EA12	O	EMIF address pin 12
15	EA11	O	EMIF address pin 11	16	EA10	O	EMIF address pin 10
17	EA9	O	EMIF address pin 9	18	EA8	O	EMIF address pin 8
19	EA7	O	EMIF address pin 7	20	EA6	O	EMIF address pin 6
21	5V	Vcc	5 V voltage supply pin	22	5V	Vcc	5 V voltage supply pin
23	EA5	O	EMIF address pin 5	24	EA4	O	EMIF address pin 4
25	EA3	O	EMIF address pin 3	26	EA2	O	EMIF address pin 2
27	BE3#	O	EMIF byte enable 3	28	BE2#	O	EMIF byte enable 2
29	BE1#	O	EMIF byte enable 1	30	BE0#	O	EMIF byte enable 0
31	GND	Vss	System ground	32	GND	Vss	System ground
33	ED31	I/O	EMIF data pin 31	34	ED30	I/O	EMIF data pin 30
35	ED29	I/O	EMIF data pin 29	36	ED28	I/O	EMIF data pin 28
37	ED27	I/O	EMIF data pin 27	38	ED26	I/O	EMIF data pin 26
39	ED25	I/O	EMIF data pin 25	40	ED24	I/O	EMIF data pin 24
41	3.3V	Vcc	3.3 V voltage supply pin	42	3.3V	Vcc	3.3 V voltage supply pin
43	ED23	I/O	EMIF data pin 23	44	ED22	I/O	EMIF data pin 22
45	ED21	I/O	EMIF data pin 21	46	ED20	I/O	EMIF data pin 20
47	ED19	I/O	EMIF data pin 19	48	ED18	I/O	EMIF data pin 18
49	ED17	I/O	EMIF data pin 17	50	ED16	I/O	EMIF data pin 16
51	GND	Vss	System ground	52	GND	Vss	System ground
53	ED15	I/O	EMIF data pin 15	54	ED14	I/O	EMIF data pin 14
55	ED13	I/O	EMIF data pin 13	56	ED12	I/O	EMIF data pin 12
57	ED11	I/O	EMIF data pin 11	58	ED10	I/O	EMIF data pin 10
59	ED9	I/O	EMIF data pin 9	60	ED8	I/O	EMIF data pin 8
61	GND	Vss	System ground	62	GND	Vss	System ground
63	ED7	I/O	EMIF data pin 7	64	ED6	I/O	EMIF data pin 6
65	ED5	I/O	EMIF data pin 5	66	ED4	I/O	EMIF data pin 4
67	ED3	I/O	EMIF data pin 3	68	ED2	I/O	EMIF data pin 2
69	ED1	I/O	EMIF data pin 1	70	ED0	I/O	EMIF data pin 0
71	GND	Vss	System ground	72	GND	Vss	System ground
73	ARE#	O	EMIF async read enable	74	AWE#	O	EMIF async write enable
75	AOE#	O	EMIF async output enable	76	ARDY	I	EMIF asynchronous ready
77	N/C	—	No connect	78	CE1#	O	Chip enable 1
79	GND	Vss	System ground	80	GND	Vss	System ground

```
//EMIF.LCD.c Core C program. Displays spectrum to LCDs through EMIF
#define IOPORT 0xA1111111              //EMIF address
int *ioport = (int *)IOPORT;          //pointer to get data out
int input, output;                    //temp storage
void set_LCD_characters();            //prototypes
void send_LCD_characters();
void init_LCD();
void LCD_PUT_CMD(int data);
void LCD_PUT_CHAR(int data);
void delay();
float bandage[16];                    //holds FFT array after downsizing
short k=0, j=0;
int toprow[16] = {0, 0, 0, 0, 0, 0, 0, 0, 0, 0, 0, 0, 0, 0, 0, 0};
int botrow[16] = {0, 0, 0, 0, 0, 0, 0, 0, 0, 0, 0, 0, 0, 0, 0, 0};
short rowselect = 1;                  //start on top row
short colselect = 0;                  //start on left of LCD
#define LCD_CTRL_INIT 0x38            //initialization for LCD
#define LCD_CTRL_OFF 0x08
#define LCD_CTRL_ON 0x0C
#define LCD_AUTOINC 0x06
#define LCD_ON 0x0C
#define LCD_FIRST_LINE 0x80
#define LCD_SECOND_LINE 0xC0          //address of second line
main()
{
  ..
  init_LCD();                         //init LCD
  while(1)                            //infinite loop
    {
      for(k=0; k<16; k++){            //for 16 bands
              float sum = 0;          //temp storage
              for(j=0; j<8; j++)      //for 8 samples per band
                  sum += x1[8*k+j];   //sum up samples
              bandage[k] = (sum/8);   //take average
      }
      set_LCD_characters();           //set up character arrays
      send_LCD_characters();          //put them on LCD
    }                                 //end of infinite loop
}                                     //end of main
interrupt void c_int11()              //ISR
{
  output_sample(bandage[buffercount/16]); //out from iobuffer
  ..
}
void set_LCD_characters()            //to fill arrays with characters
{
  int n = 0;                          //temp index variable
  for (n=0; n<16; n++)
    {
      if(bandage[n] > 40000)          //first threshold
        {
          toprow[n] = 0xFF;           //block character
          botrow[n] = 0xFF;
        }
```

FIGURE 10.12. Core C program using a C-coded FFT function to display the spectrum to LCDs through EMIF (EMIF_LCD.c).

```
    else if(bandage[n] > 20000)       //second threshold
      {
       toprow[n] = 0x20;              //blank space
       botrow[n] = 0xFF;
      }
    else                             //below second threshold
      {
       toprow[n] = 0x20;
       botrow[n] = 0x20;
      }
   }
}
void send_LCD_characters()
{
 int m=0;
 LCD_PUT_CMD(LCD_FIRST_LINE);        //start address
 for (m=0; m<16; m++)                //display top row
   LCD_PUT_CHAR(toprow[m]);
 LCD_PUT_CMD(LCD_SECOND_LINE);       //second line
 for (m=0; m<16; m++)                //display bottom row
   LCD_PUT_CHAR(botrow[m]);
}
void init_LCD()
{
 LCD_PUT_CMD(LCD_CTRL_INIT);         //put command
 LCD_PUT_CMD(LCD_CTRL_OFF);          //off display
 LCD_PUT_CMD(LCD_CTRL_ON);           //turn on
 LCD_PUT_CMD(0x01);                  //clear display
 LCD_PUT_CMD(LCD_AUTOINC);           //set address mode
 LCD_PUT_CMD(LCD_CTRL_ON);           //set it
}
void LCD_PUT_CMD(int data)
{
 *ioport = (data & 0x000000FF);      //RS=0, RW=0
 delay();
 *ioport = (data | 0x20000000);      //bring enable line high
 delay();
 *ioport = (data & 0x000000FF);      //bring enable line low
 delay();
}
void LCD_PUT_CHAR(int data)
{
 *ioport = ((data & 0x000000FF)| 0x80000000);   //RS=1, RW=0
 *ioport = ((data & 0x000000FF)| 0xA0000000);   //enable high
 *ioport = ((data & 0x000000FF)| 0x80000000);   //enable Low
 delay();
}
void delay()                                //create 1 ms delay
{
 int q=0, junk=2;
 for (q=0; q<8000; q++)
     junk = junk*junk;
}
```

FIGURE 10.12. (*Continued*)

points and samples at 32 kHz to allow a frequency display range from 0 to 16 kHz. The second component of the program is associated with the EMIF-LCD.

LCD Component

Since the LCD is 16 characters wide, each character is chosen to correspond to one band. The FFT range can then be decomposed linearly into sixteen 1-kHz bands, with each band being determined in a nested "for loop." The 256-point FFT is then decomposed into 16 bands with eight samples per band. The average of the samples is taken and placed into an array of size 16. Using thresholds, this array is then parsed to determine which character (blank or filled) is to be displayed on the LCD.

Each LCD character has two different states, either fully on or fully off (four states total). These characters are then placed in arrays, one array for the top row of the LCD and one for the bottom row. These arrays are accessed by the function that writes data to the appropriate LCD. Two functions are used to transfer data to the LCD:

1. The first function, LCD_PUT_CMD, is used primarily by an initialization function ($init\_LCD$). It masks the proper data bits and configures the control lines. The LCD has setup and hold times that must be achieved for proper operation. The LCD_PUT_CMD function sets the control lines, with delays to ensure that there are no timing glitches, and then pulses the enable control line. Clocking the data into the LCD occurs during the falling edge of the enable line.

2. The second function, LCD_PUT_CHAR, sends the characters to the LCD and requires different control signals. The cursor address is autoincremented so that a character is sent to the proper position on the LCD.

With only one port to use, the two functions LCD_PUT_CHAR and LCD_PUT_CMD include bitwise AND and OR operations to mask and set only certain bits.

The delay function creates a 1-ms delay to meet the timing requirements (setup and hold times) of the LCD for proper operation.

EMIF-LCD Pins Description

Table 10.4 displays information of the LCD pins and the EMIF connector. EMIF pins information on connector J4 is shown in Table 10.3 (associated with the previous project) and contained in the file $c6713\_dsk\_schem.pdf$, included with CCS. The least significant data pins (ED0–ED7) for the characters are selected, and the three most significant data pins (ED29–ED31) for the control lines are selected. The first six pins on the LCD are used for power and control signals. To enable the data for output through the EMIF bus, pin 75 of the External Peripheral Interface connector J3 (not J4) is to be connected to ground (see also the previous project).

TABLE 10.4   EMIF-LCD Pin Connections

LCD PinNumber	Name	Function	DSK (EMIF) Pin Connection J4
1	Vss	Ground	Gnd
2	Vdd	Supply	+5V
3	Vee	Contrast	Gnd
4	RS	Register select	ED31
5	R/W	Read/write	ED30
6	E	Enable	ED29
7	D0	Data bit 0	ED0
8	D1	Data bit 1	ED1
9	D2	Data bit 2	ED2
10	D3	Data bit 3	ED3
11	D4	Data bit 4	ED4
12	D5	Data bit 5	ED5
13	D6	Data bit 6	ED6
14	D7	Data bit 7	ED7

Build this project as **EMIF_LCD**. Use either an input signal from a signal generator or an input audio signal. Verify the graphical frequency display on the LCDs.

Some possible improvements to this project include:

1. More thresholds so that more levels of frequency intensities can be represented. More than four thresholds would better illustrate the frequency intensity.
2. The bands can be displayed logarithmically instead of linearly. A logarithmic display would allow for a wider range of frequencies. An up-sampling scheme would then be used.

10.7 TIME–FREQUENCY ANALYSIS OF SIGNALS WITH SPECTROGRAM

This project makes use of the short time Fourier transform (STFT) for the analysis of signals, resulting in a spectrogram plot [33, 34]. A spectrogram is a plot of the frequencies that make up a particular signal. The magnitude of the frequency at a particular time is represented by the colors in the graph. This plot of frequency versus time provides information on the changing frequency content of a signal over time.

The spectrogram is the square of the absolute value of the STFT of a signal. The STFT looks at a nonstationary signal as small blocks in time and takes the Fourier transform of each block to obtain the frequency content of the signal at that time. This involves multiplying the signal with a moving window to observe smaller segments of the signal and taking the Fourier transform of the product. The use of a sliding window and its size needs to be determined. A large window size (length) can be chosen to enhance the frequency resolution, but at the expense of the time resolution, and vice versa. The window increment, which represents the distance between successive windows, also needs to be determined.

A spectrogram can be more useful than a plot of the spectrum since there can be a different spectrum for each time. The spectrogram is plotted as frequency versus time as a three-dimensional plot. Consider a musical scale consisting of eight musical notes representing the C scale major: C, D, E, F, G, A, B, C with the following sinusoidal frequencies: 262, 294, 330, . . . , 523, respectively, starting with the middle C at a frequency of 262 Hz. The subsequent C is one octave higher at 523 Hz, which represents a doubling in frequency. A spectrogram plot of frequency versus time would identify each note as it is played.

Time–frequency analysis techniques include the STFT, Gabor expansion, and energy distribution-based techniques such as the Wigner–Ville distribution. These techniques are used to study the behavior of nonstationary signals such as music and speech signals.

The files for this project are in the folder **spectrogram** (with separate subfolders). The spectrogram project is decomposed into three separate sections (versions), all of which make use of MATLAB's function `imagesc` to plot the spectrogram:

1. Simulation using MATLAB to read a .wav file and plot its spectrogram
2. RTDX with MATLAB and use of a C-coded FFT function
3. RTDX with Visual C++ and a radix-4 optimized FFT function

10.7.1 Simulation Using MATLAB

This is a simulated version using MATLAB. Figure 10.13a shows the MATLAB file `spectrogram.m` that plots a spectrogram, using the function *wavread* to read a .wav file `chirp.wav` that is a swept sinusoidal signal. MATLAB's FFT function is also used, as well as the function `imagesc`, to find the spectrogram of the input .wav file.

Run the MATLAB program and verify Figure 10.13b as the spectrogram of a chirp signal. It illustrates the increase in frequency of the swept sinusoidal signal over time. You can readily test other .wav files on the CD.

10.7.2 Spectrogram with RTDX Using MATLAB

This version of the project makes use of RTDX with MATLAB for transferring data from the DSK to the PC host. Section 9.1 introduces the use of a configuration (.cdb) file and Section 9.2 illustrates RTDX with MATLAB.

%**Spectrogram.m** Reads .wav file,plots spectrogram using STFT with MATLAB

```
[x,fs,bits] = wavread('chirp.wav'); %read .wav file
N = length(x);
t=(0:N-1)/fs;
set(0,'DefaultAxesColorOrder',[0 0 0],...
    'DefaultAxesLineStyleOrder','-|-.|--|:');
figure(1); plot(t,x);                    %plots time-domain signal
xlabel('Time (sec)'); ylabel('Amplitude'); title('Waveform of signal');
M=256;  B=floor(N/M);                    %divide signal->blocks of M samples
x_mat=reshape(x(1:M*B),[M B]);           %reshape vector into MxB matrix
win=hamming(M);                          %Hamming window before FFT
win_mat=repmat(win,[1 B]);
x_fft=fft(x_mat.*win_mat);               %perform FFT
y=abs(x_fft(1:M/2,:));                    %want positive freq and mag info
t=(1:B)*(M/fs);                          %values for time and freq axes
f=((0:M-1)/(M-1))*(fs/2);
figure(2);
imagesc(t,f,dB(y));                      %plot spectrogram
colormap(jet);    colorbar;   set(gca,'ydir','normal');
xlabel('Time (sec)');  ylabel('Frequency (Hz)');  title('Spectrogram');
```
(a)

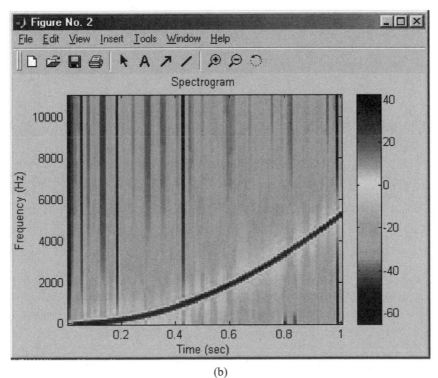

(b)

FIGURE 10.13. Spectrogram simulation with MATLAB: (a) MATLAB program to read and find the spectrogram of an input .wav file and (b) spectrogram plot of an input chirp signal.

Figure 10.14a shows the core source program `spectrogram_rtdx_mtl.c` that runs on the DSK and can readily be completed using the program `FFT128c.c` in Chapter 6 (the complete executable file is on the CD). It calls the C-coded FFT function used in Chapter 6 and enables an RTDX output channel to write/send the resulting FFT data to the PC running MATLAB for finding the spectrogram. A total of $N/2$ (128 points) are sent (in lieu of 256) for better resolution (continuity). The (`.cdb`) configuration file is used to set interrupt INT11, as in Section 9.1. From this configuration file, select *Input/Output* → *RTDX*. Right-click on properties and change the RTDX buffer size to 8200. Within CCS, select *Tools* → *RTDX* → *Configure* to set the host buffer size to 2048 (from 1024).

An input signal is read in blocks of 256 samples. Each block of data is then multiplied with a Hamming window of length 256 points. The FFT of the windowed data is calculated and squared. Half of the resulting FFT of each block of 256 points is then transferred to the PC running MATLAB to find the spectrogram. Build this project as **spectrogram_rtdx_mtl**. Within CCS, select *Tools* → *RTDX* → *Configure*.

```
//Partial_Spectrogram_rtdx_mtl.c Core program for Time-Frequency
//analysis with spectrogram using RTDX-MATLAB
. . . See FFT256c.c
#include <rtdx.h>                         //RTDX support file
#include "hamming.cof"                    //Hamming window coefficients
RTDX_CreateOutputChannel(ochan);         //create output channel C6x->PC

main()
{
//. . . calculate twiddle constants
 comm_intr();                             //init DSK, codec, McBSP
 while(!RTDX_isOutputEnabled(&ochan))     //wait for PC to enable RTDX
     puts("\n\n Waiting . . . ");         //while waiting
 while(1)                                 //infinite loop
  {
   . . .
   for (i = 0 ; i < PTS ; i++)            //swap buffers
    {
      samples[i].real=h[i]*iobuffer[i];   //multiply by Hamming coeffs
      iobuffer[i] = x1[i];                //process frame to iobuffer
    }
   . . . use FFT magnitude squared
   RTDX_write(&ochan,x1,sizeof(x1)/2);    //send 128 samples to PC
  }                                        //end of infinite loop
}                                          //end of main
interrupt void c_int11()                   //ISR
{. . . as in FFT256c.c }
```

(a)

FIGURE 10.14. Spectrogram using RTDX with MATLAB: (a) core program to calculate FFT and transfer FFT data from the DSK to the PC; (b) spectrogram plot of an external chirp input signal; and (c) spectrogram plot of a 500-Hz square wave input signal.

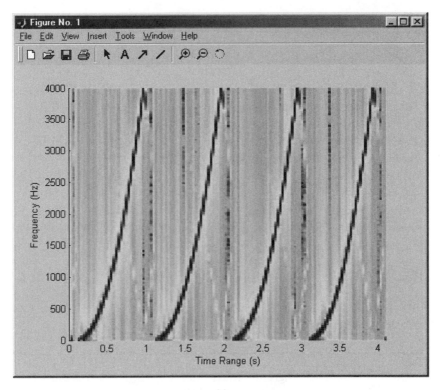

(b)

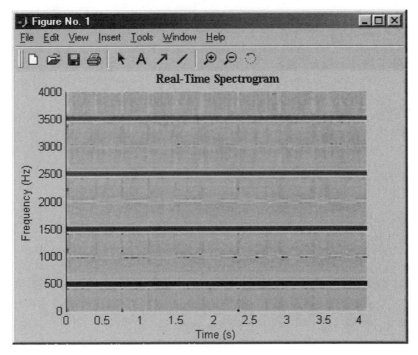

(c)

FIGURE 10.14. (*Continued*)

Open MATLAB, select the appropriate path, and run `spectrogram_rtdx.m` (on the CD). Within MATLAB, CCS will enable RTDX and will load and run the COFF (`.out`) executable file. Then MATLAB will plot the resulting spectrogram of an input signal. Input/play `Chirp.wav` (output of a soundcard as input to the DSK) and verify the spectrogram of this input signal plotted by MATLAB, as shown in Figure 10.14b. For a chirp input signal, the transfer of 128 points (in lieu of 256) yields a better spectrogram.

For a faster and accurate plot, delete the commands within the MATLAB file that include the labels (x and y axes, and title) in the spectrogram plot.

Use a 500-Hz square wave as input and verify the spectrogram plot shown in Figure 10.14c. A darker red strip is formed at the 500-Hz fundamental frequency, and lighter red strips at the other harmonics of 1500, 2500, and 3500 Hz. For this type of input, you may choose to transfer the entire block of 256-point FFT data at each time.

You can extend this project version using TI's optimized FFT function (see Chapter 6).

10.7.3 Spectrogram with RTDX Using Visual C++

This project is also tested using RTDX with Visual C++ for data transfer from the DSK to the PC host. The program `spectrogram_rtdx_r4.c` (on the CD) implements a 256-point radix-4 FFT using TI's optimized FFT function and the associated support files for digit reversal. See also the two radix-4 FFT examples in Chapter 7 and Section 10.4. As with the MATLAB version for RTDX, only 128 points are transferred at a time.

Change the buffer size to 8200 within the (`.cdb`) file, as with the previous MATLAB version. Within CCS, change the host buffer size from 1024 to 2048. Enable RTDX (there is no MATLAB file for doing so). Load/run the `.out` file.

The Visual C++ support files are on the CD. Access/run the VC++ application file `vc_spectrogram.exe`. You should get the Visual C++ dialog message in Figure 10.15 until MATLAB plots the spectrogram of a real-time input signal. Input/play

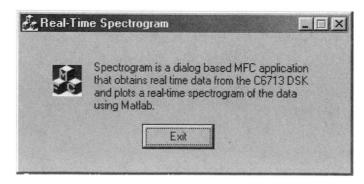

FIGURE 10.15. Visual C++ dialog message for a spectrogram.

the (.wav) chirp signal and verify that the results are identical to those achieved with the spectrogram in Figure 10.14b, being continuously updated within MATLAB. The file vc_spectrogramdlg.cpp contains the MATLAB commands for plotting the spectrogram. However, MATLAB is not used in this version to provide the RTDX link.

As in Section 10.7.2, you can obtain a fast and accurate plot by deleting the commands for including the title and the labels within the spectrogram plot. These commands are in the file vc_spectrogramdlg.cpp.

You can extend this project version using the radix-2 FFT (in lieu of the radix-4). Chapter 6 includes several examples based on the radix-2 FFT.

10.8 AUDIO EFFECTS (ECHO AND REVERB, HARMONICS, AND DISTORTION)

This project illustrates various audio effects such as distortion, echo and reverb, and harmonics [35]. Figure 10.16 shows the core program soundboard.c (virtually complete) that implements this project. The overall program flow consists of pre-amplification, distortion, echo/reverb, harmonics, and postamplification. Preamp and postamp are included to avoid overdriving the output. A sampling rate of 16 kHz is chosen, and a total of 10 sliders are used for the overall control. The slider gel file is on the CD in the folder soundboard.

The distortion effect is the simplest to implement. It requires overamplifying each sample and clipping it at maximum and minimum values. The acquired input sample is amplified based on whether it is positive or negative. The amplification polynomial used for the distortion component is used to amplify the signal in a nonlinear fashion. The result is scaled by a distortion magnitude controlled by a slider, then clipped so as not to overdrive the output.

The resulting output is processed for an echo/reverb effect (see Examples 2.4 and 2.5 on echo effects). The length of the echo is controlled by changing the buffer size where the samples are stored. A dynamic change of the echo length leads to a reverb effect. A fading effect with a decaying echo is obtained with a slider.

The third effect is harmonics boost. A harmonics buffer is used for this effect. Two main loop sections are created to produce two separate sets of harmonics. The larger (outer) loop combines the input with samples from the harmonics buffer at twice the input frequency. The smaller (inner) loop produces the next harmonics at four times the input frequency. The magnitudes of the harmonics are controlled with a slider.

These effects were tested successfully using the input from a keyboard with the keyboard output to a speaker. The audio output is sent to both channels of the codec (see Example 2.9), using the stereo capability of the onboard codec. The executable and gel files are included in the folder **soundboard**.

A drum effect section is included in the program for expanding the project. The use of external memory must be considered when applying many effects.

```
//Soundboard.c  Core C program for sound effects
union {Uint32 uint; short channel[2];} AIC23_data;
union {Uint32 uint; short channel[2];} AIC23_input;
short EchoLengthB = 8000;            //echo delay
short EchoBuffer[8000];              //create buffer
short echo_type = 1;                 //to select echo or delay
short Direction = 1;                 //1->longer echo,-1->shorter
short EchoMin=0,EchoMax=0;           //shortest/longest echo time
short DistMag=0,DistortionVar=0,VolSlider=100,PreAmp=100,DistAmp=10;
short HarmBuffer[3001];              //buffer
short HarmLength=3000;               //delay of harmonics
float output2;
short DrumOn=0,iDrum=0,sDrum=0;      //turn drum sound when = 1
int   DrumDelay=0,tempo=40000;       //delay counter/drum tempo
short ampDrum=40;                    //volume of drum sound
..                                   //addtl casting
interrupt void c_int11()             //ISR
{
AIC23_input.uint = input_sample();  //newest input data
input=(short)(AIC23_input.channel[RIGHT]+AIC23_input.channel[LEFT])/2;
input = input*.0001*PreAmp*PreAmp;
output=input;
output2=input;                       //distortion section
if (output2>0)
output2=0.0035*DistMag*DistMag*DistMag*((12.35975*(float)input)
     - (0.359375*(float)input*(float)input));
else   output2 =0.0035*DistMag*DistMag*DistMag*(12.35975*(float)input
     + 0.359375*(float)input*(float)input);
output2/=(DistMag+1)*(DistMag+1)*(DistMag+1);
if (output2 > 32000.0)  output2 = 32000.0 ;
else if (output2 < -32000.0 )  output2 = -32000.0;
output= (output*(1/(DistMag+1))+output2); //overall volume slider
input = output;                      //echo/reverb section
iEcho++;                             //increment buffer count
if (iEcho >= EchoLengthB) iEcho = 0;       //if end of buffer reinit
output=input + 0.025*EchoAmplitude*EchoBuffer[iEcho];//newest+oldest
if(echo_type==1) EchoBuffer[iEcho] = output; //for decaying echo
else EchoBuffer[iEcho]=input;        //for single echo (delay)
EchoLengthB += Direction;            //alter the echo length
if(EchoLengthB<EchoMin+100){Direction=1;} //echo delay is shortest->
if(EchoLengthB>EchoMax){Direction=-1;}   //longer,if longest->shorter
input=output;                        //output echo->harmonics gen
if(HarmBool==1) {                    //everyother sample...
 HarmBool=0;                         //switch the count
 HarmBuffer[iHarm]=input;            //store sample in buffer
 if(HarmBool2==1){                   //everyother sample...
  HarmBool2=0;                       //switch the count
  HarmBuffer[uHarm] += SecHarmAmp*.025*input;//store sample in buffer
 }
 else{HarmBool2=1; uHarm++;          //or just switch the count,
     if(uHarm>HarmLength) uHarm=0;   //and increment the pointer
 }
}
```

FIGURE 10.16. Core C program to obtain various audio effects (soundboard.c).

```
else{HarmBool=1; iHarm++;                  //or just switch the count
if(iHarm>HarmLength) iHarm=0;}             //and increment the pointer
output=input+HarmAmp*0.0125*HarmBuffer[jHarm];//add harmonics to output
jHarm++;                                   //and increment the pointer
if(jHarm>HarmLength) jHarm=0;              //reinit when maxed out
DrumDelay--;                               //decrement delay counter
if(DrumDelay<1) {                          //drum section
     DrumDelay=50000-Tempo;                //if time for drumbeat
     DrumOn=1;                             //turn it on
}
if(0){                                     //if drum is on
 output=output+(kick[iDrum])*.05*(ampDrum);//play next sample
 if((sDrum%2)==1) {iDrum++;}               //but play at Fs/2
 sDrum++;                                  //incr sample number
 if(iDrum>2500){iDrum=0; DrumOn=0;}        //drum off if last sample
}
output = output*.0001*VolSlider*VolSlider;
AIC23_data.channel[LEFT]=output;
AIC23_data.channel[RIGHT]=AIC23_data.channel[LEFT];
output_sample(AIC23_data.uint);           //output to both channels
}
main()          //init DSK,codec,McBSP and while(1) infinite loop
```

FIGURE 10.16. (*Continued*)

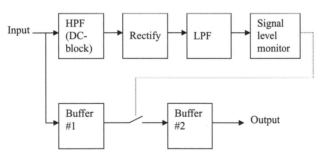

FIGURE 10.17. Block diagram for the detection of a voice signal from a microphone and playback of that signal in the reverse direction.

10.9 VOICE DETECTION AND REVERSE PLAYBACK

This project detects a voice signal from a microphone, then plays it back in the reverse direction. Figure 10.17 shows the block diagram that implements this project. All the necessary files are in the folder detect_play. Two circular buffers are used: an input buffer to hold 80,000 samples (10 seconds of data) continuously being updated and an output buffer to play back the input voice signal in the reverse direction. The signal level is monitored, and its envelope is tracked to determine whether or not a voice signal is present.

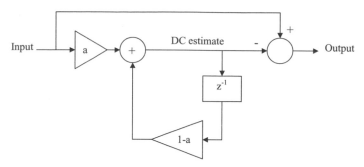

FIGURE 10.18. DC blocking first order IIR highpass filter for voice signal detection and reverse playback.

When a voice signal appears and subsequently dies out, the signal-level monitor sends a command to start the playback of the accumulated voice signal, specifying the duration of the signal in samples. The stored data are transferred from the input buffer to the output buffer for playback. Playback stops when one reaches the end of the entire signal detected.

The signal-level monitoring scheme includes rectification and filtering (using a simple first order IIR filter). An indicator specifies when the signal reaches an upper threshold. When the signal drops below a low threshold, the time difference between the start and end is calculated. If this time difference is less than a specified duration, the program continues into a no-signal state (if noise only). Otherwise, if it is more than a specified duration, a signal-detected mode is activated.

Figure 10.18 shows the DC blocking filter as a first-order IIR highpass filter. The coefficient a is much smaller than 1 (for a long time constant). The estimate of the DC filter is stored as a 32-bit integer.

The lowpass filter for the envelope detection is also implemented as a first order IIR filter, similar to the DC blocking filter except that the output is returned directly rather than being subtracted from the input. The filter coefficient a is larger for this filter to achieve a short time contant.

Build and test this project as **detect_play**.

10.10 PHASE SHIFT KEYING—BPSK ENCODING AND DECODING WITH PLL

See also the two projects on binary phase shift keying (BPSK) and modulation schemes in Sections 10.11 and 10.12. This project is decomposed into smaller mini-projects as background for the final project. The final project is the transmission of an encoded BPSK signal with voice as input and the reception (demodulation) of this signal with phase-locked loop (PLL) support on a second DSK. All the files associated with these projects are located in separate subfolders within the folder **PSK**.

10.10.1 BPSK Single-Board Transmitter/Receiver Simulation

BPSK is a digital modulation technique that separates bits by shifting the carrier 180 degrees. A carrier frequency signal is chosen that is known by both the transmitter and the receiver. Each bit is encoded as a phase shift in the carrier at some predetermined period. When a 0 is sent, the carrier is transmitted with no phase shift, and when a 1 is sent, the carrier is phase shifted by 180 degrees [36–39].

CCS Component

The necessary files for this project are on the CD in **BPSK_sim** within the folder **PSK**. Figure 10.19 shows the C source program BPSK_sim.c that modulates a bit stream of 10 bits set in the program. Since there is no carrier synchronization, demodulation is performed by the same program on the same DSK board.

Build this project as **BPSK_sim**. Connect the DSK output to the input to verify the demodulation of the transmitted sequence. Run the program. The demodulator program prints the demodulated sequence within CCS. Verify that it is the same as the sequence set in the array encodeSeq to be encoded.

The array buffer stores the entire received vector that can be plotted within CCS. Select *View→Graph→Time/Frequency*. Use buffer as the address, 190 as the acquisition and display size, 8000 as the sample rate, and a 16-bit signed integer format. Figure 10.20a shows the CCS plot of the received sequence: {1, 0, 1, 1, 0, 0, 0, 1, 0, 1} as set in the program. Note that when the received sequence changes from a 0 to a 1 or from a 1 to a 0, a change of phase is indicated in the positive and negative y axis, respectively. Change the sequence to be encoded in the program to {0, 1, 0, 0, 1, 1, 1, 0, 1, 0} and verify the CCS plot in Figure 10.20b.

MATLAB Component

The MATLAB program BPSK_sim.m is also included on the CD. It simulates the modulation and demodulation of a random bit stream. Run this MATLAB file and verify the plots in Figures 10.21a and 10.21b for signal-to-noise ratios (SNRs) of 0.5 and 5.0, respectively. They display the transmitted and received waveforms of a random bit stream. The SNR can be changed in the program. The MATLAB program also displays the decision regions and detection, as shown in Figures 10.22a and 10.22b, for SNRs of 0.5 and 5.0, respectively. With small values of SNR, the received signals fall outside the appropriate decision regions, resulting in errors in detection. The received signal is noisier, resulting in some false detection. This occurs when the correlator produces an incorrect phase for the incoming symbol. Correct detections are marked with blue ×'s and incorrect detections with red circles. For larger values of SNR, there are no false detections and the correlated signals lie well within the detection region.

```
//BPSK.c BPSK Modulator/Demod. DSK Output sequence --> Input
#include "dsk6713_aic23.h"          //codec-DSK support file
#include <math.h>
#include <stdio.h>
Uint32 fs=DSK6713_AIC23_FREQ_16KHZ; //set sampling rate
#define PI 3.1415926
#define N 16                        //# samples per symbol
#define MAX_DATA_LENGTH 10          //size of mod/demod vector
#define STABILIZE_LEN 10000         //# samples for stabilization
float phi_1[N];                     //basis function
short r[N] = {0};                   //received signal
int rNum=0,    beginDemod=0;        //# of received samples/demod flag
short encSeqNum=0,  decSeqNum=0;    //# encoded/decoded bits
short encSymbolVal=0,decSymbolVal=0;//encoder/decoder symbol index
short encodeSeq[MAX_DATA_LENGTH]={1,0,1,1,0,0,0,1,0,1};//encoded seq
short decodeSeq[MAX_DATA_LENGTH];   //decoded sequence
short sigAmp[2] = {-10000, 10000};  //signal amplitude
short buffer[N*(MAX_DATA_LENGTH+3)];//received vector for debugging
short buflen=0,   stabilizeOutput=0;
interrupt void c_int11()            //interrupt service routine
{
 int i,  outval= 0;
 short X = 0;
 if(stabilizeOutput++ < STABILIZE_LEN) //delay start to Stabilize
 {
  r[0] = input_sample();
  output_sample(0);
  return;
  }
 if(encSeqNum < MAX_DATA_LENGTH)     //modulate data sequence
 {
  outval = (int) sigAmp[encodeSeq[encSeqNum]]*phi_1[encSymbolVal++];
  if(encSymbolVal>=N) {encSeqNum++;   encSymbolVal=0; }
  output_sample(outval);
 }
 else output_sample(0);             //0 if MAX_DATA_LENGTH exceeded
 r[rNum++] = (short) input_sample();//input signal
 buffer[buflen++] = r[rNum - 1];
 if(beginDemod)                     //demod received signal
 {
  if(decSeqNum<2 && rNum==N)  {     //account for delay in signal
      decSeqNum ++;    rNum = 0; }
  if(rNum == N)                     //synchronize to symbol length
  {
   rNum = 0;
   for(i=0; i<N; i++)               //correlate with basis function
           X += r[i]*phi_1[i];
   decodeSeq[decSeqNum-2] = (X >= 0) ? 1: 0;    //do detection
   if(++decSeqNum == MAX_DATA_LENGTH+2) //print received sequence
```

FIGURE 10.19. C program that modulates a sequence of 10 numbers to illustrate BPSK, using a single DSK for modulation and demodulation (BPSK.c).

```
    {
      for(i=0;  i<decSeqNum-2;  i++)
              printf("Received Value: %d\n", decodeSeq[i]);
      exit(0);
      }
    }
  }
  else  { beginDemod = 1; rNum = 0; }
}
void main()
{
    int i; comm_intr();                    //init DSK, codec, McBSP
    for(i=0; i<=N; i++)
      phi_1[i] = sin(2*PI*i/N);            //basis function
    while(1);                              //infinite loop
}
```

FIGURE 10.19. (*Continued*)

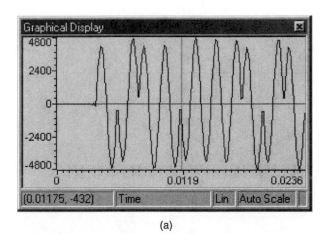

(a)

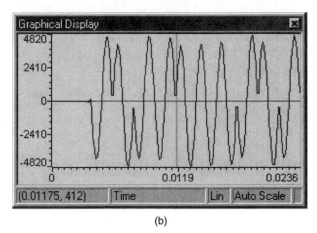

(b)

FIGURE 10.20. CCS plot of a received sequence, representing a BPSK modulated signal:
(a) sequence of {1, 0, 1, 1, 0, 0, 0, 1, 0, 1} and (b) sequence of {0, 1, 0, 0, 1, 1, 1, 0, 1, 0}.

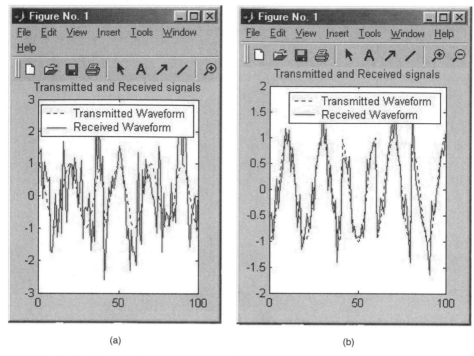

(a)   (b)

FIGURE 10.21. MATLAB plots simulating the modulation of a random bit stream showing the transmitted and received waveforms for (a) SNR = 0.5 and (b) SNR = 5.0.

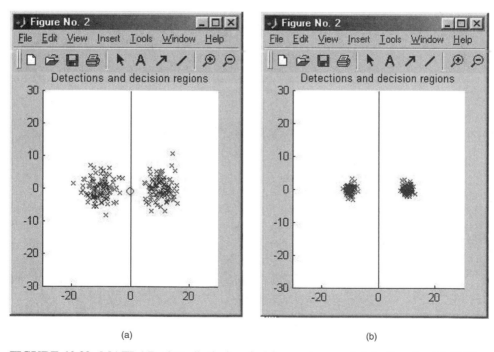

(a)   (b)

FIGURE 10.22. MATLAB plots displaying decision regions and detection for (a) SNR = 0.5 and (b) SNR = 5.0.

10.10.2 BPSK Transmitter/Voice Encoder with Real-Time Input

CCS Component

Figure 10.23 shows the C source program bpsk_ReIn.c that implements a transmitter/voice encoder with a real-time input signal. You can use your voice as input from a microphone connected to the mic input.

Build this project as **BPSK_ReIn**. All the necessary files for this project are on the CD in **BPSK_ReIn** within the folder **PSK**. Use voice as input to the DSK, with the DSK output to a scope. Verify that a representative segment of the encoded BPSK output signal from the DSK is as shown in Figure 10.24.

```
//BPSK_ReIn.c Illustrates transmitter/voice encoder with Real IN
#include "dsk6713_aic23.h"              //codec-DSK support file
#include <math.h>
Uint32 fs=DSK6713_AIC23_FREQ_32KHZ; //set sampling rate
#define NUMSAMP 4                       //# samples per symbol
#define MAX_DATA_LENGTH 10              //size of mod/demod vector
short encSeqNum=0, encSymbolVal=0;  //# encoded bits/symbol index
short sin_table[NUMSAMP]={0,10000,0,-10000};
short sample_data;   short bits[16]={0};   short outval=1;

interrupt void c_int11()                //interrupt service routine
{
 int i;
 short j=0;
 sample_data=(short)input_sample(); //input sample
 if(encSeqNum == 32)                    //decimate 32kHz to 1kHZ
 {
  encSeqNum = 0;
  if((sample_data>1000)||(sample_data<-1000)) {//above noise threshold
  for(i=0;i<8;i++) bits[i]=(sample_data&(1<<i))?1:-1;} //8sig bits
  else {for(i=0;i<8;i++) bits[i]=0;} //get next bit
 }
 outval = (short) bits[j];
 output_sample(outval*sin_table[encSymbolVal++]);//output next sample
 if(encSymbolVal>=NUMSAMP) {encSymbolVal=0; j++;} //reset encSymbolVal
 encSeqNum++;
 if (j==8)    j=0;                      //start next sample
}
void main()
{comm_intr();     while(1);}            //init DSK/infinite loop
```

FIGURE 10.23. C program to illustrate a transmitter/voice encoder using a real-time input signal (bpsk_ReIn.c).[20]

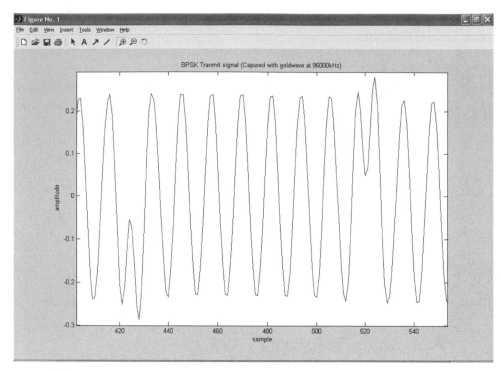

FIGURE 10.24. Plot of encoded DSK output using voice as input to the DSK.

MATLAB Component

The corresponding MATLAB file for this project `bpsk_ReIn.m` is on the CD. Verify the resulting MATLAB plots in Figure 10.25. The upper graph shows the received waveform signal segment. A `.wav` file is used to model the input signal being encoded as a BPSK signal. The plots show successive samples being encoded and decoded. The `.wav` sample is decimated to 1 kHz, converted to a bit stream, and then modulated to a BPSK signal that is then plotted. The upper graph shows which amplitude of the voice signal is being modulated into a BPSK signal. Note that as the circle moves along the received waveform in the upper graph, the corresponding BPSK signal and transmitted bits are displayed in the lower graph and are continuously encoded (updated).

10.10.3 Phase-Locked Loop

This project is a PLL receiver. In BPSK, the receiver must be able to lock onto the phase of a received signal in order to distinguish between 1's and 0's. A sinusoid of

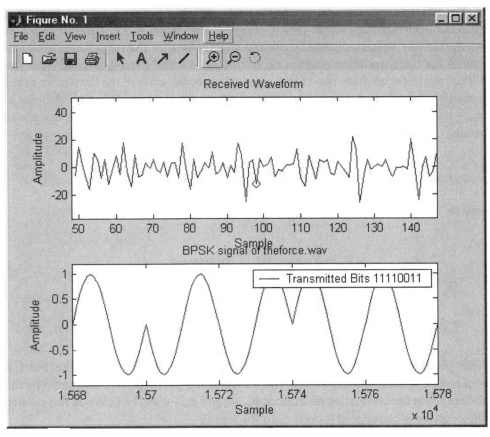

FIGURE 10.25. MATLAB plots of an encoded voice signal (lower graph) and received segment (upper graph).

1 kHz, with varying phase, is used as the real-time input to the DSK. This input signal has eight unique phase shifts. The real-time output signal is the phase of the received signal. Two DSKs are required to implement this project.

To determine the phase of an incoming sinusoid, the maximum of the correlation coefficient is calculated between the received sinusoid and a sinusoid offset by a phase estimate. The correlation coefficient, Y, between two sinusoids is given by

$$Y = \int_{0}^{2\pi} \sin(t\omega + \phi_{\text{carrier}})\sin(t\omega + \phi_{\text{est}})$$

The received sine wave has a phase of ϕ_{carrier}, and an estimate of the phase is ϕ_{est}. The correlation coefficient has a maximum Value when ϕ_{carrier} and ϕ_{est} are equal.

```
//BPSK_demod.c PLL demodulator. Input from 1st DSK
#include "dsk6713_aic23.h"          //codec-DSK support file
#include <math.h>
Uint32 fs=DSK6713_AIC23_FREQ_16KHZ; //set sampling rate
#define NUMSAMP 16                  //# samples per symbol
#define PI   3.1415926
short sample_data;                  //input sample
short ri=0,    r[10000]={0};        //buffer index/received data
short r_symbol[NUMSAMP];            //buffer to receive one period
short SBind=0,   phiBind=0;         //symbol/phi buffer index
float phiBuf[1000] = {0};           //buffer to view phi estimates
float y1, y2,   damp=1;             //correlation vectors,damping
float phi = PI;                     //phase estimate

interrupt void c_int11()            //interrupt service routine
{
  int i,   max=1;
  sample_data=(short)input_sample(); //receive sample
  r[ri++] = sample_data;
  r_symbol[SBind++] = sample_data;   //put sample in symbol buffer
  if(ri >= 10000)   ri = 0;          //reset buffer index
  if(SBind == NUMSAMP)               //after one period is received
  {                                  //then perform phi estimate
    SBind = 0;                       //reset buffer index
    y1 = 0, y2 = 0;
    for(i=0; i<NUMSAMP; i++)         //correlate received symbol
    {
      y1 += r_symbol[i]*sin(2*PI*i/NUMSAMP + phi - 0.1);
      y2 += r_symbol[i]*sin(2*PI*i/NUMSAMP + phi + 0.1);
      if(r_symbol[i] > max)          max = r_symbol[i];
    }
    y1=y1/max;        y2=y2/max;     //normalize correlation coefs
    phi = phi + 0.4*(y2 - y1)*phi;   //determine new estimate for phi
    if(phi < 1)             phi=phi+2*PI; //normalize phi
    if(phi >(2*PI+1))    phi=phi-2*PI;
    phiBuf[phiBind++]=phi;           //put phi in buffer for viewing
    if(phiBind >= 1000)  phiBind = 0; //reset buffer index
  }
output_sample(phi);
}
void main()
{
  comm_intr();    while(1);          //init DSK/infinite loop
}
```

FIGURE 10.28. C program implementing a PLL demodulator (bpsk_demod.c).

Figure 10.29b shows a CCS plot of the PLL output buffer that receives only one period of the sine wave. Use a starting address of r_symbol, an acquisition and display size of 16, and a 16-bit signed integer (not a 32-bit float, as for phiBuf).

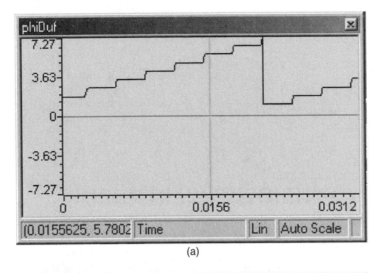

(a)

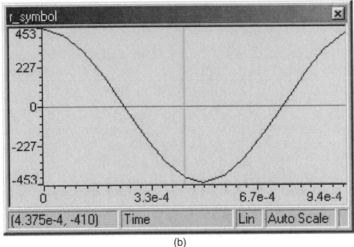

(b)

FIGURE 10.29. CCS plot of a PLL demodulator: (a) output showing eight different amplitudes and (b) output buffer that receives only one period.

10.10.4 BPSK Transmitter and Receiver with PLL

The support files for this project are in the subfolders **transmitter** and **receiver**. This project is the final product and includes the demodulation of a transmitted BPSK signal. It uses two DSKs: one to transmit a BPSK signal and the other to demodulate it. The transmitter.c program shown in Figure 10.30 uses the stereo capability of the AIC23 codec to transmit a 12-kHz carrier signal through the right channel and the BPSK encoded voice signal through the left channel. In this case, you can use a stereo cable that connects the output of the first DSK running the transmitter program to the input of the second DSK running receiver.c. Use voice

```c
//transmitter.c Transmits voice as a BPSK signal
#include "dsk6713_aic23.h"           //codec-DSK support file
#include <math.h>
#include "lp1500.cof"                //1500 Hz coeff lowpass filter
Uint32 fs=DSK6713_AIC23_FREQ_48KHZ;  //set sampling rate
#define NUMSAMP    4                 //# samples per Symbol
#define MAX_DATA_LENGTH 10           //size of Mod/Demod vector
#define NUM_BITS 8                   //number of bits per sample
#define SYNC_INTERVAL 100            //interval between sync bits
short encSeqNum   = 8;               //number of encoded bits
short encSymbolVal = 0;              //encoder symbol index
short sin_table[NUMSAMP]={0,1000,0,-1000}; //for carrier
short bits[8];                       //holds encoded sample
short sampleBuffer[2000];            //to view sample
short sIndex = 0;                    //index sampleBuffer
short syncSequence[8]={1,1,1,-1,1,-1,-1,1};//synchronization sequence
short outval=1;                      //bit value to be encoded
short encodeVal = 0;                 //filtered input value
int yn = 0;                          //init filter's output
short gain=10;                       //gain on output
short syncTimer = 0;                 //tracks time between syncs
#define LEFT   0                     //setup left/right channel
#define RIGHT 1
union {Uint32 uint; short channel[2];} AIC23_data;

interrupt void c_int11()             //interrupt service routine
{
 int i;
 short sample_data;
 sample_data = input_sample();
 yn = fircircfunc(sample_data,h,N); //asm func passing to A4,B4,A6
 if(encSymbolVal >= NUMSAMP)        //increment through waveform
 {
  encSymbolVal = 0;
  encSeqNum++;
 }
 if(encSeqNum == NUM_BITS)          //when all 8 bits sent
 {                                  //get a new sample
  encSeqNum = 0;
  if(syncTimer++ >= SYNC_INTERVAL)  //determine whether
  {                                 //to send sync sequence
   syncTimer = 0;
   for(i=0; i<8; i++)               //put sync sequence in bit
      bits[i] = syncSequence[i];
  }
  else
  {                                 //get the bits
   encodeVal = (short) (yn >> 15);
   for(i=8; i<16; i++)              //encode input sequence
      bits[i-8]=(encodeVal&(1<<i)) ? 1 : -1; //shift
  }
  sampleBuffer[sIndex++] = encodeVal;
  if(sIndex >= 2000)     sIndex = 0;
 }
 outval = (short) bits[encSeqNum];
 AIC23_data.channel[RIGHT]=gain*sin_table[encSymbolVal];//carrier
 AIC23_data.channel[LEFT]=gain*outval*sin_table[encSymbolVal++];//data
 output_sample(AIC23_data.uint);    //output to both channels
}
void main(){
 comm_intr();    while(1); }        //init,infinite loop
```

FIGURE 10.30. C program for BPSK transmission (transmitter.c).

as input. Verify the successful reception (demodulation) of the transmitted BPSK signal, with the receiver output connected to a speaker.

See Example 4.12 for the use of an FIR filter function implemented in ASM code. For this project, $N = 8$, so that the size of the circular buffer is 512 bytes (a 16-bit value occupies two memory locations).

The input is lowpass-filtered, decimated, and converted to an 8-bit stream. The bit stream is then modulated as a BPSK signal, and four output samples are generated for each bit. Each sample of the voice is a 16-bit integer. Because of sampling rate limitations, only the most significant 8 bits are used for transmission. This yields a resolution of 256 sample levels for the amplitude of the voice, which results in some degradation in the fidelity of the received signal.

The procedure is to sample the voice, get the most significant 8 bits, then transmit one period of a sine wave for each bit. Each period of a sine wave is constructed by outputting to the D/A converter four values of the sine wave. Therefore, for one voice sample, 30 output samples are necessary. This is a severe limitation since the maximum sampling rate is 96 kHz. The maximum sampling rate of the voice that we can implement is then 96 kHz/32, or 3 kHz.

The receiver uses eight samples to determine the phase of the phase-locked loop component allowing for a 48-kHz sampling rate by the transmitter. It can be verified that the receiver's voice bandwidth is approximately 3 kHz. To reconstruct a byte, the receiver must know where the frame starts for each byte. The transmitter periodically sends a synchronization sequence that is 1 byte long. This occurs once every 100 bytes.

To achieve frame synchronization, a synchronization sequence is sent periodically by the transmitter. This sequence is 8 bits long and is detected by the receiver by correlating the incoming bits with the expected sequence. A trigger variable looks over the previously received 8 bits and counts the number of bits that match the synchronization sequence. If the trigger variable is equal to 8, then the synchronization sequence was detected. With 8 bits in the synchronization sequence, there are 256 possible values, so that there is a 1/256 possibility that the sequence will occur randomly. This is too high a probability, and since we are receiving bits at 12 kHz (96 kHz/eight samples per bit), we would expect the sequence to occur randomly about 47 times a second (12 kHz/256). To lower this rate, we make sure that successive synchronization sequences are separated by the expected interval before declaring that the sequence has actually been received. When a correlation is detected, the frame index is reset to zero.

Since the receiver is reconstructing voice samples at a rate of 64 kHz, it needs to interpolate received voice samples to provide the DAC with a sample every time the interrupt routine is invoked. The receiver uses Newton's Forward interpolation with a third-degree polynomial to interpolate the sample values [39]. The generic expansion follows for points f_0 through f_n:

$$p(x) = f_0 + u\Delta_1 f_1 + [u(u-1)/2!]\Delta_2 f_2 + [u(u-1)(u-2)/3!]\Delta_3 f_3 + \cdots$$
$$+ [u(u-1)(u-2)\cdots(u-n+1)/n!]\Delta_n f_n + \cdots$$

where $u = [(x - x_i)/(x_{i+1} - x_i)]$ and f_i is the value of the function $f(n)$ at x_i. To interpolate, based on three points, this equation becomes

$$p(x) = f_0 + u(f_1 - f_0) + [u(u-1)/2](f_2 - 2f_1 + f_0)$$

Interpolating the output values significantly increases the quality of the output voice.

Possible improvements include the following:

1. At least a quadrature phase shift keying (QPSK) scheme can be used for the transmitter/receiver to allow much higher data rates across the channel.

2. Noise can be added to the system to increase the practicality of the project.

3. In addition to a phase estimator, a frequency estimator can be added to the receiver. Channels can sometimes introduce frequency distortion into a signal, and this would help the correlator to decode the modulated sequence.

10.11 BINARY PHASE SHIFT KEYING

This miniproject implements BPSK (see also Section 10.10). Two separate boards are used, one to modulate a signal simulating the transmitter component and the other to demodulate the received signal, simulating the receiver component.

Modulation

The modulation scheme transmits binary data using the polar nonreturn to zero (NRZ), ±1 V for the input data. The input is multiplied by a carrier signal with a frequency of $f_c = 8\,\text{kHz}$. For input data with values of ±1 V, the amplitude of the carrier remains the same, but not the phase. An input of +1 V yields a carrier output with a zero-phase shift, while an input of −1 V yields a carrier output that has been shifted by 180°.

A 100-Hz square wave with an amplitude of ±1 V is chosen as the input data. Using a threshold detector at 0 V, it is determined from the input whether the output signal carrier is a positive or a negative cosine. An 8-kHz cosine as the carrier is generated using a 4-point lookup table, sampling at 32 kHz. If the sampled data are greater than zero, then the output carrier is the generated cosine multiplied by +1; if the sampled data are less than zero, then the output carrier is the generated cosine multiplied by −1. Whenever the input signal switches from +1 to −1, or vice versa, the phase of the cosine wave is scaled by 180°. This change in phase looks like an M or a W on an oscilloscope. Figure 10.31 shows the core of the C source code

```
//BPSK_modulate.c Core program for BPSK modulation
. . .
short cos_table[4] = {1000,0,-1000,0};
interrupt void c_int11()
{
   input_data  = ((short)input_sample());
   if(input_data>0) bpsk_signal = cos_table[i++];
   else bpsk_signal = -1*cos_table[i++];
   output_sample(bpsk_signal);
   if(i > 3) i=0;
}
void main()
{ comm_intr(); while(1); }
```

FIGURE 10.31. Core C program for BPSK modulation (`bpsk_modulate.c`).

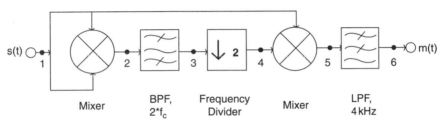

FIGURE 10.32. Carrier recovery block diagram for BPSK demodulation.

`bpsk_modulate.c` for the modulation scheme. Build the modulation component of the project. Verify that the output is an 8-kHz sinusoidal waveform, which becomes the input to the second DSK.

Demodulation

The second DSK simulates a pozar as a carrier recovery to demodulate the received signal. Demodulation can occur regardless of the input phase. The carrier recovery scheme is shown in Figure 10.32 and consists of a mixer, a bandpass filter centered at 16 kHz, a frequency divider by 2, a second mixer, and a lowpass filter with a cutoff frequency of 4 kHz. The output at each node is (with an input $m(t) = \pm 1\,\text{V}$, $f_m = 100\,\text{Hz}$):

Node 1: $s(t) = m(t)\cos(2\pi f_c t + \theta)$

Node 2: $m^2(t)\cos^2(2\pi f_c t + \theta) = \frac{1}{2} + \frac{1}{2}\cos[2(2\pi f_c t + \theta)]$

Node 3: $\frac{1}{2}\cos[2(2\pi f_c t + \theta)]$

Node 4: $\frac{1}{2}\cos(2\pi f_c t + \theta)$

Node 5: $\frac{1}{2}\cos(2\pi f_c t + \theta) m(t) \cos(2\pi f_c t + \theta) = \frac{m(t)}{4}\{\cos[2(2\pi f_c t + \theta)] + 1\}$

Node 6: $\frac{m(t)}{4}$

For the demodulator, the sampling frequency is set at 48 kHz (in lieu of 32 kHz) to prevent aliasing and allow for the use of a bandpass filter at node 2, since the output of the first mixer is at 16 kHz.

The signal at node 1 is the output of the modulator: a cosine wave (with an M or W) due to any phase shift. At node 2, it is a 16-kHz signal with a DC component. At node 3, the signal is filtered by a 30th order least squares FIR bandpass filter centered at 16 kHz. The FIR filter uses a least squares design with MATLAB's SPTool. The 16-kHz filtered signal is downsampled (decimated) to obtain an 8-kHz signal at node 4. The downsampling is achieved by setting every other input value to zero. The last stage of demodulation uses a product detector—a combination of a mixer and a lowpass filter—to recover the original binary input. The mixer multiplies the 8-kHz signal with the original input signal. This yields two signals: one at twice the carrier frequency and the other as a DC component with the original $m(t)$ input signal. This signal is then lowpass filtered to yield the original binary signal, regardless of the input phase. The lowpass filter is a 30th order Kaiser FIR filter, also designed with MATLAB's SPTool. The output at node 6 is then a 100-Hz square wave, the same as the modulator input signal. Figure 10.33 shows the core of the C source program *bpsk_demodulate.c* for the demodulator.

Verify that the original input signal to the modulator is recovered as the output from the demodulator. Experiment with different sampling rates, filter characteristics, and carrier frequencies to reduce the occasional output noise.

10.12 MODULATION SCHEMES—PAM AND PSK

This project implements both pulse amplitude modulation and phase shift keying schemes. See also the projects in Sections 10.10 and 10.11. The files for this project are included in the folder **modulation_schemes**.

10.12.1 Pulse Amplitude Modulation

In pulse amplitude modulation (PAM), the amplitude of the pulse conveys the information. The information symbols are transmitted at discrete and uniformly spaced time intervals. They are mapped to a train of pulses in the form of a carrier signal. The amplitude of these pulses represents a one-to-one mapping of the infor-

```
//BPSK_demodulate.c   Core C program for BPSK demodulation
...
double mixer_out, pd;
interrupt void c_int11()
{
 input_signal=((short)input_sample()/10);
 mixer_out = input_signal*input_signal;
 dly[0] = mixer_out;
 ..
 filter_output = (yn >> 15);            //output of 16 kHz BP filter
 x = 0;                                 //init downsampled value
 if (flag == 0)                         //discard input sample value
   flag = 1;                            //don't discard at next sampling
 else {
   x = filter_output;                   //downsampled value is input value
   flag = 0;
 }
 pd = x * input_signal;                 //product detector
 dly2[0] = ((short)pd);                 //for 4 kHz LP filter
 ..
 m = (yn2 >> 15);                       //output of LP filter
 output_sample(m);
 return;
}
void main()
{ comm_intr(); while(1); }
```

FIGURE 10.33. Core C program for BPSK demodulation (`bpsk_demodulate.c`).

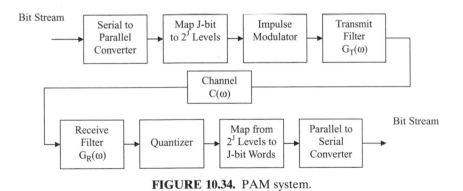

FIGURE 10.34. PAM system.

mation symbols to the respective levels. For example, in binary PAM, bit 1 is represented by a pulse with amplitude A and bit 0 by $-A$.

At the receiver, the information is recovered by obtaining the amplitude of each
pulse. The pulse amplitudes are then mapped back to the information symbol. Figure
10.34 shows the block diagram of a typical PAM system. This is a simplified version

without the introduction of adaptive equalizers or symbol clock recovery, which takes into account the effects of the channel. The incoming bit stream (output of the DSK) is parsed into J-bit words, with different lengths of parsing, resulting in different numbers of levels. For example, there are eight levels when $J = 3$. These levels are equidistant from each other on a constellation diagram and symmetric around the zero level, as shown in Figure 10.35. The eight constellation points represent the levels, with each level coded by a sequence of 3 bits. Tables 10.5–10.7 show the mapping levels.

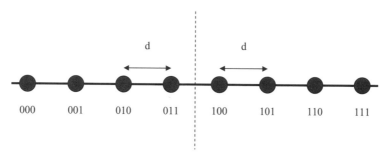

FIGURE 10.35. Constellation diagram of an eight-level PAM.

TABLE 10.5   Four-Level PAM Lookup Table for Mapping

Symbol Block	Level (in hex)
0000	0x7FFF
0101	0x2AAA
1010	−0x2AAB
1111	−0x8000

TABLE 10.6   Eight-Level PAM Lookup Table for Mapping

Symbol Block	Level (in hex)
000	0x7FFF
001	0x5B6D
010	0x36DB
011	0x1249
100	−0x1249
101	−0x36DB
110	−0x5B6D
111	−0x7FFF

TABLE 10.7   Sixteen-Level PAM Lookup Table for Mapping

Symbol Block	Level (in hex)
0000	0x7FFF
0001	0x6EEE
0010	0x5DDD
0011	0x4CCC
0100	0x3BBB
0101	0x2AAA
0110	0x1999
0111	0x0888
1000	–0x0889
1001	–0x199A
1010	–0x2AAB
1011	–0x3BBC
1100	–0x4CCD
1101	–0x5DDE
1110	–0x6EEF
1111	–0x8000

Transmitter/Receiver Algorithm

An input sample is composed of 16 bits. Depending on the type of PAM, an appropriate masking is used. The same transmitter and receiver implementations apply to four-level and eight-level PAM with differences in masking, shifting, and lookup tables (see Tables 10.5–10.7). For the 8-PAM, the LSB of the input sample is discarded so that the remaining number of bits (15) is an integer multiple of 3, which does not have a noticeable effect on the modulated waveform and on the recovered voice.

Consider the specific case of a 16-PAM. In order to achieve the desired symbol rate, the input sample is decomposed into segments 4 bits long. Each input sample is composed of four segments. Parsing the input sample is achieved through the use of masking and shifting. The first symbol block is obtained with masking of the four least significant bits by *anding* the input sample with 0x000F. The second symbol block is obtained through shifting the original input sample by four to the right and masking the four LSBs. These steps are repeated until the end of the input sample length and produce four symbol blocks. Assume that the input sample is 0xA52E. In this case, 1110 (after masking the four LSBs) is mapped to –0x6EEF, as shown in Table 10.7. Each symbol block is composed of 4 bits mapped into the 16 uniformly spaced levels between –0x8000 and 0x7FFF. The spacing between each level is 0x1111, selected for uniform spacing. The selected level is then transmitted as a square wave. The period of the square wave is achieved by outputting the same level many times to ensure a smooth-looking square wave at the output of the transmitter.

TABLE 10.9   Input and Output Scheme for Voice Scrambler

	Period 1	Period 2	Period 3	Period 4	Period 5	Period 6	Period 7	Period 8
Input	Sample 1	X	Sample 2	X	Sample 3	X	Sample 4	X
Output	X	X	X	X	Sample 1	Sample 2	Sample 3	Sample 4

FIGURE 10.40. Hard-decision decoding setup.

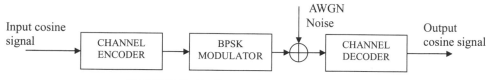

FIGURE 10.41. Soft decision decoding setup.

frequency noise in the output. Note that the scrambling scheme uses bit manipulation that requires no external synchronization between the scrambling transmitter and the unscrambling receiver.

The (complete) executable file for the IIR and scrambling implementations is on the CD as `minimicro.out`, and the unscrambling executable file is on the CD as `minimicrob.out`. These executable files can be used first to test the different implementations for IIR filtering and the scrambling/unscrambling scheme. The appropriate support files are included in the folder **IIR_ctrl**.

DIP switch values 6 to 15 yield no output and can be used for expanding this project to implement additional IIR or FIR filters and/or another scrambling scheme. RTDX can be used to pass the designed coefficients (see the FIR project incorporating RTDX and Chapter 9).

10.14 CONVOLUTIONAL ENCODING AND VITERBI DECODING

Channel coding schemes widely used in communication systems mostly consist of the convolutional encoding and Viterbi decoding algorithms to reduce the bit errors on noisy channels. This project implements a 3-output, 1-input, 2-shift register (3,1,2) convolutional encoder used for channel encoding and a channel decoder employing soft decision and basic Viterbi decoding techniques.

Soft Decision and Basic Viterbi Decoding
The system setups are used for soft decision and Viterbi decoding techniques. In Figures 10.40 and 10.41, the channel encoder represents a (3,1,2) convolutional

encoding algorithm, and the channel decoder represents the Viterbi decoding algorithm.

In the Viterbi decoding setup shown in Figure 10.40, a cosine signal is the input to the channel encoder algorithm. The encoded output is stored in a buffer. The elements of this buffer provide the input to the channel decoder algorithm that decodes it and returns the original cosine signal. Both the encoder and decoder outputs are displayed within CCS.

In the soft decision decoding setup shown in Figure 10.41, a cosine signal is given as input to the channel encoder algorithm. The binary output of the channel encoder is modulated using the BPSK technique, whereby the 0 output of the channel encoder is translated into −1 and the 1 output is translated into +1. Additive white Gaussian noise (AWGN) is generated and added to the modulated output. The signal that is corrupted by the additive noise is fed to the channel decoder. Both the encoder and decoder outputs are displayed within CCS. The variance of AWGN is varied, and the decoder's performance is observed.

(3,1,2) Convolutional Encoder

Convolutional coding provides error correction capability by adding redundancy bits to the information bits. The convolutional encoding is usually implemented by the shift register method and associated combinatorial logic that performs modulo-two addition, an XOR operation. A block diagram of the implemented (3,1,2) convolutional encoder is shown in Figure 10.42, where u is the input, $v(1)$, $v(2)$, $v(3)$ are the outputs, and A, B are the shift registers. The outputs are

$$v(1) = u$$

$$v(2) = u \oplus b$$

$$v(3) = u \oplus a \oplus b$$

where a and b are the contents of the shift registers A and B, respectively. Initially the contents of the shift registers are 0's. The shift registers go through four different

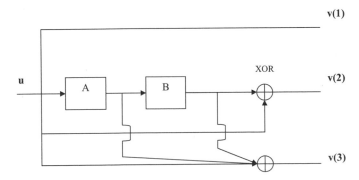

FIGURE 10.42. A (3,1,2) convolutional encoder.

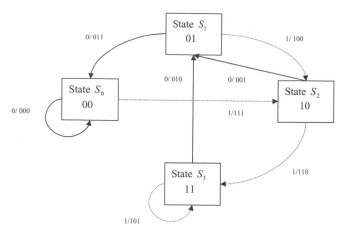

FIGURE 10.43. State diagram for encoding.

states, depending on the input (0 or 1) received. Once all the input bits are processed, the contents of the shift registers are again reset to zero by feeding two 0's (since we have two shift registers) at the input.

State Diagram

The basic state diagram of the encoder is shown in Figure 10.43, where S_0, S_1, S_2, and S_3 represent the different states of the shift registers. Furthermore, m/xyz indicates that on receiving an input bit m, the output of the encoder is xyz; that is, if $u = m => v(1) = x$, $v(2) = y$, $v(3) = z$ for that particular state of shift registers A and B. The arrows indicate the state changes on receiving the inputs.

Trellis Diagram

The corresponding trellis diagram for the state diagram is shown in Figure 10.44. The four possible states of the encoder are shown as four rows of horizontal dots. There is one column of four dots for the initial state of the encoder and one for each time instant during the message. The solid lines connecting the dots in the diagram represent state transitions when the input bit is a 0. The dotted lines represent transitions when the input bit is a 1. For this encoding scheme, each encoding state at time n is linked to two states at time $n + 1$. The Viterbi algorithm is used for decoding the trellis-coded information bits by expanding the trellis over the received symbols. The Viterbi algorithm reduces the computational load by taking advantage of the special structure of the trellis codes.

Modulation and AWGN for Soft Decision

In the soft decision decoding setup, the 1/0 output of the convolutional encoder is mapped into an antipodal baseband signaling scheme (BPSK) by translating 0's to −1's and 1's to +1's. This can be accomplished by performing the operation

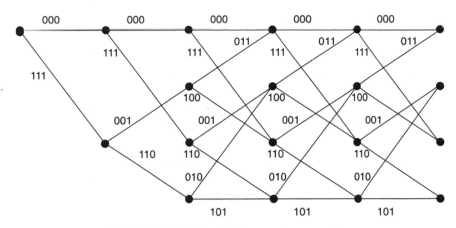

FIGURE 10.44. Trellis diagram for encoding.

$y = 2x - 1$ on each convolutional encoder output symbol, where x is the encoder output symbol and y is the output of the BPSK modulator.

AWGN is added to this modulated signal to create the effect of channel noise. AWGN is a noise whose voltage distribution over time has characteristics that can be described using a Gaussian distribution, that is, a bell curve. This voltage distribution has zero mean and a standard deviation that is a function of the SNR of the received signal. The standard deviation of this noise can be varied to obtain signals with different SNRs at the decoder input.

A zero-mean Gaussian noise with standard deviation σ can be generated as follows. In order to obtain Gaussian random numbers, we take advantage of the relationships between uniform, Rayleigh, and Gaussian distributions. C only provides a uniform random number generator, `rand( )`. Given a uniform random variable U, a Rayleigh random variable R can be obtained using

$$R = \sqrt{2\sigma^2 \ln(1/(1-U))} = \sigma\sqrt{2\ln(1/(1-U))}$$

where σ^2 is the variance of the Rayleigh random variable. Given R and a second uniform random variable V, a Gaussian random variable G can be obtained using

$$G = R\cos V$$

Viterbi Decoding Algorithm

The Viterbi decoding algorithm uses the trellis diagram to perform the decoding. The basic cycle repeated by the algorithm at each stage into the trellis is:

1. *Add*: At each cycle of decoding, the branch metrics enumerating from the nodes (states) of the previous stage are computed. These branch metrics are added to the previously accumulated and saved path metrics.

2. *Compare*: The path metrics leading to each of the encoder's states are compared.

3. *Select*: The highest-likelihood path (survivor) leading to each of the encoder's states is selected, and the lower-likelihood paths are discarded.

A metric is a measure of the "distance" between what is received and all of the possible channel symbols that could have been received. The metrics for the soft decision and the basic Viterbi decoding techniques are computed using different methods. For basic Viterbi decoding, the metric used is the Hamming distance, which specifies the number of bits by which two symbols differ. For the soft decision technique, the metric used is the Euclidean distance between the signal points in a signal constellation. More details of the decoding algorithm are presented elsewhere [40, 41].

Implementation

Build this project as **viterbi**. The complete C source program and the executable (.out) files are included on the CD in the folder Viterbi. Several functions are included in the program to perform convolutional encoding and BPSK modulation, add white Gaussian noise, and implement the Viterbi decoding algorithm (the more extensive function).

The following time-domain graphs can be viewed within CCS—input, encoder output, and decoder output—using the addresses *input, enc_output*, and *dec_output*, respectively. For the graphs, use an acquisition buffer size of 128, a sampling frequency of 8000, a 16-bit signed integer for both input and decoder output, and a 32-bit float for the encoder output.

Three gel files are used (included on the CD):

1. *Input.gel*: to select one of the following three input signals: *cos666* (default), *cos666 + cos1500*, and *cos666 + cos2200*, where 666 represents a 666-Hz cosine.

2. *Technique.gel*: to select between soft decision and basic Viterbi decoding.

3. *Noise.gel*: to select a suitable standard deviation for AWGN. One of five different values (0, 0.3, 0.4, 2.0, 3.0) of the standard deviation of the AWGN can be selected.

Results

The following results are obtained:

Case 1: input = cosine 666 Hz, using soft decision

Case 2: input = cosine 666 Hz, standard deviation $\sigma = 0.4$

Case 3: input = cosine 666 Hz, standard deviation $\sigma = 3.0$

Case 4: input = cosine (666 + 1500) Hz, using basic Viterbi decoding (noise level 0)

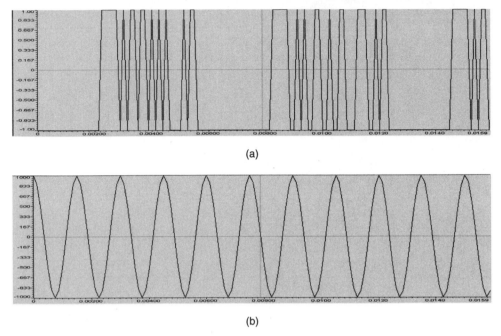

(a)

(b)

FIGURE 10.45. CCS plots of output using case 1: (a) convolutional encoder varying between +1/−1 and (b) Viterbi decoder.

With the default settings, the encoded output will appear between the +1 and −1 voltage levels, as shown in Figure 10.45a. The output of the Viterbi decoder is shown in Figure 10.45b. With an increase in the noise level, slight variations will be observed around the +1 and −1 voltage levels at the encoder output. These variations will increase with an increase in noise level. It can be observed from the decoder outputs that it is able to recover the original cosine signal. With the noise level set at 0, 0.3, or 0.4 using the *noise.gel* slider, the decoder is still able to recover the original cosine signal, even though there is some degradation in the corresponding encoder output, as shown in Figure 10.46. With further increase in the noise level with $\sigma = 3.0$, the decoder output is degraded, as shown in Figure 10.47.

Figure 10.48 illustrates case 4 using cosine (666 + 1500) as input. With the *technique.gel* slider selected for Viterbi decoding, the encoder output appears between the 0 and 1 voltage levels, as shown in Figure 10.48b, since the input is of plain binary form. The decoded output is the restored input cosine signal shown in Figure 10.48c. There is no additive noise added in this case.

This project can be extended for real-time input and output signals.

Illustration of the Viterbi Decoding Algorithm

Much of the material introduced here can be found in Ref. 41. To illustrate the Viterbi decoding algorithm, consider the basic Viterbi symbol inputs. Each time a

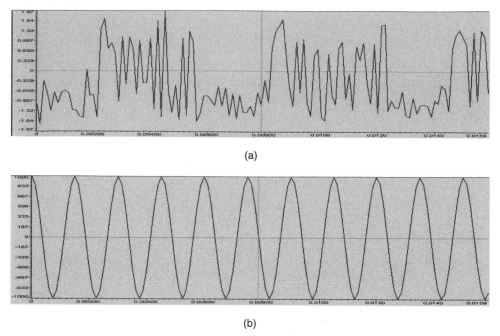

(a)

(b)

FIGURE 10.46. CCS plots of output using case 2: (a) convolutional encoder with AWGN ($\sigma = 0.4$) and (b) Viterbi decoder.

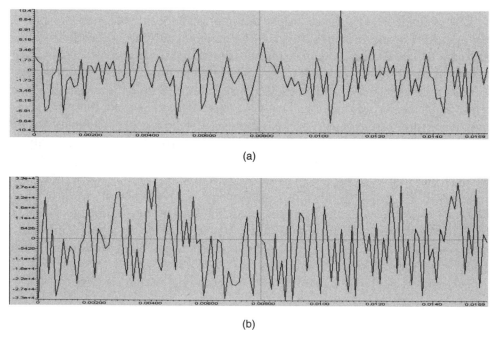

(a)

(b)

FIGURE 10.47. CCS plots of output using case 3: (a) convolutional encoder with AWGN ($\sigma = 0.3$) and (b) Viterbi decoder.

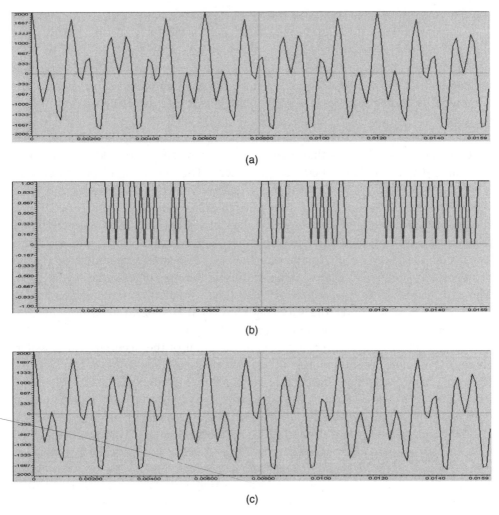

FIGURE 10.48. CCS plots using case 4: (a) input to convolutional encoder; (b) output from convolutional encoder (between 0 and 1); and (c) output from a Viterbi decoder.

triad of channel symbols is received, a metric is computed to measure the "distance" between what is received and all of the possible channel symbol triads that could have been received. Going from $t = 0$ to $t = 1$, there are only two possible channel symbol triads that could have been received: 000 and 111. This is because the convolutional encoder was initialized to the all-0's state, and given one input bit = 1 or 0, there are only two states to transition to and two possible outputs of the encoder: 000 and 111.

The metric used is the Hamming distance between the received channel symbol triad and the possible channel symbol triad. The Hamming distance is computed by simply counting how many bits are different between the received channel symbol triad and the possible channel symbol triad. The results can only be zero, one, two, or three. The Hamming distance (or other metric) values computed at each time instant, for the paths between the states at the previous time instant and the states at the current time instant, are called *branch metrics*. For the first time instant, these results are saved as *accumulated error metric* values associated with states. From the second time instant on, the accumulated error metrics are computed by adding the previous accumulated error metrics to the current branch metrics.

Consider that at $t = 1$, 000 is received at the input of the decoder.. The only possible channel symbol triads that could have been received are 000 and 111. The Hamming distance between 000 and 000 is zero. The Hamming distance between 000 and 111 is three. Therefore, the branch metric value for the branch from State 00 to State 00 is zero, and for the branch from State 00 to State 10 it is two. Since the previous accumulated error metric values are equal to zero, the accumulated metric values for State 00 and for State 10 are equal to the branch metric values. The accumulated error metric values for the other two states are undefined (in the program, this undefined value is initialized to be the maximum value for integer). The path history table is updated for every time instant. This table, which has an entry for each state, stores the surviving path for that state at each time instant. These results at $t = 1$ are shown in Figure 10.49a.

Consider that at $t = 2$, 110 is received at the input of the decoder. The possible channel symbol triads that could have been received in going from $t = 1$ to $t = 2$ are 000 going from State 00 to State 00, 111 going from State 00 to State 10, 001 going from State 10 to State 01, and 110 going from State 10 to State 11. The Hamming distance is two between 000 and 110, one between 111 and 110, three between 001 and 110, and zero between 110 and 110. These branch metric values are added to the previous accumulated error metric values associated with each state that we came from to get to the current states. At $t = 1$, we can only be at State 00 or State 10. The accumulated error metric values associated with those states were 0 and 2, respectively. The calculation of the accumulated error metric associated with each state at $t = 2$ is shown in Figure 10.49b.

Consider that at $t = 3$, 010 is received. There are now two different ways that we can get from each of the four states that were valid at $t = 2$ to the four states that are valid at $t = 3$. To handle that, we compare the accumulated error metrics associated with each branch and discard the larger one of each pair of branches leading into a given state. If the members of a pair of accumulated error metrics going into a particular state are equal, that value is saved. The operation of adding the previously accumulated error metrics to the new branch metrics, comparing the results, and selecting the smaller accumulated error metric to be retained for the next time instant is called the *add-compare-select* operation. The path history for a state is also updated by selecting the path corresponding to the smallest path metric for that state. This can be found by adding the current selected path transition to the path history of its previous state. The result for $t = 3$ follows.

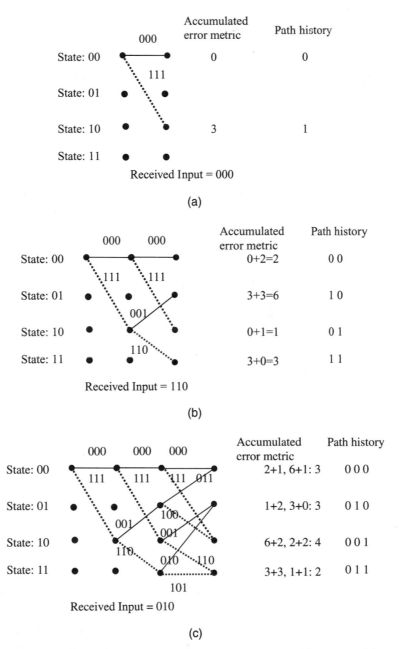

FIGURE 10.49. Trellis diagrams to illustrate Viterbi decoding: (a) $t = 1$; (b) $t = 2$; and (c) $t = 3$.

At $t = 3$, the decoder has reached its steady state; that is, it is possible to have eight possible state transitions. For every other time instant from now on, the same process gets repeated until the end of input is reached. The last two inputs that are received in a Viterbi decoder are also considered special cases. At the convolutional encoder, when the end of input is reached, we input two trailing zeros in order to reset the shift register states to zero. As a consequence of this, in a Viterbi decoder, in the last but one time instant, the only possible states in the Viterbi decoder are State 00 and State 01. Therefore, the expected inputs are 000, 011, 001, and 010. And for the last time instant, the only possible state is 00. Therefore, the expected inputs are only 000 and 011. This case is illustrated in Figure 10.49c.

In the program, it is assumed that the decoder has a memory of only 16, meaning that at any one time, the path history can store only 16 paths. As soon as the first 16 channel symbol triads are read, the path history becomes full. The path history in this source code is an array named *path_history*. Each variable of this array maintains the path history for a particular state, with each bit in the variable storing a selected path with the rightmost bit storing the most recent path. Therefore, before processing the 17th channel symbol triad, the minimum branch metric state is found, and the leftmost bit in the path history of this state is output into a variable *dec_output*. For every other time instant afterward, this process is repeated and the leftmost bit of the selected *path_history* variable is output to *dec_output*. On completing the decoding algorithm, *dec_output* contains the desired decoder output.

A variable named *output_table* lists the output symbols for every input at a particular state, as shown in the following table:

	Output Symbols If:	
Current State	Input = 0	Input = 1
00	000	111
01	011	100
10	001	110
11	010	101

The soft decision Viterbi algorithm functions in a similar fashion, except that the metric is computed in a different way. The metric is specified using the Euclidean distance between the signal points in a signal constellation. In the soft decision algorithm, the output of the encoder is sent in the form of BPSK-modulated symbols, that is, 0 is sent as -1 and 1 is sent as $+1$. Before this distance is found, BPSK modulation is performed on the possible channel symbol triad. Assume that a channel symbol triad containing $\{a1, a2, a3\}$ is received, and the expected input channel symbol triad is 001. After BPSK modulation, it can be written as $\{b1, b2, b3\}$, where $b1 = -1, b2 = -1$, and $b3 = +1$. Then, the distance between these two channel symbols is found using

$$distance = abs(b1 - a1) + abs(b2 - a2) + abs(b3 - a3)$$

10.15 SPEECH SYNTHESIS USING LINEAR PREDICTION OF SPEECH SIGNALS

Speech synthesis is based on the reproduction of human intelligible speech through artificial means [42–45]. Examples of speech synthesis technology include *text-to-speech* systems. The creation of synthetic speech covers a range of processes; and even though they are often lumped under the general term *text-to-speech*, a lot of work has been done to generate speech from sequences of the speech sounds. This would be a speech-sound (phoneme) to audio waveform synthesis, rather than going from text to phonemes (speech sounds) and then to sound. One of the first practical applications of speech synthesis was a speaking clock. It used optical storage for phrases and words (noun, verb, etc.), concatenated to form complete sentences. This led to a series of innovative products such as vocoders, speech toys, and so on. Advances in the understanding of the speech production mechanism in humans, coupled with similar advances in DSP, have had an impact on speech synthesis techniques. Perhaps the most singular factors that started a new era in this field were the computer processing and storage technologies. While speech and language were already important parts of daily life before the invention of the computer, the equipment and technology that developed over the last several years have made it possible to produce machines that speak, read, and even carry out dialogs. A number of vendors provide both recognition and speech technology. Some of the latest applications of speech synthesis are in cellular phones, security networks, and robotics.

There are different methods of speech synthesis based on the source. In a text-to-speech system, the source is a text string of characters read by the program to generate voice. Another approach is to associate intelligence in the program so that it can generate voice without external excitation. One of the earliest techniques was *Formant synthesis*. This method was limited in its ability to represent voice with high fidelity due to its inherent drawback of representing phonemes by three frequencies. This method and several analog technologies that followed were replaced by digital methods. Some early digital technologies were RELP (residue excited) and VELP (voice excited). These were replaced by new technologies, such as LPC (linear predictive coding), CELP (code excited), and PSOLA (pitch synchronous overlap-add). These technologies have been used extensively to generate artificial voice.

Linear Predictive Coding

Most methods that are used for analyzing speech start by transforming acoustic data into spectral form by performing short time Fourier analysis of the speech wave. Although this type of spectral analysis is a well-known technique for studying signals, its application to speech signal suffers from limitations due to the nonstationary and quasiperiodic properties of the speech wave. As a result, methods based on spectral analysis often do not provide a sufficiently accurate description of speech articulation. Linear predictive coding (LPC) represents the speech waveform directly in terms of time-varying parameters related to the transfer function

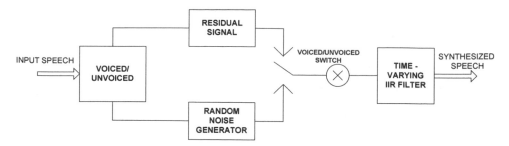

FIGURE 10.50. Diagram of the speech synthesis process.

of the vocal tract and the characteristics of the source function. It uses the knowledge that any speech can be represented by certain types of parametric information, including the filter coefficients (that model the vocal tract) and the excitation signal (that maps the source signals). The implementation of LPC reduces to the calculation of the filter coefficients and excitation signals, making it suitable for digital implementation.

Speech sounds are produced as a result of acoustical excitation of the human vocal tract. During production of the voiced sounds, the vocal chord is excited by a series of nearly periodic pulses generated by the vocal cords. In unvoiced sounds, excitation is provided by the air passing turbulently through constrictions in the tract. A simple model of the vocal tract is a discrete time-varying linear filter. Figure 10.50 is a diagram of the LPC speech synthesis. To reproduce the voice signal, the following are required:

1. An excitation signal
2. The LPC filter coefficients

The excitation mechanism can be approximated using a residual signal generator (for voiced signals) or a white Gaussian noise generator (for unvoiced signals) with adjustable amplitudes and periods. The linear predictor P, a transversal filter with p delays of one sample interval each, forms a weighted sum of past samples as the input of the predictor. The output of the predictor at the nth sampling instant is given by

$$s_n = \sum_{k=1}^{p} a_k \cdot (s_m) + \delta_n$$

where $m = n - k$ and δ_n represents the nth excitation sample.

Implementation

The input to the program is a sampled array of input speech using an 8-kHz sampling rate. The samples are stored in a header file. The length of the input speech

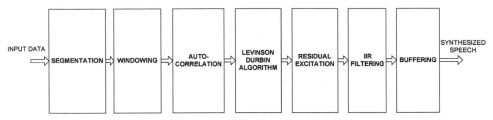

FIGURE 10.51. Speech synthesis algorithm with various modules.

array is 10,000 samples, translating into approximately 1.25 seconds of speech. The input array is segmented into a large number of frames, each 80 B long with an overlap of 40 B for each frame. Each frame is then passed to the following modules: windowing, autocorrelation, LPC, residual, IIR, and accumulate. External memory is utilized. A block diagram of the LPC speech synthesis algorithm with the various modules is shown in Figure 10.51.

1. *Segmentation.* This module separates the input voice into overlapping segments. The length of the segment is such that the speech segment appears stationary as well as quasiperiodic. The overlap provides a smooth transition between consecutive speech frames.
2. *Windowing.* The speech waveform is decomposed into smaller frames using the Hamming window. This suppresses the sidelobes in the frequency domain.
3. *Levinson–Durbin algorithm.* To calculate the LPC coefficients, the autocorrelation matrix of the speech frame is required. From this matrix, the LPC coefficients can be obtained using

$$r(i) = \sum_{k=1}^{p} a_k \cdot r(|i-k|)$$

where $r(i)$ and ak represent the autocorrelation array and the coefficients, respectively.
4. *Residual signal.* For synthesis of the artificial voice, the excitation is given by the residual signal, which is obtained by passing the input speech frame through an FIR filter. It serves as an excitation signal for both voiced and unvoiced signals. This limits the algorithm due to the energy and frequency calculations required for making decisions about voiced/unvoiced excitation since, even for an unvoiced excitation that has a random signal as its source, the same principle of residue signal can still be used. This is because, in the case of unvoiced excitation, even the residue signal obtained will be random.

5. *Speech synthesis.* With the representation of the speech frame in the form of the LPC filter coefficients and the excitation signal, speech can be synthesized. This is done by passing the excitation signal (the residual signal) through an IIR filter. The residual signal generation and the speech synthesis modules imitate the vocal chord and the vocal tract of the speech production system in humans.

6. *Accumulation and buffering.* Since speech is segmented at the beginning, the synthesized voice needs to be concatenated. This is performed by the accumulation and buffering module.

7. *Output.* When the entire synthesized speech segment is obtained, it is played. During playback, the data are downsampled to 4 kHz to restore the intelligibility of the speech.

Implementation

The complete support files are on the CD in the folder speech_syn. Generate a .wav file of the speech sample to be synthesized. For example, include *goaway.wav* in the MATLAB file input_read.m. The MATLAB file samples it for 8 kHz and stores the input samples array in the header file input.h. Include this generated header file in the main C source program speech.c. Build this project as **speech_syn**. Run the MATLAB program input_read.m to generate the two header files input.h (containing the input samples) and hamming.h (for the Hamming coefficients). Load/run speech_syn.out and verify the synthesized speech "go away" from a speaker connected to the DSK output. Three other .wav files are included in the folder and can be tested readily.

Results

Speech is synthesized for the following: "Go away," "Hello, professor," "Good evening," and "Vacation." The synthesized output voice is found to have considerable fidelity to the original speech. The voice/unvoiced speech phonemes are reproduced with considerable accuracy. This project can be improved with a larger buffer size for the samples and noise suppression filters. There is noise after each time the sentence is played. A speech recognition algorithm can be implemented in conjunction with the speech synthesis to facilitate a dialog.

10.16 AUTOMATIC SPEAKER RECOGNITION

This project implements an automatic speaker recognition system [46–50]. *Speaker recognition* refers to the concept of recognizing a speaker by his/her voice or speech samples. This is different from speech recognition. In automatic speaker recognition, an algorithm generates a hypothesis concerning the speaker's identity or authenticity. The speaker's voice can be used for ID and to gain access to services such as banking, voice mail, and so on.

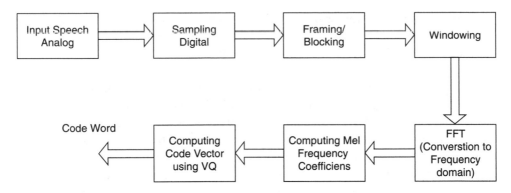

(Spelling error-Computing-not computung)

FIGURE 10.52. Steps for speaker recognition implementation.

Speaker recognition systems contain two main modules: *feature extraction* and *classification*.

1. Feature extraction is a process that extracts a small amount of data from the voice signal that can be used to represent each speaker. This module converts a speech waveform to some type of parametric representation for further analysis and processing. Short-time spectral analysis is the most common way to characterize a speech signal. The Mel-frequency cepstrum coefficients (MFCCs) are used to parametrically represent the speech signal for the speaker recognition task. The steps in this process are shown in Figure 10.52:

   (a) Block the speech signal into frames, each consisting of a fixed number of samples.

   (b) Window each frame to minimize the signal discontinuities at the beginning and end of the frame.

   (c) Use FFT to convert each frame from time to frequency domain.

   (d) Convert the resulting spectrum into a Mel-frequency scale.

   (e) Convert the Mel spectrum back to the time domain.

2. Classification consists of models for each speaker and a decision logic necessary to render a decision. This module classifies extracted features according to the individual speakers whose voices have been stored. The recorded voice patterns of the speakers are used to derive a classification algorithm. Vector quantization (VQ) is used. This is a process of mapping vectors from a large vector space to a finite number of regions in that space. Each region is called a *cluster* and can be represented by its center, called a *codeword*. The collection of all clusters is a *codebook*. In the training phase, a speaker-specific VQ codebook is generated for each known speaker by clustering his/her training acoustic vectors. The distance from a vector to the closest codeword of a codebook is called a *VQ distortion*. In the recognition phase, an input utterance of an

unknown voice is vector-quantized using each trained codebook, and the total VQ distortion is computed. The speaker corresponding to the VQ codebook with the smallest total distortion is identified.

Speaker recognition can be classified with identification and verification. *Speaker identification* is the process of determining which registered speaker provides a given utterance. *Speaker verification* is the process of accepting or rejecting the identity claim of a speaker. This project implements only the speaker identification (ID) process. The speaker ID process can be further subdivided into *closed set* and *open set*. The *closed set* speaker ID problem refers to a case where the speaker is known *a priori* to belong to a set of M speakers. In the *open set* case, the speaker may be out of the set and, hence, a "none of the above" category is necessary. In this project, only the simpler closed set speaker ID is used.

Speaker ID systems can be either *text-independent* or *text-dependent*. In the *text-independent* case, there is no restriction on the sentence or phrase to be spoken, whereas in the *text-dependent* case, the input sentence or phrase is indexed for each speaker. The text-dependent system, implemented in this project, is commonly found in speaker verification systems in which a person's password is critical for verifying his/her identity.

In the *training phase*, the feature vectors are used to create a model for each speaker. During the *testing phase*, when the test feature vector is used, a number will be associated with each speaker model indicating the degree of match with that speaker's model. This is done for a set of feature vectors, and the derived numbers can be used to find a likelihood score for each speaker's model. For the speaker ID problem, the feature vectors of the test utterance are passed through all the speakers' models and the scores are calculated. The model having the best score gives the speaker's identity (which is the decision component).

This project uses MFCC for feature extraction, VQ for classification/training, and the Euclidean distance between MFCC and the trained vectors (from VQ) for speaker ID. Much of this project was implemented with MATLAB [47].

Mel-Frequency Cepstrum Coefficients

MFCCs are based on the known variation of the human ear's critical bandwidths. A Mel-frequency scale is used with a linear frequency spacing below 1000 Hz and a logarithmic spacing above that level. The steps used to obtain the MFCCs follow.

1. *Level detection.* The start of an input speech signal is identified based on a prestored threshold value. It is captured after it starts and is passed on to the framing stage.
2. *Frame blocking.* The continuous speech signal is blocked into frames of N samples, with adjacent frames being separated by M ($M < N$). The first frame consists of the first N samples. The second frame begins M samples after the

first frame and overlaps it by $N - M$ samples. Each frame consists of 256 samples of speech signal, and the subsequent frame starts from the 100th sample of the previous frame. Thus, each frame overlaps with two other subsequent frames. This technique is called *framing*. The speech sample in one frame is considered to be stationary.

3. *Windowing.* After framing, windowing is applied to prevent spectral leakage. A Hamming window with 256 coefficients is used.

4. *Fast Fourier transform.* The FFT converts the time-domain speech signal into a frequency domain to yield a complex signal. Speech is a real signal, but its FFT has both real and imaginary components.

5. *Power spectrum calculation.* The power of the frequency domain is calculated by summing the square of the real and imaginary components of the signal to yield a real signal. The second half of the samples in the frame are ignored since they are symmetric to the first half (the speech signal being real).

6. *Mel-frequency wrapping.* Triangular filters are designed using the Mel-frequency scale with a bank of filters to approximate the human ear. The power signal is then applied to this bank of filters to determine the frequency content across each filter. Twenty filters are chosen, uniformly spaced in the Mel-frequency scale between 0 and 4 kHz. The Mel-frequency spectrum is computed by multiplying the signal spectrum with a set of triangular filters designed using the Mel scale. For a given frequency f, the mel of the frequency is given by

$$B(f) = [1125 \ln(1 + f/700)] \text{ mels}$$

If m is the mel, then the corresponding frequency is

$$B^{-1}(m) = [700 \exp(m/1125) - 700] \text{ Hz}$$

The frequency edge of each filter is computed by substituting the corresponding mel. Once the edge frequencies and the center frequencies of the filter are found, boundary points are computed to determine the transfer function of the filter.

7. *Mel-frequency cepstrum coefficients.* The log mel spectrum is converted back to time. The discrete cosine transform (DCT) of the log of the signal yields the MFCCs.

Speaker Training—VQ

VQ is a process of mapping vectors from a large vector space to a finite number of regions in that space. Each region is called a *cluster* and can be represented by its center, the codeword. As noted earlier, a codebook is the collection of all the clusters. An example of a one-dimensional VQ has every number less than –2 approximated by –3; every number between –2 and 0 approximated by –1; every number between 0 and 2 approximated by +1; and every number greater than 2 approximated by +3. These approximate values are uniquely represented by 2 bits, yielding a one-dimensional, 2-bit VQ. An example of a two-dimensional VQ consists of 16 regions and 16 stars, each of which can be uniquely represented by 4 bits (a two-dimensional 4-bit VQ). Each pair of numbers that fall into a region are approximated by a star associated with that region. The stars are called *codevectors*, and the regions are called *encoding regions*. The set of all the codevectors is called the *codebook*, and the set of all encoding regions is called the *partition* of the space.

Speaker Identification (Using Euclidean Distances)

After computing the MFCCs, the speaker is identified using a set of trained vectors (samples of registered speakers) in an array. To identify the speaker, the Euclidean distance between the trained vectors and the MFCCs is computed for each trained vector. The trained vector that produces the smallest Euclidean distance is identified as the speaker.

Implementation

The design is first tested with MATLAB. A total of eight speech samples from eight different people (eight speakers, labeled S1 to S8) are used to test this project. Each speaker utters the same single digit, *zero*, once in a training session (then also in a testing session). A digit is often used for testing in speaker recognition systems because of its applicability to many security applications. This project was implemented on the C6711 DSK and can be transported to the C6713 DSK. Of the eight speakers, the system identified six correctly (a 75% identification rate). The identification rate can be improved by adding more vectors to the training codewords. The performance of the system may be improved by using two-dimensional or four-dimensional VQ (training header file would be $8 \times 20 \times 4$) or by changing the quantization method to dynamic time wrapping or hidden Markov modeling. A readme file to test this project is on the CD in the folder **speaker_recognition**, along with all the appropriate support files. These support files include several modules for framing and windowing, power spectrum, threshold detection, VQ, and the Mel-frequency spectrum.

10.17 μ-LAW FOR SPEECH COMPANDING

An analog input such as speech is converted into digital form and compressed into 8-bit data. μ-Law *encoding* is a nonuniform quantizing logarithmic compression scheme for audio signals. It is used in the United States to compress a signal into a logarithmic scale when coding for transmission. It is widely used in the telecommunications field because it improves the SNR without increasing the amount of data.

The dynamic range increases, while the number of bits for quantization remains the same. Typically, μ-law compressed speech is carried in 8-bit samples. It carries more information about smaller signals than about larger signals. It is based on the observation that many signals are statistically more likely to be near a low-signal level than a high-signal level. As a result, there are more quantization points closer to the low level.

A lookup table with 256 values is used to obtain the quantization levels from 0 to 7. The table consists of a 16×16 set of numbers: Two 0's, two 1's, four 2's, eight 3's, sixteen 4's, thirty-two 5's, sixty-four 6's, and one hundred twenty-eight 7's. More higher-level signals are represented by 7 (from the lookup table). Three exponent bits are used to represent the levels from 0 to 7, 4 mantissa bits are used to represent the next four significant bits, and 1 bit is used for the sign bit.

The 16-bit input data are converted from linear to 8-bit μ-law (simulated for transmission), then converted back from μ-law to 16-bit linear (simulated as receiving), and then output to the codec.

From the 16-bit sample signal, the eight MSBs are used to choose a quantization level from the lookup table of 256 values. The quantization is from 0 to 7 so that 0 and 1 range across 2 values, . . . , 2 ranges across 4 values, 3 ranges across 8 values, . . . , and 7 ranges across 128 values. This is a logarithmic companding scheme.

Build this project as **Mulaw**. The C source file for this project, Mulaw.c, is included on the CD.

10.18 SB-ADPCM ENCODER/DECODER: IMPLEMENTATION OF G.722 AUDIO CODING

An audio signal is sampled at 16 kHz, transmitted at a rate of 64 kbits/s, and reconstructed at the receiving end [51, 52].

Encoder

The subband adaptive differential pulse code-modulated (SB-ADPCM) encoder consists of a transmit quadrature mirror filter that splits the input signal into a low frequency band, 0 to 4 kHz, and a high frequency band, 4 to 8 kHz. The low and high frequency signals are encoded separately by dynamically quantizing an adaptive predictor's output error. The low and high encoder error signals are encoded with 6 and 2 bits, respectively. As long as the error signal is small, a negligible amount of overall quantization noise and good performance can be obtained. The low and high band bits are multiplexed, and the result is 8 bits sampled at 8 kHz for a bit rate of 64 kbits/s. Figure 10.53 shows a block diagram of an SB-ADPCM encoder.

Transmit Quadrature Mirror Filter

The transmit quadrature mirror filter (QMF) takes a 16-bit audio signal sampled at 16 kHz and separates it into a low band and a high band. The filter coefficients

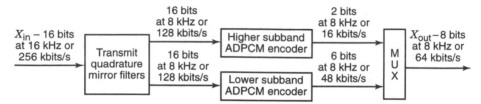

FIGURE 10.53. Block diagram of the ADPCM encoder.

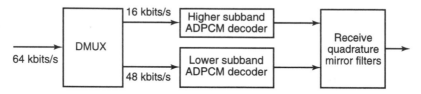

FIGURE 10.54. Block diagram of the ADPCM encoder.

represent a 4-kHz lowpass filter. The sampled signal is separated into odd and even samples, with the effect of aliasing the signals from 4 to 8 kHz. This aliasing causes the high frequency odd samples to be 180° out of phase with the high frequency even samples. The low frequency even and odd samples are in phase. When the odd and even samples are added after being filtered, the low frequency signals constructively add, while the high frequency signals cancel each other, producing a low band signal sampled at 8 kHz.

The low subband encoder converts the low frequencies from the QMF into an error signal that is quantized to 6 bits.

Decoder

The decoder decomposes a 64-kbits/s signal into two signals to form the inputs to the lower and higher SB-ADPCM decoder, as shown in Figure 10.54. The receive QMF consists of two digital filters to interpolate the lower and higher subband ADPCM decoders from 8 to 16 kHz and produce output at a rate of 16 kHz. In the higher SB-ADPCM decoder, adding the quantized difference signal to the signal estimate produces the reconstructed signal.

Components of the ADPCM decoder include an inverse adaptive quantizer, quantizer adaptation, adaptive prediction, predicted value computation, and reconstructed signal computation. With input from a CD player, the DSK reconstructed output signal sound quality was good. Buffered input and reconstructed output data also confirmed successful results from the decoder.

Build this project as **G722**. The support files (encoder and decoder functions, etc.) to implement this project are included on the CD in the folder G722.

10.19 ENCRYPTION USING THE DATA ENCRYPTION STANDARD ALGORITHM

Cryptography is the art of communicating with secret data. In voice communication, cryptography refers to the encrypting and decrypting of voice data through a possibly insecure data line. The goal is to prevent anyone who does not have a "key" from receiving and understanding a transmitted message.

The data encryption standard (DES) is an algorithm that was formerly considered to be the most popular method for private key encryption. DES is still appropriate for moderately secured communication. However, with current computational power, one would be able to break (decrypt) the 56-bit key in a relatively short period of time. As a result, for very secure communication, the DES algorithm has been modified into the triple-DES or (AES) standards. DES is a very popular private-key encryption algorithm and was an industry standard until 1998, after which it was replaced by triple-DES and AES, two slightly more complex algorithms derived from DES [53–56]. Triple-DES increases the size of the key and the data blocks used in this project, essentially performing the same algorithm three times before sending the ciphered data. AES encryption, known as the *Rijndael algorithm*, is the new standard formally implemented by the National Institute of Standards and Technology (NIST) for data encryption in high-level security communications.

DES is a bit-manipulation technique with a 64-bit block cipher that uses an effective key of 56 bits. It is an iterated Feistel-type cipher with 16 rounds. The general model of DES has three main components for (see Figure 10.55): (1) initial permutation; (2) encryption—the core iteration/f-function (16 rounds); and (3) final permutation. X and Y are the input and output data streams in 64-bit block segments, respectively, and $K1$ through $K16$ are distinct keys used in the encryption algorithm. The initial permutation is based on the predefined Table 10.10. The value at each position is used to scramble the input before the encryption routine. For example, the 58th bit of data is moved into the first position of a 64-bit array, the 50th bit into position 2, and so on. The input stream is permutated using a nonrepetitive random table of 64 integers (1–64) that corresponds to a new position of each bit in the 64-bit data block. The final permutation is the reverse of the initial permutation to reorder the samples into the correct original formation. The initial permutation is

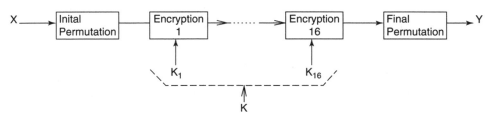

FIGURE 10.55. DES model.

TABLE 10.10 Initial Permutation

			IP				
58	50	42	34	26	18	10	2
60	52	44	36	28	20	12	4
62	54	46	38	30	22	14	6
64	56	48	40	32	24	16	8
57	49	41	33	25	17	9	1
59	51	43	35	27	19	11	3
61	53	45	37	29	21	13	5
63	55	47	39	31	23	15	7

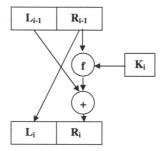

FIGURE 10.56. Encryption process—one round.

followed by the actual encryption. The permutated 64-bit block is divided into a left and a right block of 32 bits each. Sixteen rounds take place, each undergoing a similar procedure, as illustrated in Figure 10.56. The right block is placed into the left block of the next round, and the left block is combined with an encoded version of the right block and placed into the right block of the next round, or

$$L_i = R_{i-1}$$
$$R_i = L_{i-1} \oplus f(R_{i-1}, k_i)$$

where L_{i-1} and R_{i-1} are the left and right blocks, respectively, each with 32 bits, and k_i is the distinct key for the particular round of encryption. The original key is sent through a key scheduler that alters the key for each round of encryption. The left block is not utilized until the very end, when it is XORed with the encrypted right block.

The f-function operating on a 32-bit quantity expands these 32 bits into 48 bits using the expansion table (see Table 10.11). This expansion table performs a permutation while duplicating 16 of the bits (the rightmost two columns). For example, the first integer is 32, so that the first bit in the output block will be bit 32; the second integer is 1, so that the second bit in the output block will be bit 1; and so on.

TABLE 10.11   Expansion of 32 bits to 48

32	1	2	3	4	5
4	5	6	7	8	9
8	9	10	11	12	13
12	13	14	15	16	17
16	17	18	19	20	21
20	21	22	23	24	25
24	25	26	27	28	29
28	29	30	31	32	1

TABLE 10.12   S-Box Example, S_1

14	4	13	1	2	15	11	8	3	10	6	12	5	9	0	7
0	15	7	4	14	2	13	1	10	6	12	11	9	5	3	8
4	1	14	8	13	6	2	11	15	12	9	7	3	10	5	0
15	12	8	2	4	9	1	7	5	11	3	14	10	0	6	14

TABLE 10.13   P-Box

16	7	20	21	29	12	28	17	1	15	23	26	5	18	31	10
2	8	24	14	32	27	3	9	19	13	30	6	22	11	4	25

The 48-bit key transformations are XORed with these expanded data, and the results are used as the input to eight different S-boxes. Each S-box takes 6 consecutive bits and outputs only 4 bits. The 4 output bits are taken directly from the numbers found in a corresponding S-box table. This process is similar to that of a decoder where the 6 bits act as a table address and the output is a binary representation of the value at that address. The zeroth and fifth bits determine the row of the S-box, and the first through fourth bits determine which column the number is located in. For example, 110100 points to the third row (10) and 10th column (1010). The first 6 bits of data correspond to the first of eight S-box tables, shown in Table 10.12. The 32 bits of output from the S-boxes are permutated according to the P-box shown in Table 10.13, and then output from the f-function shown in Figure 10.57. For example, from Table 10.13, bits 1 and 2 from the input block will be moved to bits 16 and 7 in the output, respectively. After the 16 rounds of encryption, a final permutation occurs, which reverses the initial permutation, yielding an encrypted data signal.

The signal output from the encryption algorithm is not decipherable by the human ear even if the signal is filtered in any way. For testing purposes, the first three onboard switches were utilized: sw0 for selecting different keys; sw1 to enable encryption only, or both encryption and decryption; and sw2 as an on/off switch (a loop program).

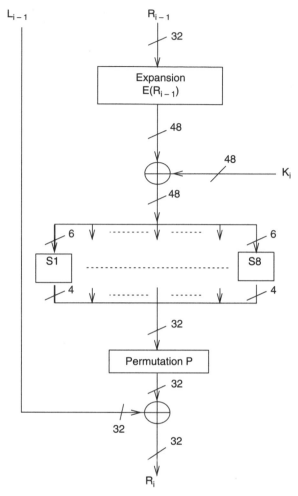

FIGURE 10.57. Core f-function of DES.

This project was successfully implemented on the C6711 DSK with a different onboard codec and can be transported to a C6713 DSK. All the necessary files are in the folder **encryption**. The sections of code associated with the onboard switches need to be modified so that the corresponding available library support functions are utilized. The highest level of compiler optimization (-o3) was utilized in building this project.

10.20 PHASE-LOCKED LOOP

The PLL project implements a software-based linear PLL. The basic PLL causes a particular system to track another PLL. It consists of a phase detector, a loop filter, and a voltage-controlled oscillator. The software PLL is more versatile. However, it is limited by the range in frequency that can be covered, since the PLL function must be executed at least once every period of the input signal [57–59].

Initially, the PLL was tested using MATLAB, then ported to the C6x using C. The PLL locks to a sine wave, generated either internally within the program or from an external source. Output signals are viewed on a scope or on a PC using RTDX.

Figure 10.58 shows a block diagram of the linear PLL implemented in two versions:

1. Using an external input source, with the output of the digitally controlled oscillator (DCO) to an oscilloscope
2. Using RTDX with an input sine wave generated from a lookup table and various signals viewed using Excel

The phase detector, from Figure 10.58, multiplies the input sine wave by the square wave output of the DCO. The sum and difference frequencies of the two inputs to the phase detector produce an output with a high and a low frequency component, respectively. The low frequency component is used to control the loop,

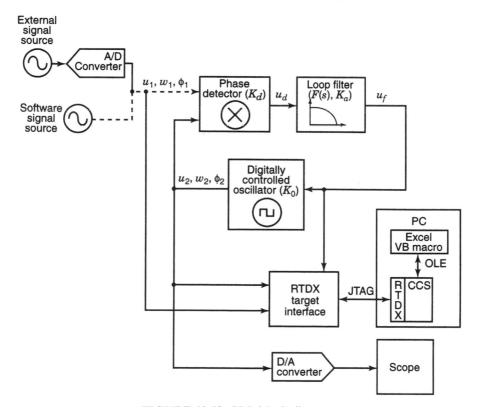

FIGURE 10.58. PLL block diagram.

while the high frequency component is filtered out. When the PLL is locked, the two inputs to the phase detector are at the same frequency but with a quadrature (90°) relationship.

The loop filter is a lowpass filter that passes the low frequency output component of the phase detector while it attenuates the undesired high frequency component. The loop filter is implemented as a single-pole IIR filter with a zero to improve the loop's dynamics and stability. The scaled output of the loop filter represents the instantaneous incremental phase step the DCO is to take. The DCO outputs a square wave as a Walsh function: +1 for phase between 0 and π and −1 for phase between $-\pi$ and 0, with an incremental phase proportional to the number at its input.

RTDX for Real-Time Data Transfer

The RTDX feature was used to transfer data to the PC host using a sine wave from a lookup table as input. A single output channel was created to pass to CCS the input signal, the output of both the loop filter and the DCO, and time stamps. CCS buffers these data so that they can be accessed by other applications on the PC host. CCS has an interface that allows PC applications to access buffered RTDX data. Visual Basic Excel was used to display the results on the PC monitor. Chapter 9 introduced RTDX with several examples using different schemes.

This project was implemented on the C6211 DSK and can be transported to the C6713 DSK. All the necessary files, including the MATLAB file to test the project, are on the CD in the folder **PLL**.

10.21 MISCELLANEOUS PROJECTS

The following projects can also be used as a source of ideas to implement other projects.

10.21.1 Multirate Filter

With multirate processing, a filter can be realized with fewer coefficients than with an equivalent single-rate approach. Possible applications include a controlled noise source and background noise synthesis.

Introduction

Multirate processing uses more than one sampling frequency to perform a desired processing operation. The two basic operations are *decimation*, which is a sampling-rate reduction, and *interpolation*, which is a sampling-rate increase. Decimation techniques have been used in filtering. Multirate decimators can reduce the computational requirements of the filter. Interpolation can be used to obtain a sampling-rate increase. For example, a sampling-rate increase by a factor of K can be achieved by padding $K - 1$ zeros between pairs of consecutive input samples x_i and x_{i+1}. We can also obtain a noninteger sampling-rate increase or decrease by cascading the decimation process with the interpolation process. For example, if a net sampling-rate increase of 1.5 is desired, we would interpolate by a factor of 3, padding (adding) two zeros between each input sample, and then decimate with the interpolated input samples shifted by 2 before each calculation. Decimating or interpolating over several stages generally results in better efficiency [60–67].

Design Considerations

A binary random signal is fed into a bank of filters that are used to shape the output spectrum. The functional block diagram of the multirate filter is shown in Figure 10.59. The frequency range is divided into 10 octave bands, with each band $\frac{1}{3}$-octave controllable. The control of each octave band is achieved with three filters. The coefficients of these filters are combined to yield a composite filter with one set of coefficients for each octave. Only three unique sets of filter coefficients (low, middle, and high) are required, because the center frequency and the bandwidth are proportional to the sampling frequency. Each of the $\frac{1}{3}$-octave filters has a bandwidth of approximately 23% of its center frequency, a stopband rejection of greater than 45 dB, with an amplitude that can be controlled individually. This control provides the capability of shaping an output pseudorandom noise spectrum. The sampling rate of the output is chosen to be 16,384 Hz. Forty-one coefficients are used for the highest $\frac{1}{3}$-octave filter to achieve these requirements. The middle $\frac{1}{3}$-octave filter coefficients were used as BP41.cof in Chapter 4.

In order to meet the filter specifications in each region with a *constant* sampling rate, the number of filter coefficients must be doubled from one octave filter to the next lower one. As a result, the lowest-octave filter would require 41×2^9 coefficients. With 10 filters ranging from 41 to 41×2^9 coefficients, the computational requirements would be considerable. To reduce these computational requirements, a multirate approach is used, as shown in Figure 10.59.

The noise generator is a software-based implementation of a maximal length sequence technique used for generating pseudorandom numbers. This pseudorandom noise generator was implemented in Example 3.3. The output of the noise generator provides uncorrelated noise input to each of the 10 sets of bandpass filters. The noise generation example in Chapter 3 uses the process shown in Figure 10.60.

Because each $\frac{1}{3}$-octave filter can be scaled individually, a total of 30 levels can be controlled. The output of each octave bandpass filter (except the last one) becomes the input to an interpolation lowpass filter, using a 2:1 interpolation factor. The ripple in the output spectrum is minimized by having each adjacent $\frac{1}{3}$-octave filter with crossover frequencies at the 3-dB points.

The center frequency and bandwidth of each filter are determined by the sampling rate. The sampling rate of the highest-octave filter is processed at 16,384 samples per second (you can use a sampling rate of 16 kHz, 48 kHz, etc.), and each successively lower-octave band is processed at half the rate of the next higher band.

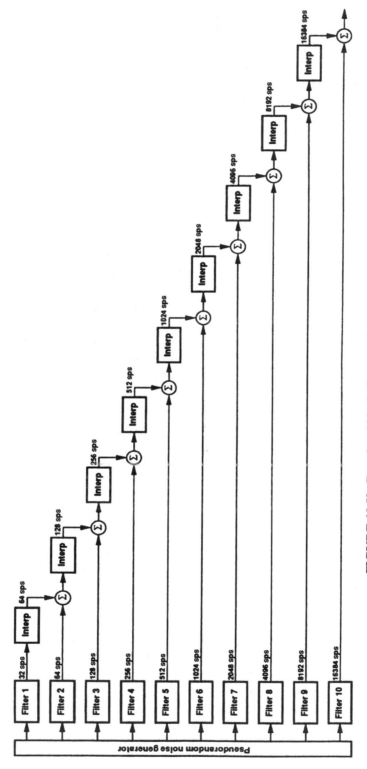

FIGURE 10.59. Functional block diagram of a 10-band multirate filter.

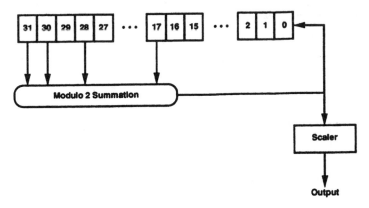

FIGURE 10.60. A 32-bit pseudorandom generator.

Only three separate sets of 41 coefficients are used for the lower, middle, and higher $\frac{1}{3}$-octave bands. For each octave band, the coefficients are combined as follows:

$$H_{ij} = (H_{lj})(L_{3i-2}) + (H_{mj})(L_{3i-1}) + (H_{hj})(L_{3i})$$

where $i = 1, 2, \ldots, 10$ bands and $j = 0, 1, \ldots, 40$ coefficients; $L_1, L_2, \ldots, L_{30}$ represent the level of each $\frac{1}{3}$-octave band filter; and H_{lj}, H_{mj}, H_{hj} represent the jth coefficient of the lower, middle, and higher $\frac{1}{3}$-octave band FIR filter. For example, for the first band ($i = 1$),

$$H_0 = (H_{l0})(L_1) + (H_{m0})(L_2) + (H_{h0})(L_3)$$
$$H_1 = (H_{l1})(L_1) + (H_{m1})(L_2) + (H_{h1})(L_3)$$
$$\vdots$$
$$H_{40} = (H_{l40})(L_1) + (H_{m40})(L_2) + (H_{h40})(L_3)$$

and for band 10 ($i = 10$),

$$H_0 = (H_{l0})(L_{28}) + (H_{m0})(L_{29}) + (H_{h0})(L_{30})$$
$$H_1 = (H_{l1})(L_{28}) + (H_{m1})(L_{29}) + (H_{h1})(L_{30})$$
$$\vdots$$
$$H_{40} = (H_{l40})(L_{28}) + (H_{m40})(L_{29}) + (H_{h40})(L_{30})$$

For an efficient design with the multirate technique, lower-octave bands are processed at a lower sampling rate, then interpolated up to a higher sampling rate, by a factor of 2, to be summed with the next higher octave band filter output, as shown in Figure 10.59. Each interpolation filter is a 21-coefficient FIR lowpass filter, with a cutoff frequency of approximately one-fourth of the sampling rate. For each input, the interpolation filter provides two outputs, or

$$y_1 = x_0 I_0 + 0I_1 + x_1 I_2 + 0I_3 + \cdots + x_{10} I_{20}$$
$$y_2 = 0I_0 + x_0 I_1 + 0I_2 + x_1 I_3 + \cdots + x_9 I_{19}$$

where y_1 and y_2 are the first and second interpolated outputs, respectively, x_n are the filter inputs, and I_n are the interpolation filter coefficients. The interpolator is processed in two sections to provide the data-rate increase by a factor of 2.

For the multirate filter, the approximate number of multiplication operations (with accumulation) per second is

$$\begin{aligned} \text{MAC/S} &= (41 + 21)(32 + 64 + 128 + 256 + 512 + 1024 + 2048 + 4096 + 8192) \\ &\quad + (41)(16,384) \\ &\approx 1.686 \times 10^6 \end{aligned}$$

The approximate number of multiplications/accumulation per second for an equivalent single-rate filter is then

$$\text{MAC/S} = F_s \times 41(1 + 2 + 2^2 + 2^3 + \cdots + 2^9) = 687 \times 10^6$$

which would considerably increase the processing time requirements.

A brief description (recipe) of the main processing follows, for the first time through (using three buffers B_1, B_2, B_3).

Band 1
1. Run the bandpass filter and obtain one output sample.
2. Run the lowpass interpolation filter twice and obtain two outputs. The interpolator provides two sample outputs for each input sample.
3. Store in buffer B_2, size 512, at locations 1 and 2 (in memory).

Band 2
1. Run the bandpass filter two times and sum with the two previous outputs stored in B_2 from band 1.
2. Store the summed values in B_2 at the same locations 1 and 2 (again).
3. Pass the sample in B_2 at location 1 to the interpolation filter twice and obtain two outputs.
4. Store these two outputs in buffer B_3, size 256, at locations 1 and 2.
5. Pass the sample in B_2 at location 2 to the interpolation filter twice and obtain two outputs.
6. Store these two outputs in buffer B_3 at locations 3 and 4.

Band 3

1. Run the bandpass filter four times and sum with the previous four outputs stored in B_3 from band 2.
2. Store the summed values in B_3 at locations 1 through 4.
3. Pass the sample in B_3 at location 1 to the interpolation filter twice and obtain two outputs.
4. Store these two outputs in buffer B_2 at locations 1 and 2.
5. Pass the sample in B_3 at location 2 to the interpolation filter twice and obtain two outputs.
6. Store these two outputs in buffer B_2 at locations 3 and 4.
7. Repeat steps 3 and 4 for the other two samples at locations 3 and 4 in B_3. For each of these samples, obtain two outputs, and store each set of two outputs in buffer B_2 at locations 5 through 8.

Band 10

1. Run the bandpass filter 512 times and sum with the previous 512 outputs stored in B_2 from band 9.
2. Store the summed values in B_2 at locations 1 through 512.

No interpolation is required for band 10. After all the bands are processed, wait for the output buffer B_1, size 512, to be empty. Then switch the buffers B_1 and B_2—the last working buffer with the last output buffer. The main processing is then repeated.

The multirate filter was implemented on the C25 processor using 9 bands and on the C30 processor using 10 bands [8] and can be transported to the C6x. Using a total of 30 different levels, any specific $\frac{1}{3}$-octave filter can be turned on or off. For example, all the filter bands can be turned on except bands 2 and 5. Figure 10.61 shows the frequency response of the three $\frac{1}{3}$-octave filters of band 9 implemented on the C30. Note that if a sampling rate of 8 kHz is set (for the highest band), the middle $\frac{1}{3}$-octave band 1 filter would have a center frequency of 4 Hz (one-fourth of the equivalent sampling rate for band 1).

10.21.2 Acoustic Direction Tracker

This project uses two microphones to capture an audio signal. From the delay associated with the signal reaching one of the microphones before the other, a relative angle where the source is located can be determined. A signal radiated at a distance from its source can be considered to have a plane wavefront, as shown in Figure 10.62. This allows the use of equally spaced sensors (many microphones can be used as acoustical sensors) in a line to ascertain the angle at which the signal is radiating. Since one microphone is closer to the source than the other, the signal received by

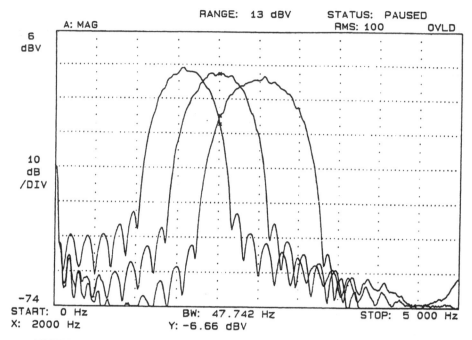

RANGE: 13 dBV   STATUS: PAUSED
A: MAG   RMS: 100   OVLD

FIGURE 10.61. Frequency response of the three $\frac{1}{3}$-octave filters of band 9.

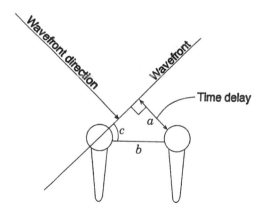

FIGURE 10.62. Signal reception with two microphones.

the more distant microphone is delayed in time. This time shift corresponds to the angle where the source is located and the relative distance between the microphones and the source. The angle $c = \arcsin(a/b)$, where the distance a is the product of the speed of sound and the time delay (phase/frequency).

Figure 10.63 shows a block diagram of the acoustic signal tracker. Two 128-point arrays of data are obtained, cross-correlating the first signal with the second and then the second signal with the first. The resulting cross-correlation data are decom-

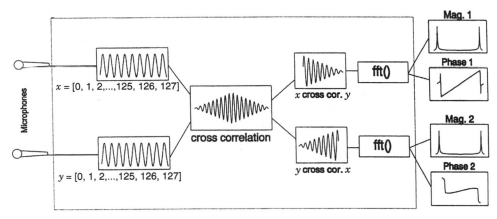

FIGURE 10.63. Block diagram of an acoustic signal tracker.

posed into two halves, each transformed using a 128-point FFT. The resulting phase is the phase difference of the two signals.

This project was implemented on the C30 [17] and can be transported to the C6713 processor. To test this project, a speaker was positioned a few feet from the two microphones, which are separated by 1 foot. The speaker receives a 1-kHz signal from a function generator. A track of the source speaker is plotted over time on the PC monitor. Plots of the cross-correlation and the magnitude of the cross-correlation of the two microphone signals were also displayed on the PC monitor.

10.21.3 Neural Network for Signal Recognition

The goal of this project is to recognize a signal. The FFT of a signal becomes the input to a neural network that is trained to recognize the signal using the back-propagation learning rule.

Design and Implementation

The neural network consists of three layers with a total of 90 nodes: 64 input nodes in the first layer, 24 nodes in the middle or hidden layer, and 2 output nodes in the third layer. The 64 points as input to the neural network are obtained by retaining half of the 128 points resulting from a 128-point FFT of the signal to be recognized. In recent years, many books and articles on neural networks have been published [68, 69]. Neural network products are now available from many vendors.

Many different rules have been described in the literature for training a neural network. The back-error propagation is one of the most widely used for a wide range of applications. Given a set of input, the network is trained to give a desired response. If the network gives the wrong answer, then it is corrected by adjusting its parameters so that the error is reduced. During this correction process, one starts

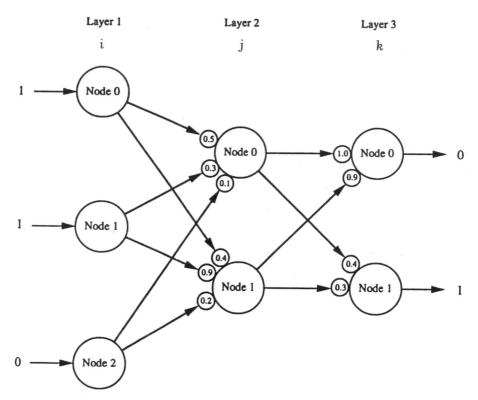

FIGURE 10.64. Three-layer neural network with seven nodes.

with the output nodes and propagation is backward to the input nodes (back propagation). Then the propagation process is repeated.

To illustrate the procedure for training a neural network using the back-propagation rule, consider a simple three-layer network with seven nodes, as shown in Figure 10.64. The input layer consists of three nodes, and the hidden layer and output layer each consist of two nodes. Given the following set of inputs—input No. 1 = 1 into node 0, input No. 2 = 1 into node 1, and input No. 3 = 0 into node 2—the network is to be trained to yield the desired output 0 at node 0 and 1 at node 1. Let the subscripts i, j, k be associated with the first, second, and third layers, respectively. A set of random weights are initially chosen, as shown in Figure 10.64. For example, the weight $w_{11} = 0.9$ represents the weight value associated with node 1 in layer 1 and node 1 in the middle or hidden layer 2. The weighted sum of the input value is

$$s_j = \sum_{i=0}^{2} w_{ji} x_i$$

where $j = 0, 1$ and $i = 0, 1, 2$. Then

$$s_0 = w_{00} x_0 + w_{01} x_1 + w_{02} x_2 = (0.5)(1) + (0.3)(1) + (0.1)(0) = 0.8$$

Similarly, $s_1 = 1.3$. A function of the resulting weighted sum $f(s_j)$ is next computed. This transfer function f of a processing element must be differentiable. For this project, f is chosen as the hyperbolic tangent function tanh. Other functions, such as the unit step function or the smoother sigmoid function, also can be used. The output of the transfer function associated with the nodes in the middle layer is

$$x_j = f(s_j) = \tanh(s_j), \quad j = 0, 1$$

The output of node 0 in the hidden layer then becomes

$$x_0 = \tanh(0.8) = 0.664$$

Similarly, $x_1 = 0.862$. The weighted sum at each node in layer 3 is

$$s_k = \sum_{j=0}^{1} w_{kj} x_j, \quad k = 0, 1$$

to yield

$$s_0 = w_{00}x_0 + w_{01}x_1 = (1.0)(0.664) + (0.9)(0.862) = 1.44$$

Similarly, $s_1 = 0.524$. The output of the transfer function is associated with the output layer, and replacing j by k,

$$x_k = f(s_k), \quad k = 0, 1$$

Then $x_0 = \tanh(1.44) = 0.894$, and $x_1 = \tanh(0.524) = 0.481$. The error in the output layer can now be found using

$$e_k = (d_k - x_k)f'(s_k)$$

where $d_k - x_k$ reflects the amount of error, and $f'(s)$ represents the derivative of $\tanh(s)$, or

$$f'(x) = (1 + f(s))(1 - f(s))$$

Then

$$e_0 = (0 - 0.894)(1 + \tanh(1.44))(1 - \tanh(1.44)) = -0.18$$

Similarly, $e_1 = 0.399$. Based on this output error, the contribution to the error by each hidden layer node is to be found. The weights are then adjusted based on this error using

$$\Delta w_{kj} = \eta e_k x_j$$

where η is the network learning rate constant, chosen as 0.3. A large value of η can cause instability, and a very small one can make the learning process much too slow. Then

$$\Delta w_{00} = (0.3)(-0.18)(0.664) = -0.036$$

Similarly, $\Delta w_{01} = -0.046$, $\Delta w_{10} = 0.08$, and $\Delta w_{11} = 0.103$. The error associated with the hidden layer is

$$e_j = f'(s_j)\sum_{k=0}^{1} e_k w_{kj}$$

Then

$$e_0 = (1+\tanh(0.8))(1-\tanh(0.8))\{(-0.18)(1.0)+(0.399)(0.4)\} = -0.011$$

Similarly, $e_1 = -0.011$. Changing the weights between layers i and j,

$$\Delta w_{ji} = \eta e_j x_i$$

Then

$$\Delta w_{00} = (0.3)(-0.011)(1) = -0.0033$$

Similarly, $\Delta w_{01} = -0.0033$, $\Delta w_{02} = 0$, $\Delta w_{10} = -0.0033$, $\Delta w_{11} = -0.0033$, and $\Delta w_{12} = 0$. This gives an indication of by how much to change the original set of weights chosen. For example, the new set of coefficients becomes

$$w_{00} = w_{00} + \Delta w_{00} = 0.5 - 0.0033 = 0.4967$$

and $w_{01} = 0.2967$, $w_{02} = 0.1$, and so on.

This new set of weights represents only the values after one complete cycle. These weight values can be verified using a training program for this project. For this procedure of training the network, readjusting the weights is continuously repeated until the output values converge to the set of desired output values. For this project, the training program is such that the training process can be halted by the user, who can still use the resulting weights.

This project was implemented on the C30 and can be transported to the C6x. Two sets of inputs were chosen: a sinusoidal and a square wave input. The FFT (128-point) of each input signal is captured and stored in a file, with a total of 4800 points: 200 vectors, each with 64 features (retaining one-half of the 128 points).

Another program scales each set of data (sine and square wave) so that the values are between 0 and 1.

To demonstrate this project, two output values for each node are displayed on the PC screen. Values of +1 for node 0 and −1 for node 1 indicate that a sinusoidal input is recognized, and values of −1 for node 0 and +1 for node 1 indicate that a square wave input is recognized.

This project was successful but was implemented for only the two sets of chosen data. Much work remains to be done, such as training more complex sets of data and examining the effects of different training rules based on the different signals to be recognized.

10.21.4 Adaptive Temporal Attenuator

An adaptive temporal attenuator (ATA) suppresses undesired narrowband signals to achieve a maximum signal-to-interference ratio. Figure 10.65 shows a block diagram of the ATA. The input is passed through delay elements, and the outputs from selected delay elements are scaled by weights. The output is

$$y[k] = \mathbf{m}^{\mathrm{T}} \cdot \mathbf{r}[k] = \sum_{i=0}^{N-1} (\mathbf{m}_i \cdot \mathbf{r}[k-i])$$

where $\mathbf{m}$ is a weight vector, $\mathbf{r}$ a vector of delayed samples selected from the input signal, and N the number of samples in $\mathbf{m}$ and $\mathbf{r}$. The adaptive algorithm computes the weights based on the correlation matrix and a direction vector:

$$\mathbf{C}[k, \delta = 0] \cdot \mathbf{m}[k] = \lambda \mathbf{D}$$

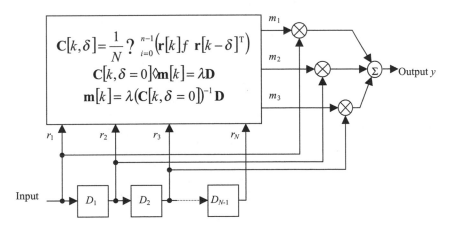

FIGURE 10.65. Block diagram of an adaptive temporal attenuator.

where $\mathbf{C}$ is a correlation matrix, $\mathbf{D}$ a direction vector, and λ a scale factor. The correlation matrix $\mathbf{C}$ is computed as an average of the signal correlation over several samples:

$$\mathbf{C}[k, \delta] = \frac{1}{N_{AV}} \sum_{i=0}^{n-1} (\mathbf{r}[k] \otimes \mathbf{r}[k-\delta]^{T})$$

where N_{AV} is the number of samples included in the average. The direction vector $\mathbf{D}$ indicates the signal desired:

$$\mathbf{D} = [1 \quad \exp(j\omega_T \tau)] \quad \cdots \quad \exp[j\omega_T (N-1)\tau]^{T}$$

where ω_T is the angular frequency of the signal desired, τ the delay between samples that create the output, and N the order of the correlation matrix.

This procedure minimizes the undesired-to-desired ratio (UDR) [70]. UDR is defined as the ratio of the total signal power to the power of the signal desired, or

$$\mathrm{UDR} = \frac{P_{total}}{P_d} = \frac{\mathbf{m}[k]^{T} \cdot \mathbf{C}[k, 0] \cdot \mathbf{m}[k]}{P_d (\mathbf{m}[k]^{T} \cdot \mathbf{D})^2} = \frac{1}{P_d (\mathbf{m}[k]^{T} \cdot \mathbf{D})}$$

where P_d is the power of the signal desired.

MATLAB is used to simulate the ATA, then ported to the C6x for real-time implementation. Figure 10.66 shows the test setup using a fixed desired signal of 1416 Hz and an undesired signal of 1784 Hz (which can be varied). From MATLAB, and optimal value of τ is found to minimize UDR. This is confirmed in real time, since for that value of τ (varying τ with a GEL file), the undesired signal (initially displayed from an HP3561A analyzer) is greatly attenuated.

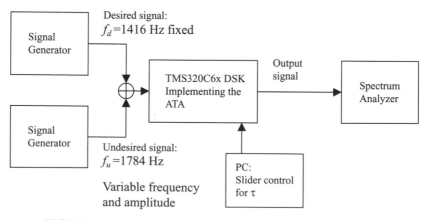

FIGURE 10.66. Test setup for an adaptive temporal attenuator.

10.21.5 FSK Modem

This project implements a digital modulator/demodulator. It generates 8-ary FSK carrier tones. The following steps are performed in the program.

1. The sampled data are acquired as input.
2. The 6 MSBs are separated into two 3-bit samples.
3. The most significant portion of the sample data selects an FSK tone.
4. The FSK tone is sent to a demodulator.
5. The FSK tone is windowed using the Hanning window function.
6. DFT (16-point) results are obtained for the windowed FSK tone.
7. DFT results are sent to the function that selects the frequency with the highest amplitude, corresponding to the upper 3 bits of the sampled data.
8. The process is repeated for the lower 3 bits of the sampled data.
9. The bits are combined and sent to the codec.
10. The gel program allows for an option to interpolate or upsample the reconstructed data for a smoother output waveform.

10.21.6 Image Processing

This project implements various schemes used in image processing:

1. *Edge detection:* for enhancing edges in an image using Sobe's edge detection
2. *Median filtering:* nonlinear filter for removing noise spikes in an image
3. *Histogram equalization:* to make use of the image spectrum
4. *Unsharp masking:* spatial filter to sharpen the image, emphasizing its high frequency components
5. *Point detection:* for emphasizing single-point features in the image

A major issue was using/loading the images as .h files in lieu of using real-time images (due to the course's one-semester time constraint). During the course of this project, the following evolved: a code example for additive noise with a Gaussian distribution, with adjustable variance and mean, and a code example of histogram transformation to map the distribution of one set of numbers to a different distribution (used in image processing).

10.21.7 Filter Design and Implementation Using a Modified Prony's Method

This project designs and implements a filter based on a modified Prony's method [71–74]. The method is based on the correlation property of the filter's representation and does not require computation of any derivatives or an initial guess of the coefficient vector. The filter's coefficients are calculated recursively to obtain the filter's impulse response.

10.21.8 PID Controller

Both nonadaptive and adaptive controllers using the proportional, integral, and derivative (PID) control algorithm have been implemented [17, 75, 76].

10.21.9 Four-Channel Multiplexer for Fast Data Acquisition

A four-channel multiplexer module was designed and built for this project, implemented in C [8]. It includes an 8-bit flash ADC, a FIFO, a MUX, and a crystal oscillator (2 or 20 MHz). An input is acquired through one of the four channels. The FFT of the input signal is displayed in real time on the PC monitor.

10.21.10 Video Line Rate Analysis

This project is discussed in Refs. 8 and 77 and implemented using C and C30 code. It analyzes a video signal at the horizontal (line) rate. Interactive algorithms commonly used in image processing for filtering, averaging, and edge enhancement using C code are utilized for this analysis. The source of the video signal is a charge-coupled device (CCD) camera as input to a module designed and built for this project. This module includes flip-flops, logic gates, and a clock. Displays on the PC monitor illustrate various effects on one horizontal video line signal from either a 500-kHz or a 3-MHz IIR lowpass filter and from an edge enhancement algorithm.

ACKNOWLEDGMENTS

I owe a special debt to all the students who have made this chapter possible. They include students from Roger Williams University, the University of Massachusetts–Dartmouth, and the Worcester Polytechnic Institute (WPI) who have contributed to my general background in real-time DSP applications over the last 20 years: in particular, the undergraduate and graduate students at WPI who have recently taken my two courses on real-time DSP. Many projects and mini-projects from these students are included in this chapter. A special thanks to the following students: N. Alsindi, E. Boron, A. Buchholz, J. Chapman, G. Colangelo, J. Coyne, H. Daempfling, T. Daly, D. Debiasio, A. Dupont, J. Elbin, J. Gaudette, E. Harvey, K. Krishna, M. Lande, M. Lauer, E. Laurendo, R. Lemdiasov, M. Marcantonio, A. Nadkarni, S. Narayanan, A. Navalekar, A. Obi, P. Phadnis, J. Quartararo, V. Rangan, D. Sebastian, M. Seward, D. Tulsiani, and K. Yuksel.

REFERENCES

1. R. Chassaing, *DSP Applications Using C and the TMS320C6x DSK*, Wiley, Hoboken, NJ, 2002.

2. J. H. McClellan, R. W. Schafer, and M. A. Yoder, *DSP First: A Multimedia Approach*, Prentice Hall, Upper Saddle River, NJ, 1998.

3. N. Kehtarnavaz and M. Keramat, *DSP System Design Using the TMS320C6000*, Prentice Hall, Upper Saddle River, NJ, 2001.

4. N. Dahnoun, *DSP Implementation Using the TMS320C6x Processors*, Prentice Hall, Upper Saddle River, NJ, 2000.

5. S. Tretter, *Communication System Design Using DSP Algorithms—With Laboratory Experiments for the TMS320C6701 and TMS320C6711*, Kluwer Academic, Boston, 2003.

6. M. Morrow, T. Welch, C. Cameron, and G. York, Teaching real-time beamforming with the C6211 DSK and MATLAB, *Proceedings of the Texas Instruments DSPS Fest Annual Conference*, 2000.

7. R. Chassaing, *Digital Signal Processing Laboratory Experiments Using C and the TMS320C31 DSK*, Wiley, Hoboken, NJ, 1999.

8. R. Chassaing, *Digital Signal Processing with C and the TMS320C30*, Wiley, Hoboken, NJ, 1992.

9. C. Marven and G. Ewers, *A Simple Approach to Digital Signal Processing*, Wiley, Hoboken, NJ, 1996.

10. J. Chen and H. V. Sorensen, *A Digital Signal Processing Laboratory Using the TMS320C30*, Prentice Hall, Upper Saddle River, NJ, 1997.

11. S. A. Tretter, *Communication System Design Using DSP Algorithms*, Plenum Press, New York, 1995.

12. R. Chassaing et al., Student projects on digital signal processing with the TMS320C30, *Proceedings of the 1995 ASEE Annual Conference*, June 1995.

13. J. Tang, Real-time noise reduction using the TMS320C31 digital signal processing starter kit, *Proceedings of the 2000 ASEE Annual Conference*, 2000.

14. C. Wright, T. Welch III, M. Morrow, and W. J. Gomes III, Teaching real-world DSP using MATLAB and the TMS320C31 DSK, *Proceedings of the 1999 ASEE Annual Conference*, 1999.

15. J. W. Goode and S. A. McClellan, Real-time demonstrations of quantization and prediction using the C31 DSK, *Proceedings of the 1998 ASEE Annual Conference*, 1998.

16. R. Chassaing and B. Bitler, Signal processing chips and applications, *The Electrical Engineering Handbook*, CRC Press, Boca Raton, FL, 1997.

17. R. Chassaing et al., Digital signal processing with C and the TMS320C30: Senior projects, *Proceedings of the 3rd Annual TMS320 Educators Conference*, Texas Instruments, Dallas, TX, 1993.

18. R. Chassaing et al., Student projects on applications in digital signal processing with C and the TMS320C30, *Proceedings of the 2nd Annual TMS320 Educators Conference*, Texas Instruments, Dallas, TX, 1992.

19. R. Chassaing, TMS320 in a digital signal processing lab, *Proceedings of the TMS320 Educators Conference*, Texas Instruments, Dallas, TX, 1991.

20. P. Papamichalis, Ed., *Digital Signal Processing Applications with the TMS320 Family: Theory, Algorithms, and Implementations*, Vols. 2 and 3, Texas Instruments, Dallas, TX, 1989, 1990.

21. *Digital Signal Processing Applications with the TMS320C30 Evaluation Module: Selected Application Notes*, Texas Instruments, Dallas, TX, 1991.

22. R. Chassaing and D. W. Horning, *Digital Signal Processing with the TMS320C25*, Wiley, Hoboken, NJ, 1990.

23. I. Ahmed, Ed., *Digital Control Applications with the TMS320 Family*, Texas Instruments, Dallas, TX, 1991.

24. A. Bateman and W. Yates, *Digital Signal Processing Design*, Computer Science Press, New York, 1991.

25. Y. Dote, *Servo Motor and Motion Control Using Digital Signal Processors*, Prentice Hall, Upper Saddle River, NJ, 1990.

26. R. Chassaing, A senior project course in digital signal processing with the TMS320, *IEEE Transactions on Education*, Vol. 32, pp. 139–145, 1989.

27. R. Chassaing, Applications in digital signal processing with the TMS320 digital signal processor in an undergraduate laboratory, *Proceedings of the 1987 ASEE Annual Conference*, June 1987.

28. K. S. Lin, Ed., *Digital Signal Processing Applications with the TMS320 Family: Theory, Algorithms, and Implementations*, Vol. 1, Prentice Hall, Upper Saddle River, NJ, 1988.

29. G. Goertzel, An algorithm for the evaluation of finite trigonometric series, *American Mathematics Monthly*, Vol. 65, Jan. 1958.

30. ScenixSemiconductors available at `http://www.electronicsweekly.com/_toolkits/system/feature3.asp`.

31. A. Si, Implementing DTMF detection using the Silicon Laboratories Data Access Arrangement (DAA), Scenix Semiconductors, Sept. 1999.

32. `www.gamedev.net/reference/programming/features/beatdetection/`.

33. S. Qian, *Introduction to Time-Frequency and Wavelet Transform*, Prentice-Hall, Upper Saddle River, NJ, 2002.

34. B. Boashah, *Time-Frequency Signal Analysis: Methods and Applications*, Wiley Halsted Press, Hoboken, NJ, 1992.

35. U. Zoler, *Digital Audio Signal*, Wiley, Chichester, England, 1995.

36. J. Proakis and M. Salehi, *Communication Systems Engineering*, Prentice-Hall, Upper Saddle River, NJ, 1994.

37. S. Haykin, *Communication Systems*, Wiley, Hoboken, NJ, 2001.

38. B. Sklar, *Digital Communications: Fundamentals and Applications*, Prentice Hall, Upper Saddle River, NJ, 2001.

39. http://www.physics.gmu.edu/~amin/phys251/Topics/NumAnalysis/ Approximation/polynomialInterp.html.

40. S. Lin and D. J. Costello, *Error Control Coding, Fundamentals and Applications*, Prentice Hall, Upper Saddle River, NJ, 1983.

41. C. Fleming, A tutorial on convolutional encoding with viterbi decoding. Available at http://home.netcom.com/~chip.f/viterbi/algrthms2.html.

42. J. Flanagan and L. Rabiner, *Speech Synthesis*, Dowden, Hutchinson & Ross, Stroudsburg, PA, 1973.

43. R. Rabiner and R. W. Schafer, *Digital Signal Processing of Speech Signals*, Prentice Hall, Englewood Cliffs, NJ, 1978.

44. R. Deller, J. G. Proakis, and J. H. Hansen, *Discrete-Time Processing of Signals*, Macmillan, New York, 1993.

45. B. Gold and N. Morgan, *Speech and Audio Signal Processing*, Wiley, Hoboken, NJ, 2000.

46. R. P. Ramachandran and R. J. Mammoce, Eds., *Modern Methods of Speech Processing*, Kluwer Academic, Boston, 1995.

47. M. N. Do, An automatic speaker recognition system, Audio Visual Communications Lab, Swiss Federal Institute of Technology, Lausanne.

48. X. Huang et al., *Spoken Language Processing*, Prentice Hall, Upper Saddle River, NJ, 2001.

49. R. P. Ramchandran and Peter Kabal, Joint solution for formant and speech predictors in speech processing, *Proceedings of the IEEE International Conference on Acoustics, Speech, Signal Processing*, pp. 315–318, Apr. 1988.

50. L. B. Rabiner and B. H. Juang, *Fundamentals of Speech Recognition*, Prentice Hall, Upper Saddle River, NJ, 1993.

51. *ITU-T Recommendation G.722 Audio Coding with 64 kbits/s*.

52. P. M. Embree, *C Algorithms for Real-Time DSP*, Prentice Hall, Upper Saddle River, NJ, 1995.

53. ECB Mode (Native DES), Frame Technology, 1994. Available at http://www.cs.nps. navy.mil/curricula/tracks/security/notes/chap04_38.html.

54. S. Hallyn, *DES: The Data Encryption Standard*, last modified June 27, 1996. Available at http://www.cs.wm.edu/~hallyn/des.

55. N. Nicolicim, *Data Encryption Standard (DES) History of DES*, McMaster University, lecture notes, October 9, 2001. Available at www.ece.mcmaster.ca/faculty/ nicolici/coe4oi4/2001/lecture10.pdf.

56. B. Sunar, interview and lecture notes. Available at http://www.ece.wpi.edu/~sunar.

57. Roland E. Best, *Phase-Locked Loops Design, Simulation, and Applications*, 4th ed., McGraw-Hill, New York, 1999.

58. W. Li and J. Meiners, *Introduction to Phase Locked Loop System Modeling*, SLTT015, Texas Instruments, Dallas, TX, May 2000.

59. J. P. Hein and J. W. Scott, Z-domain model for discrete-time PLL's, *IEEE Transactions on Circuits and Systems*, Vol. CS-35, pp. 1393–1400, Nov. 1988.

60. R. Chassaing, P. Martin, and R. Thayer, Multirate filtering using the TMS320C30 floating-point digital signal processor, *Proceedings of the 1991 ASEE Annual Conference*, June 1991.

61. R. E. Crochiere and L. R. Rabiner, *Multirate Digital Signal Processing*, Prentice Hall, Upper Saddle River, NJ, 1983.

62. R. W. Schafer and L. R. Rabiner, A digital signal processing approach to interpolation, *Proceedings of the IEEE*, Vol. 61, pp. 692–702, 1973.

63. R. E. Crochiere and L. R. Rabiner, Optimum FIR digital filter implementations for decimation, interpolation and narrow-band filtering, *IEEE Transactions on Acoustics, Speech, and Signal Processing*, Vol. ASSP-23, pp. 444–456, 1975.

64. R. E. Crochiere and L. R. Rabiner, Further considerations in the design of decimators and interpolators, *IEEE Transactions on Acoustics, Speech, and Signal Processing*, Vol. ASSP-24, pp. 296–311, 1976.

65. M. G. Bellanger, J. L. Daguet, and G. P. Lepagnol, Interpolation, extrapolation, and reduction of computation speed in digital filters, *IEEE Transactions on Acoustics, Speech, and Signal Processing*, Vol. ASSP-22, pp. 231–235, 1974.

66. R. Chassaing, W. A. Peterson, and D. W. Horning, A TMS320C25-based multirate filter, *IEEE Micro*, pp. 54–62, Oct. 1990.

67. R. Chassaing, Digital broadband noise synthesis by multirate filtering using the TMS320C25, *Proceedings of the 1988 ASEE Annual Conference*, Vol. 1, June 1988.

68. B. Widrow and R. Winter, Neural nets for adaptive filtering and adaptive pattern recognition, *Computer*, pp. 25–39, Mar. 1988.

69. D. E. Rumelhart, J. L. McClelland, and the PDP Research Group, *Parallel Distributed Processing: Explorations in the Microstructure of Cognition*, Vol. 1, MIT Press, Cambridge, MA, 1986.

70. I. Progri and W. R. Michalson, Adaptive spatial and temporal selective attenuator in the presence of mutual coupling and channel errors, *ION GPS-2000*, 2000.

71. F. Brophy and A. C. Salazar, Recursive digital filter synthesis in the time domain, *IEEE Transactions on Acoustics, Speech, and Signal Processing*, Vol. ASSP-22, 1974.

72. W. H. Press, S. A. Teukolsky, W. T. Vetterling, and B. P. Flannery, *Numerical Recipes in C: The Art of Scientific Computing*, Cambridge University Press, New York, 1992.

73. J. Borish and J. B. Angell, An efficient algorithm for measuring the impulse response using pseudorandom noise, *Journal of the Audio Engineering Society*, Vol. 31, 1983.

74. T. W. Parks and C. S. Burrus, *Digital Filter Design*, Wiley, Hoboken, NJ, 1987.

75. J. Tang, R. Chassaing, and W. J. Gomes III, Real-time adaptive PID controller using the TMS320C31 DSK, *Proceedings of the 2000 Texas Instruments DSPS Fest Conference*, 2000.

76. J. Tang and R. Chassaing, PID controller using the TMS320C31 DSK for real-time motor control, *Proceedings of the 1999 Texas Instruments DSPS Fest Conference*, 1999.

77. B. Bitler and R. Chassaing, Video line rate processing with the TMS320C30, *Proceedings of The 1992 International COnference On Signal Processing Applications and Technology (ICSPAT)*, 1992.

78. *MATLAB, The Language of Technical Computing, Version 6.3*, MathWorks, Natick, MA.

A

TMS320C6x Instruction Set

A.1 INSTRUCTIONS FOR FIXED- AND FLOATING-POINT OPERATIONS

Table A.1 shows a listing of the instructions available for the C6x processors. The instructions are grouped under the functional units used by these instructions. These instructions can be used with both fixed- and floating-point C6x processors. Some additional instructions are available for the fixed-point C64x processor [2].

A.2 INSTRUCTIONS FOR FLOATING-POINT OPERATIONS

Table A.2 shows a listing of additional instructions available with the floating-point processor C67x. These instructions handle floating-point type of operations and are grouped under the functional units used by these instructions (see also Table A.1).

REFERENCES

1. *TMS320C6000 CPU and Instruction Set*, SPRU189F, Texas Instruments, Dallas, TX, 2000.

2. *TMS320C6000 Programmer's Guide*, SPRU198G, Texas Instruments, Dallas, TX, 2002.

TABLE A.1 Instructions for Fixed- and Floating-Point Operations

.L Unit	.M Unit	.S Unit	.D Unit
ABS	MPY	ADD	ADD
ADD	MPYH	ADDK	ADDAB
ADDU	MPYHL	ADD2	ADDAH
AND	MPYHLU	AND	ADDAW
CMPEQ	MPYHSLU	B disp	LDB
CMPGT	MPYHSU	B IRP[a]	LDBU
CMPGTU	MPYHU	B NRP[a]	LDH
CMPLT	MPYHULS	B reg	LDHU
CMPLTU	MPYHUS	CLR	LDW
LMBD	MPYLH	EXT	LDB (15-bit offset)[b]
MV	MPYLHU	EXTU	LDBU (15-bit offset)[b]
NEG	MPYLSHU	MV	LDH (15-bit offset)[b]
NORM	MPYLUHS	MVC[a]	LDHU (15-bit offset)[b]
NOT	MPYSU	MVK	LDW (15-bit offset)[b]
OR	MPYU	MVKH	MV
SADD	MPYUS	MVKLH	STB
SAT	SMPY	NEG	STH
SSUB	SMPYH	NOT	STW
SUB	SMPYHL	OR	STB (15-bit offset)[b]
SUBU	SMPYLH	SET	STH (15-bit offset)[b]
SUBC		SHL	STW (15-bit offset)[b]
XOR		SHR	SUB
ZERO		SHRU	SUBAB
		SSHL	SUBAH
		SUB	SUBAW
		SUBU	ZERO
		SUB2	
		XOR	
		ZERO	

[a] S2 only.
[b] D2 only.
Source: Courtesy of Texas Instruments [1, 2].

TABLE A.2 Instructions for Floating-Point Operations

.L Unit	.M Unit	.S Unit	.D Unit
ADDDP	MPYDP	ABSDP	ADDAD
ADDSP	MPYI	ABSSP	LDDW
DPINT	MPYID	CMPEQDP	
DPSP	MPYSP	CMPEQSP	
DPTRUNC		CMPGTDP	
INTDP		CMPGTSP	
INTDPU		CMPLTDP	
INTSP		CMPLTSP	
INTSPU		RCPDP	
SPINT		RCPSP	
SPTRUNC		RSQRDP	
SUBDP		RSQRSP	
SUBSP		SPDP	

Source: Courtesy of Texas Instruments [1, 2].

B

Registers for Circular Addressing and Interrupts

A number of special-purpose registers available on the C6x processor are shown in Figures B.1 to B.8 [1].

1. Figure B.1 shows the address mode register (AMR) that is used for the circular mode of addressing. It is used to select one of eight register pointers (A4 through A7, B4 through B7) and two blocks of memories (BK0, BK1) that can be used as circular buffers.

2. Figure B.2 shows the control status register (CSR) with bit 0 for the global interrupt enable (GIE) bit.

3. Figure B.3 shows the interrupt enable register (IER).

4. Figure B.4 shows the interrupt flag register (IFR).

5. Figure B.5 shows the interrupt set register (ISR).

6. Figure B.6 shows the interrupt clear register (ICR).

7. Figure B.7 shows the interrupt service table pointer (ISTP).

8. Figure B.8 shows the serial port control register (SPCR).

In Section 3.7.2 we discuss the AMR register and in Section 3.14 the interrupt registers.

REFERENCE

1. *C6000 CPU and Instruction Set*, SPRU189F, Texas Instruments, Dallas, TX, 2000.

Digital Signal Processing and Applications with the TMS320C6713 and TMS320C6416 DSK,
Second Edition By Rulph Chassaing and Donald Reay
Copyright © 2008 John Wiley & Sons, Inc.

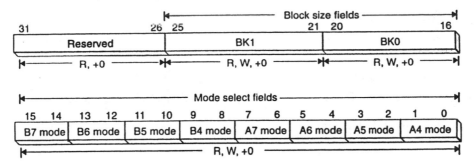

FIGURE B.1. Address mode register (AMR). (Courtesy of Texas Instruments.)

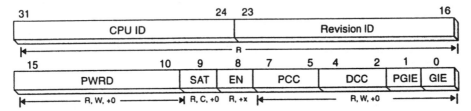

FIGURE B.2. Control status register (CSR). (Courtesy of Texas Instruments.)

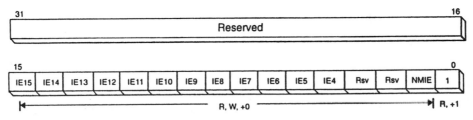

FIGURE B.3. Interrupt enable register (IER). (Courtesy of Texas Instruments.)

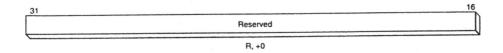

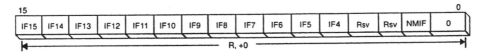

FIGURE B.4. Interrupt flag register (IFR). (Courtesy of Texas Instruments.)

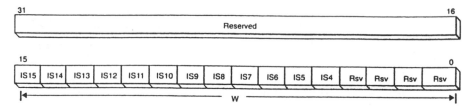

FIGURE B.5. Interrupt set register (ISR). (Courtesy of Texas Instruments.)

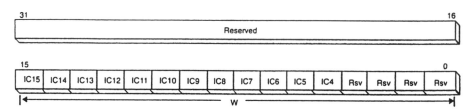

FIGURE B.6. Interrupt clear register (ICR). (Courtesy of Texas Instruments.)

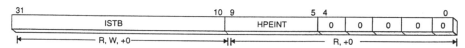

FIGURE B.7. Interrupt service table pointer (ISTP). (Courtesy of Texas Instruments.)

31		24	23	22	21	20	19	18	17	16
reserved†			$\overline{FRST}$	$\overline{GRST}$	XINTM		XSYNCERR‡	$\overline{XEMPTY}$	XRDY	$\overline{XRST}$
R, +0			RW, +0	RW, +0	RW, +0		RW, +0	R, +0	R, +0	RW, +0

15	14	13	12	10	9	8	7	6	5	4	3	2	1	0
DLB	RJUST	CLKSTP		reserved		reserved	reserved	RINTM			RSYNCERR	RFULL	RRDY	$\overline{RRST}$
RW,+0	RW, +0	RW,+0		R, +0		R, +0	R, +0	RW, +0			RW, +0	R, +0	R, +0	RW, +0§

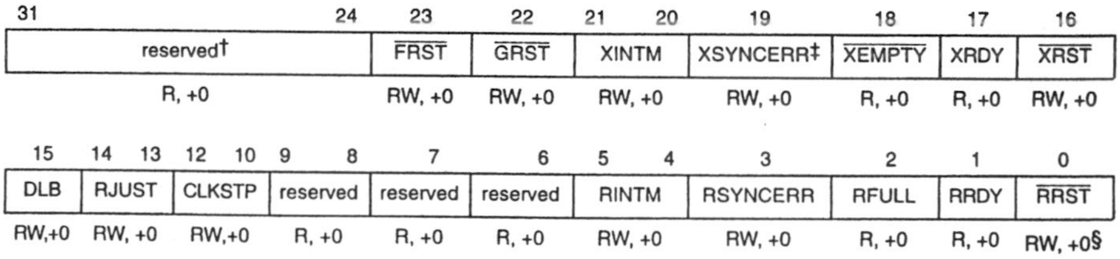

FIGURE B.8. Serial port control register (SPCR). (Courtesy of Texas Instruments.)

C

Fixed-Point Considerations

The C6713 is a floating-point processor capable of performing both integer and floating-point operations. Both the C6713 and the A1C23 codec support 2's-complement arithmetic. It is thus appropriate here to review some fixed-point concepts [1].

In a fixed-point processor, numbers are represented in integer format. In a floating-point processor, both fixed- and floating-point arithmetic can be handled. With the floating-point processor C6713, a much greater range of numbers can be represented than with a fixed-point processor.

The dynamic range of an N-bit number based on 2's-complement representation is between $-(2^{N-1})$ and $(2^{N-1} - 1)$, or between $-32,768$ and $32,767$ for a 16-bit system. By normalizing the dynamic range between -1 and 1, the range will have 2^N sections, where $2^{-(N-1)}$ is the size of each section starting at -1 up to $1 - 2^{-(N-1)}$. For a 4-bit system, there would be 16 sections, each of size $\frac{1}{8}$ from -1 to $\frac{7}{8}$.

C.1 BINARY AND TWO'S-COMPLEMENT REPRESENTATION

To make illustrations more manageable, a 4-bit system is used rather than a 32-bit word length. A 4-bit word can represent the unsigned numbers 0 through 15, as shown in Table C.1.

The 4-bit unsigned numbers represent a modulo (mod) 16 system. If 1 is added to the largest number (15), the operation wraps around to give 0 as the answer. Finite bit systems have the same modulo properties as number wheels on combina-

Digital Signal Processing and Applications with the TMS320C6713 and TMS320C6416 DSK,
Second Edition   By Rulph Chassaing and Donald Reay
Copyright © 2008 John Wiley & Sons, Inc.

TABLE C.1 Unsigned Binary Number

Binary	Decimal
0000	0
0001	1
0010	2
0011	3
.	.
.	.
.	.
1110	14
1111	15

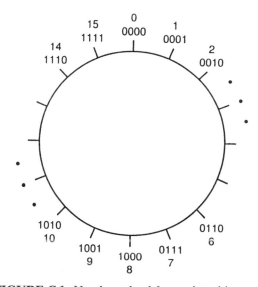

FIGURE C.1. Number wheel for unsigned integers.

tion locks. Therefore, a number wheel graphically demonstrates the addition properties of a finite bit system. Figure C.1 shows a number wheel with the numbers 0 through 15 wrapped around the outside. For any two numbers x and y in the range, the operation amounts to the following procedure:

1. Find the first number x on the wheel.
2. Step off y units in the clockwise direction, which brings you to the answer.

For example, consider the addition of the two numbers $(5 + 7)$ mod 16, which yields 12. From the number wheel, locate 5, then step 7 units in the clockwise direction to arrive at the answer, 12. As another example, $(12 + 10)$ mod 16 = 6. Starting with 12 on the number wheel, step 10 units clockwise, past zero, to 6.

Negative numbers require a different interpretation of the numbers on the wheel. If we draw a line through 8 cutting the number wheel in half, the right half will

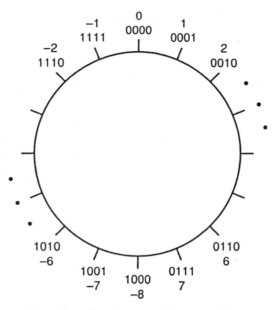

FIGURE C.2. Number wheel for signed integers.

represent the positive numbers and the left half the negative numbers, as shown in Figure C.2. This representation is the 2's-complement system. The negative numbers are the 2's complement of the positive numbers, and vice versa.

A 2's-complement binary integer,

$$B = b_{n-1} \cdots b_1 b_0$$

is equivalent to the decimal integer

$$I(B) = -b_{n-1} \times 2^{n-1} + \cdots + b_1 \times 2^1 + b_0 \times 2^0$$

where the b's are binary digits. The sign bit has a negative weight; all the others have positive weights. For example, consider the number -2:

$$1110 = -1 \times 2^3 + 1 \times 2^2 + 1 \times 2^1 + 0 \times 2^0 = -8 + 4 + 2 + 0 = -2$$

To apply the graphical technique to the operation 6 + (−2) mod 16 = 4, locate 6 on the wheel, then step off (1110) units clockwise to arrive at the answer 4.

The binary addition of these same numbers,

$$
\begin{array}{r}
0110 \\
1110 \\
\hline
10100 \\
C
\end{array}
$$

shows a carry in the most significant bit, which in the case of finite register arithmetic will be ignored. This carry corresponds to the wraparound through zero on the number wheel. The addition of these two numbers results in correct answers, by ignoring the carry in the most significant bit position, provided that the answer is in the range of representable numbers -2^{n-1} to $(2^{n-1} - 1)$ in the case of an n-bit number, or between -8 and 7 for the 4-bit number wheel example. When -7 is added to -8 in the 4-bit system, we get an answer of $+1$ instead of the correct value of -15, which is out of range. When two numbers of like sign are added to produce an answer with opposite sign, overflow has occurred. Subtraction with 2's-complement numbers is equivalent to adding the 2's complement of the number being subtracted to the other number.

C.2 FRACTIONAL FIXED-POINT REPRESENTATION

Rather than using the integer values just discussed, a fractional fixed-point number that has values between $+0.99\ldots$ and -1 can be used. To obtain the fractional n-bit number, the radix point must be moved $n - 1$ places to the left. This leaves one sign bit plus $n - 1$ fractional bits. The expression

$$F(B) = -b_0 \times 2^0 + b_1 \times 2^{-1} + b_2 \times 2^{-2} + \cdots + b_{n-1} \times 2^{-(n-1)}$$

converts a binary fraction to a decimal fraction. Again, the sign bit has a weight of negative 1 and the weights of the other bits are positive powers of 1/2. The number wheel representation for the fractional 2's-complement 4-bit numbers is shown in Figure C.3. The fractional numbers are obtained from the 2's-complement integer numbers of Figure C.2 by scaling them by 2^3. Because the number of bits in a 4-bit system is small, the range is from -1 to 0.875. For a 16-bit word, the signed integers range from $-32,768$ to $+32,767$. To get the fractional range, scale those two signed integers by 2^{-15} or 32,768, which results in a range from -1 to 0.999969 (usually taken as 1).

C.3 MULTIPLICATION

If one multiplies two n-bit numbers, the common notion is that a $2n$-bit operand will result. Although this is true for unsigned numbers, it is not so for signed numbers. As shown before, sign numbers need one sign bit with a weight of -2^{n-1}, followed by positive weights that are powers of 2. To find the number of bits needed for the result, multiply the two largest numbers together:

$$P = (-2^{n-1})(-2^{n-1}) = 2^{2n-2}$$

This number is a positive number representable in $(2n - 1)$ bits. The MSB of this result occupies the $(2n - 2)$-bit position counting from 0. Since this number is posi-

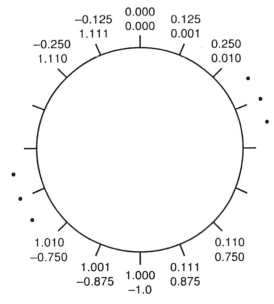

FIGURE C.3. Number wheel for fixed-point representation.

tive, its sign bit, which would show up as a negative number (a power of 2), does not appear. This is an exceptional case, which is treated as an overflow in fractional representation. Since the fractional representation requires that both operand and resultant occupy the same range, $-1 \geq$ range $< +1$, the operation $(-1) \times (-1)$ produces an unrepresentable number, $+1$.

Consider the next larger combination:

$$P = (-2^{n-1})(-2^{n-1} + 1) = 2^{2n-2} - 2^{n-1}$$

Since the second number subtracts from the first, the product will occupy up to the $(2n - 3)$-bit position, counting from 0. Thus, it is representable in $(2n - 2)$ bits. With the exceptional case ruled out, this makes the bit position $(2n - 2)$ available for the sign bit of the resultant. Therefore, $(2n - 1)$ bits are needed to support an $(n \times n)$-bit signed multiplication.

To clarify the preceding equation, consider the 4-bit case, or

$$P = (-2^3)(-2^3 + 1) = 2^6 - 2^3$$

The number 2^6 occupies bit position 6. Since the second number is negative, the summation of the two is a number that will occupy only bit positions less than bit position 6, or

$$2^6 - 2^3 = 64 - 8 = 56 = 00111000$$

Thus, bit position 6 is available for the sign bit. The 8-bit equivalent would have 2 sign bits (bits 6 and 7). The C6x supports signed and unsigned multiplies and therefore provides $2n$ bits for the product.

Consider the multiplication of two fractional 4-bit numbers, with each number consisting of 3 fractional bits and 1 sign bit. Let the product be represented by an 8-bit number. The first number is −0.5 and the second number is 0.75; the multiplication is as follows:

$$
\begin{aligned}
-0.50 &= 1.100 \\
\times 0.75 &= 0.110 \\
\hline
&\underline{11111000} \\
&\underline{111000} \\
\hline
&111.101000 \\
&\mathrm{C} \\
&= -2^1 + 2^0 + 2^{-1} + 2^{-3} = -0.375
\end{aligned}
$$

The underlined bits of the multiplicand indicate sign extension. When a negative multiplicand is added to the partial product, it must be sign-extended to the left up to the limit of the product in order to give the proper larger bit version of the same number. To demonstrate that sign extension gives the correct expanded bit number, scan around the number wheel in Figure C.2 in the counterclockwise direction from 0. Write the codes for 5-bit, 6-bit, 7-bit, . . . negative numbers. Note that they would be derived correctly by sign-extending the existing 4-bit codes; therefore, sign extension gives the correct expanded bit number. The carry-out will be ignored; however, the numbers 111.101000 (9-bit word), 11.101000 (8-bit word), and 1.101000 (7-bit word) all represent the same number: −0.375. Thus, the product of the preceding example could be represented by $(2n - 1)$ bits, or 7 bits for a 4-bit system.

When two 16-bit numbers are multiplied to produce a 32-bit result, only 31 bits are needed for the multiply operation. As a result, bit 30 is sign-extended to bit 31. The extended bits are frequently called *sign bits*.

Consider the following example: to multiply $(0101)_2$ by $(1110)_2$, which is equivalent to multiplying 5 by −2 in decimal, which would result in −10. This result is outside the dynamic range {−8, 7} of a 4-bit system. Using a Q-3 format, this corresponds to multiplying 0.625 by −0.25, yielding a result of −0.15625, which is within the fractional range.

When two Q-15 format numbers (each with a sign bit) are multiplied, the result is a Q-30 format number with one extra sign bit. The MSB is the extra sign bit. One can shift right by 15 to retain the MSBs and only one of the 2 sign bits. By shifting right by 15 (dividing by 2^{15}) to be able to store the result into a 16-bit system, this discards the 15 LSBs, thereby losing some precision. One is able to retain high precision by keeping the most significant 15 bits. With a 32-bit system, a left shift by 1 bit would suffice to get rid of the extra sign bit.

Note that when two Q-15 numbers, represented with a range of −1 to 1, are multiplied, the resulting number remains within the same range. However, the addition of two Q-15 numbers can produce a number outside this range, causing overflow. Scaling would then be required to correct this overflow.

REFERENCE

1. R. Chassaing and D. W. Horning, *Digital Signal Processing with the TMS320C25*, Wiley, Hoboken, NJ, 1990.

D

MATLAB and *Goldwave* Support Tools

This appendix gives a brief description of the use of MATLAB and *Goldwave* in support of the exercises in this book. Their use is also described at various other points in the preceding chapters.

D.1 `fdatool` FOR FIR FILTER DESIGN

MATLAB's filter design and analysis tool `fdatool` makes use of MATLAB functions, for example, `cheby1()` (see Chapter 5), that can be called from the MATLAB command line but integrates them with a graphical user interface (GUI) for the design and analysis of filters. It is invoked by typing

```
>> fdatool
```

at the MATLAB command line.

Three MATLAB functions `dsk_fir67()`, `dsk_sos_iir67()`, and `dsk_sos_iir67int()` are provided in the folder `Support` on the CD accompanying this book. These can be used in conjunction with `fdatool` to create coefficient files for use with a number of example programs.

Digital Signal Processing and Applications with the TMS320C6713 and TMS320C6416 DSK,
Second Edition   By Rulph Chassaing and Donald Reay
Copyright © 2008 John Wiley & Sons, Inc.

Example D.1: Design of FIR Bandstop Filter Using `fdatool` and `dsk_fir67()`

This example describes how the filter coefficient file `bs2700f.cof`, used in Example 4.4, was created.

Enter the `fdatool` parameters shown in Figure D.1 to design an FIR bandstop filter centered at 2700 Hz. The filter uses $N = 89$ coefficients and the Kaiser window function. Select *File → Export* and then set the parameters *Export to, Export as*, and *Variable Names* to *Workspace, Coefficients*, and *bs2700*, respectively. Click on *Export*. At the MATLAB command line, type

```
>> dsk_fir67(bs2700)
```

and enter the filename `bs2700f.cof`.

The resultant coefficient (`.cof`) file is listed in Figure D.2. This file is compatible with programs `fir.c`, `firprn.c`, `firprnbuf.c`, `adaptidfir.c`.

Figure D.3 shows Code Composer *Graphical Displays* of the coefficients and their magnitude FFT.

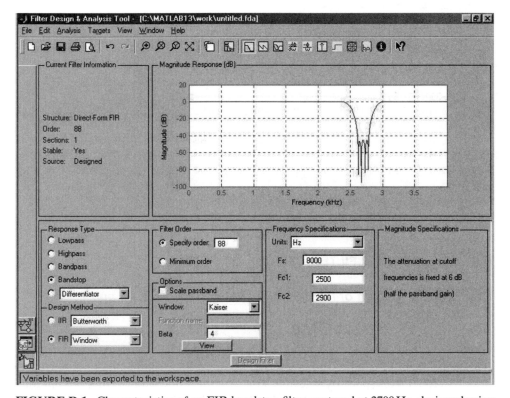

FIGURE D.1. Characteristics of an FIR bandstop filter centered at 2700 Hz, designed using `fdatool`.

```
// bs2700f.cof
// this file was generated automatically using function dsk_fir67.m

#define N 89

float h[N] = {
-4.4230E-004,7.0433E-004,-2.6120E-004,-1.7972E-004,-6.9219E-
018,2.4316E-004,
4.7954E-004,-1.7657E-003,1.5295E-003,1.3523E-003,-4.4872E-
003,3.6368E-003,
2.0597E-003,-7.4813E-003,6.1048E-003,2.2005E-003,-9.5210E-
003,7.8501E-003,
1.6112E-003,-9.1250E-003,7.2785E-003,5.9684E-004,-5.0469E-
003,2.6733E-003,
-2.6271E-017,3.1955E-003,-7.2161E-003,1.0221E-003,1.4959E-002,-
2.2570E-002,
4.8135E-003,2.8456E-002,-4.2116E-002,1.1962E-002,4.1134E-002,-
6.3159E-002,
2.2081E-002,5.0364E-002,-8.2093E-002,3.3692E-002,5.4206E-002,-
9.5274E-002,
4.4497E-002,5.1989E-002,9.0000E-001,5.1989E-002,4.4497E-002,-9.5274E-
002,
5.4206E-002,3.3692E-002,-8.2093E-002,5.0364E-002,2.2081E-002,-
6.3159E-002,
4.1134E-002,1.1962E-002,-4.2116E-002,2.8456E-002,4.8135E-003,-
2.2570E-002,
1.4959E-002,1.0221E-003,-7.2161E-003,3.1955E-003,-2.6271E-
017,2.6733E-003,
-5.0469E-003,5.9684E-004,7.2785E-003,-9.1250E-003,1.6112E-
003,7.8501E-003,
-9.5210E-003,2.2005E-003,6.1048E-003,-7.4813E-003,2.0597E-
003,3.6368E-003,
-4.4872E-003,1.3523E-003,1.5295E-003,-1.7657E-003,4.7954E-
004,2.4316E-004,
-6.9219E-018,-1.7972E-004,-2.6120E-004,7.0433E-004,-4.4230E-004
};
```

FIGURE D.2. Listing of coefficient file `bs2700f.cof`.

D.2 `fdatool` FOR IIR FILTER DESIGN

Example D.2: Design of IIR Bandstop Filter Using `fdatool` and `dsk_sos_iir67()`

Figure D.4 shows the `fdatool` window corresponding to the design of a sixth order IIR bandstop filter centered at 1800 Hz. The filter coefficients can be exported to the MATLAB workspace by selecting *File → Export* and then setting the parameters *Export to, Export as*, and *Variable Names SOS Matrix*, and *Scale Values* to *Workspace, Coefficients, SOS*, and *G*, respectively. Click on *Export*.

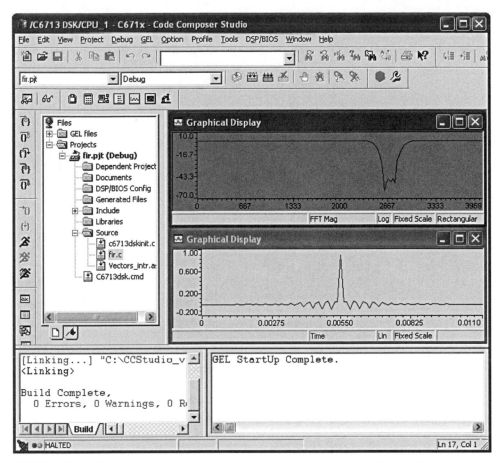

FIGURE D.3. Code Composer window showing filter coefficients read from file `bs2700f.cof` and their magnitude FFT.

At the MATLAB command line, type

```
>> dsk_sos_iir67(SOS,G)
```

and enter the filename `bs1800.cof`, or type

```
>> dsk_sos_iir67int(SOS,G)
```

and enter the filename `bs1800int.cof`.

Coefficient file `bs1800.cof` should be compatible with programs `iirsos.c`, `iirsosprn.c`, `iirsosdelta.c`, and `iirsosadapt.c`. Coefficient file `bs1800int.cof` should be compatible with program `iir.c`.

Figure D.5 shows the output produced by program `iirsosprn.c` using coefficient file `bs1800.cof`.

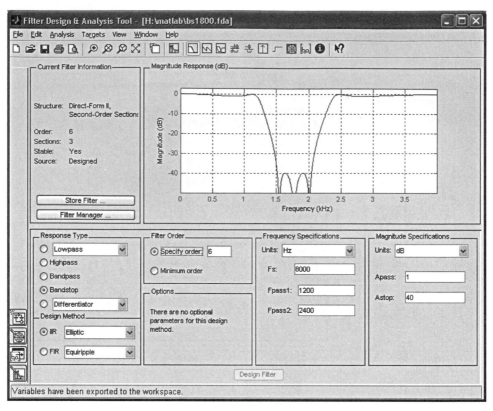

FIGURE D.4. Characteristics of a sixth order IIR bandstop filter centered at 1800 Hz using `fdatool`.

D.3 MATLAB FOR FIR FILTER DESIGN USING THE STUDENT VERSION

FIR filters can be designed using the Student Version [2] of MATLAB [1].

Example D.3: Design of FIR Filters Using the Student Version of MATLAB

Figure D.6 shows the listing of a MATLAB M-file script `mat33.m` that designs a 33-coefficient FIR bandpass filter using the Parks–McClellan algorithm based on the Remez exchange algorithm and Chebyshev approximation theory. The desired filter has a center frequency of 800 Hz (assuming a sampling frequency of 8 kHz) and its magnitude frequency response is specified in vectors `nfreq` and `mag`. Vector `nfreq` contains a set of normalized frequency points, in ascending order, in the range 0 to 1, where 1 corresponds to half the sampling frequency. Vector `mag` contains a set of gain magnitudes corresponding to the frequencies specified in vector `nfreq`.

Function `firpm()` returns the 33 coefficients of an FIR filter designed to meet the specified magnitude frequency response as closely as possible. The coefficients are

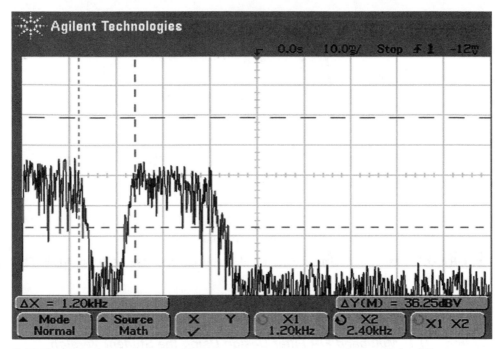

FIGURE D.5. Output generated by program `iirsosprn.c` using coefficient file `bs1800.cof`.

```
%mat33.m M-file for 33-coefficient FIR bandpass filter design
nfreq=[0 0.1 0.15 0.25 0.3 1];    % normalized frequencies
mag = [0 0 1 1 0  0];             % magnitudes at normalized
frequencies
bp33 = firpm(32,nfreq,mag);       % use Parks-McClellan
[h,w]=freqz(bp33,1,512);          % compute frequency response
plot(nfreq,mag,'b-')              % plot desired and computed
hold on                           % frequency responses
plot(w/pi,abs(h),'r')
xlabel('normalized frequency');
ylabel('magnitude');
```

FIGURE D.6. Listing of M-file `mat33.m`.

returned as vector `bp33` and both the desired magnitude frequency response (specified by vectors `nfreq` and `mag`) and the magnitude frequency response calculated using the filter coefficients `bp33` are plotted as shown in Figure D.7. Note that magnitude is plotted on a linear scale.

The filter coefficients can be exported as file `bp33f.cof`, for use by example program `fir.c`, by typing

```
>> dsk_fir67(bp33')
```

at the MATLAB command line and entering the filename `bp33f.cof`.

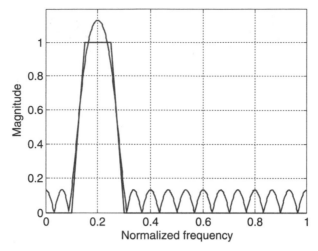

FIGURE D.7. Desired and calculated magnitude frequency responses of FIR bandpass filter designed using M-file mat33.m.

Example D.4: Multiband FIR Filter Design Using the Student Version of MATLAB

This example extends the previous one to design an FIR filter with two passbands. M-file script mat63.m is similar in structure to mat33.m but contains a different magnitude frequency response specified by vectors nfreq and mag. In addition, it produces 63 rather than 33 filter coefficients in a vector bp63. Figure D.8 shows the magnitude frequency response plot produced by mat63.m.

D.4 MATLAB FOR IIR FILTER DESIGN USING THE STUDENT VERSION

Example D.5: Design of IIR Bandstop Filter Using the Bilinear Transform in MATLAB

The analog filter having the transfer function

$$H(s) = \frac{s^2 + 347311379}{s^2 + 4324.75s + 347311379}$$ (D.1)

is a low order IIR bandstop filter centered on 3000 Hz.

Assuming either that it has been entered by typing

```
>> b = [1, 0, 347311379];
>> a = [1, 4324.75, 347311379];
```

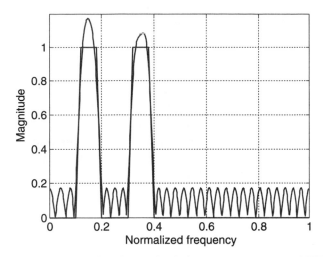

FIGURE D.8. Desired and calculated magnitude frequency responses of FIR filter designed using M-file mat63.m.

or that it has been designed using

```
>> [b,a]=cheby1(1,2,[2*pi*2550,2*pi*3450],'stop','s');
```

its frequency response can be displayed by typing

```
>> freqs(b,a)
```

at the MATLAB command line.

The bilinear transform method of creating an IIR filter based on this analog prototype can be implemented by typing

```
>>[bz,az]=bilinear(b,a,8000);
```

where 8000 specifies a sampling rate of 8 kHz, and the bilinear transform used is

$$s = \frac{2}{8000} \frac{(z-1)}{(z+1)} \tag{D.2}$$

yielding

$$H(z) = \frac{0.8971 + 0.2716z^{-1} + 0.8971z^{-2}}{1 + 0.2716z^{-1} + 0.7942z^{-2}} \tag{D.3}$$

The frequency response of this filter can be displayed by typing

```
>> freqz(bz,az)
```

and should confirm that its stopband is centered not at 3000 Hz (1.885e-4 rad/s) but at approximately 2200 Hz (1.382e-4 rad/s). This is due to the frequency warping effect, described in Chapter 5, of the bilinear transform. A digital filter having a stopband centered at 3000 Hz can be designed by prewarping the prototype analog filter design to match its gain at that frequency. In MATLAB, this is achieved by passing another parameter to function `bilinear()`. Typing

```
>> [bz,az]=bilinear(b,a,8000,3000);
```

causes the bilinear transform used to be

$$s = \frac{2\pi 3000}{\tan(\pi 3000/8000)} \frac{(z-1)}{(z+1)} \qquad (D.4)$$

yielding

$$H(z) = \frac{0.9236 + 1.2956z^{-1} + 0.9236z^{-2}}{1 + 1.2956z^{-1} + 0.8472z^{-2}} \qquad (D.5)$$

Typing

```
>> freqz(bz,az)
```

should confirm that equation (D.5) represents a bandstop filter centered at 3000 Hz as specified by the analog prototype. Either filter (equation (D.3) or equation (D.5)) can be implemented as a single direct form II, second order stage. The MATLAB vectors `bz` and `az` can be written to a `.cof` file by typing

```
>> dsk_sos_iir67([bz,az],[1;1]);
```

and then used by programs `iirsos.c`, `iirsosprn.c`, `iirsosdelta.c`, and `iirsosadapt.c`.

D.5 USING THE *GOLDWAVE* SHAREWARE UTILITY AS A VIRTUAL INSTRUMENT

Goldwave is a shareware utility software program that can turn a PC with a sound-card into a virtual instrument. It can be downloaded from the Internet [3]. One can create a function generator to generate different signals such as a sine wave and random noise. It can also be used as an oscilloscope and as a spectrum analyzer, and to record/edit a speech signal. Effects such as echo and filtering can be applied to stored sounds. Lowpass, highpass, bandpass, and bandstop filters can be implemented on a soundcard with *Goldwave* and their effects on a signal illustrated readily.

Goldwave was used to record the speech contained in files `mefsin.wav` and `corrupt.wav` used in Chapters 2 and 4 and to add the unwanted sine wave components to those recordings.

REFERENCES

1. *MATLAB, The Language of Technical Computing*, MathWorks, Natick, MA, 2003.

2. *MATLAB Student Version*, MathWorks, Natick, MA, 2000.

3. *Goldwave*, available at `www.goldwave.com`.

E

Fast Hartley Transform

Whereas complex additions and multiplications are required for an FFT, the Hartley transform [1–8] requires only real multiplications and additions. The FFT maps a real function of time into a complex function of frequency, whereas the fast Hartley transform (FHT) maps the same real-time function into a real function of frequency. The FHT can be particularly useful in cases where the phase is not a concern.

The discrete Hartley transform (DHT) of a time sequence $x(n)$ is defined as

$$H(k) = \sum_{n=0}^{N-1} x(n)\text{cas}\left(\frac{2\pi nk}{N}\right), \quad k = 0, 1, \ldots, N-1 \tag{E.1}$$

where

$$\text{cas } u = \cos u + \sin u \tag{E.2}$$

In a similar development to the FFT, (E.1) can be decomposed as

$$H(k) = \sum_{n=0}^{(N/2)-1} x(n)\text{cas}\left(\frac{2\pi nk}{N}\right) + \sum_{n=N/2}^{N-1} x(n)\text{cas}\left(\frac{2\pi nk}{N}\right) \tag{E.3}$$

Digital Signal Processing and Applications with the TMS320C6713 and TMS320C6416 DSK, Second Edition   By Rulph Chassaing and Donald Reay
Copyright © 2008 John Wiley & Sons, Inc.

Let $n = n + N/2$ in the second summation of (E.3):

$$H(k) = \sum_{n=0}^{(N/2)-1} \left\{ x(n)\text{cas}\left(\frac{2\pi nk}{N}\right) + x\left(n + \frac{N}{2}\right)\text{cas}\left(\frac{2\pi k(n + N/2)}{N}\right) \right\} \qquad \text{(E.4)}$$

Using (E.2) and the identities

$$\sin(A + B) = \sin A \cos B + \cos A \sin B$$
$$\cos(A + B) = \cos A \cos B - \sin A \sin B \qquad \text{(E.5)}$$

for odd k,

$$\begin{aligned}
\text{cas}\left(\frac{2\pi k(n + N/2)}{N}\right) &= \cos\left(\frac{2\pi nk}{N}\right)\cos(\pi k) - \sin\left(\frac{2\pi nk}{N}\right)\sin(\pi k) \\
&\quad + \sin\left(\frac{2\pi nk}{N}\right)\cos(\pi k) + \cos\left(\frac{2\pi nk}{N}\right)\sin(\pi k) \\
&= -\cos\left(\frac{2\pi nk}{N}\right) - \sin\left(\frac{2\pi nk}{N}\right) \\
&= -\text{cas}\left(\frac{2\pi nk}{N}\right) \qquad \text{(E.6)}
\end{aligned}$$

and, for even k,

$$\text{cas}\left(\frac{2\pi k(n + N/2)}{N}\right) = \cos\left(\frac{2\pi nk}{N}\right) + \sin\left(\frac{2\pi nk}{N}\right) = \text{cas}\left(\frac{2\pi nk}{N}\right) \qquad \text{(E.7)}$$

Using (E.6) and (E.7), (E.4) becomes

$$H(k) = \sum_{n=0}^{(N/2)-1} \left[x(n) + x\left(n + \frac{N}{2}\right) \right]\text{cas}\left(\frac{2\pi nk}{N}\right), \quad \text{for even } k \qquad \text{(E.8)}$$

and

$$H(k) = \sum_{n=0}^{(N/2)-1} \left[x(n) - x\left(n + \frac{N}{2}\right) \right]\text{cas}\left(\frac{2\pi nk}{N}\right), \quad \text{for odd } k \qquad \text{(E.9)}$$

Let $k = 2k$ for even k, and let $k = 2k + 1$ for odd k. Equations (E.8) and (E.9) become

$$H(2k) = \sum_{n=0}^{(N/2)-1} \left[x(n) + x\left(n + \frac{N}{2}\right) \right]\text{cas}\left(\frac{2\pi n 2k}{N}\right) \qquad \text{(E.10)}$$

$$H(2k + 1) = \sum_{n=0}^{(N/2-1)} \left[x(n) - x\left(n + \frac{N}{2}\right) \right]\text{cas}\left(\frac{2\pi n[2k + 1]}{N}\right) \qquad \text{(E.11)}$$

Furthermore, using (E.5)

$$\text{cas}\left(\frac{2\pi n(2k+1)}{N}\right) = \cos\left(\frac{2\pi n}{N}\right)\left\{\cos\left(\frac{2\pi n 2k}{N}\right) + \sin\left(\frac{2\pi n 2k}{N}\right)\right\}$$
$$+ \sin\left(\frac{2\pi n}{N}\right)\left\{\cos\left(\frac{2\pi n 2k}{N}\right) - \sin\left(\frac{2\pi n 2k}{N}\right)\right\}$$

and

$$\sin\left(\frac{2\pi k n}{N}\right) = -\sin\left(\frac{2\pi k(N-n)}{N}\right)$$
$$\cos\left(\frac{2\pi k n}{N}\right) = \cos\left(\frac{2\pi k(N-n)}{N}\right)$$

Equation (E.11) becomes

$$H(2k+1) = \sum_{n=0}^{(N/2)-1}\left\{\left[x(n) - x\left(n+\frac{N}{2}\right)\right]\cos\left(\frac{2\pi n}{N}\right)\text{cas}\left(\frac{2\pi n 2k}{N}\right)\right.$$
$$\left. + \sin\left(\frac{2\pi n}{N}\right)\text{cas}\left(\frac{2\pi 2k(N-n)}{N}\right)\right\} \tag{E.12}$$

Substituting $N/2 - n$ for n in the second summation, (E.12) becomes

$$H(2k+1) = \sum_{n=0}^{(N/2)-1}\left\{\left[x(n) - x\left(n+\frac{N}{2}\right)\right]\cos\left(\frac{2\pi n}{N}\right)\right.$$
$$\left. + \left[x\left(\frac{N}{2}-n\right) - x(N-n)\right]\sin\left(\frac{2\pi n}{N}\right)\right\}\text{cas}\left(\frac{2\pi n 2k}{N}\right) \tag{E.13}$$

Let

$$a(n) = x(n) + x\left(n+\frac{N}{2}\right)$$
$$b(n) = \left[x(n) - x\left(n+\frac{N}{2}\right)\right]\cos\left(\frac{2\pi n}{N}\right)$$
$$+ \left[x\left(\frac{N}{2}-n\right) - x(N-n)\right]\sin\left(\frac{2\pi n}{N}\right)$$

Equations (E.10) and (E.13) become

$$H(2k) = \sum_{n=0}^{(N/2)-1} a(n)\text{cas}\left(\frac{2\pi n 2k}{N}\right) \tag{E.14}$$

$$H(2k+1) = \sum_{n=0}^{(N/2)-1} b(n)\text{cas}\left(\frac{2\pi n 2K}{N}\right) \tag{E.15}$$

A more complete development of the FHT can be found in Ref. 3. We now illustrate the FHT with two exercises: an 8-point FHT and a 16-point FHT. We will then readily verify these results from the FFT exercises in Chapter 6.

Exercise E.1: Eight-Point Fast Hartley Transform

Let the rectangular sequence $x(n)$ be represented by $x(0) = x(1) = x(2) = x(3) = 1$, and $x(4) = x(5) = x(6) = x(7) = 0$. The flow graph in Figure E.1 is used to find $X(k)$. We will now use $X(k)$ instead of $H(k)$. The sequence is first permuted and the intermediate results after the first two stages are as shown in Figure E.1. The coefficients Cn and Sn are (with $N = 8$)

$$Cn = \cos(2\pi n/N)$$

$$Sn = \sin(2\pi n/N)$$

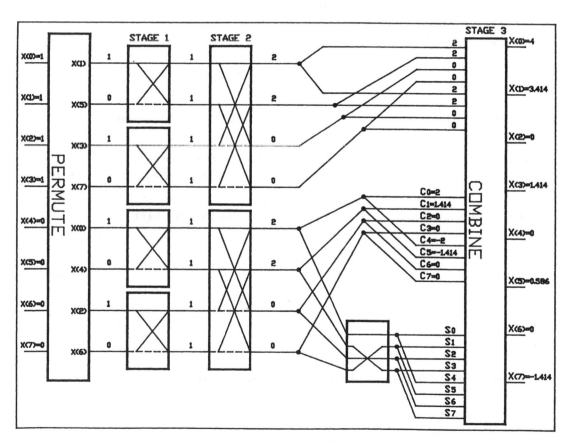

FIGURE E.1. Eight-point FHT flow graph.

The output sequence $X(k)$ after the final stage 3 is also shown in Figure E.1. For example,

$$X(0) = 2 + 2C0 + 2S0 = 2 + 2(1) + 2(0) = 4$$
$$X(1) = 2 + 2C1 + 2S1 = 2 + 1.414 + 0 = 3.41$$
$$\vdots$$
$$X(7) = 0 + 0(C7) + 2S7 = -1.414 \qquad \qquad (E.16)$$

This resulting output sequence can be verified from the $X(k)$ obtained with the FFT, using

$$\text{DHT}\{x(n)\} = \text{Re}\{\text{DFT}[x(n)]\} - \text{Im}\{\text{DFT}[x(n)]\} \qquad (E.17)$$

For example, from the eight-point FFT in Exercise 6.1, $X(1) = 1 - j2.41$, and

$$\text{Re}\{X(1)\} = 1$$
$$\text{Im}\{X(1)\} = -2.41$$

Using (E.17),

$$\text{DHT}\{x(1)\} = X(1) = 1 - (-2.41) = 3.41$$

as in (E.16). Conversely, the FFT can be obtained from the FHT using

$$\text{Re}\{\text{DFT}[x(n)]\} = \tfrac{1}{2}\{\text{DHT}[x(N-n)] + \text{DHT}[x(n)]\}$$
$$\text{Im}\{\text{DFT}[x(n)]\} = \tfrac{1}{2}\{\text{DHT}[x(N-n)] - \text{DHT}[x(n)]\} \qquad (E.18)$$

For example, using (E.18) to obtain $X(1) = 1 - j2.41$ from the FHT,

$$\text{Re}\{X(1)\} = \tfrac{1}{2}\{X(7) + X(1)\} = \tfrac{1}{2}\{-1.41 + 3.41\} = 1$$
$$\text{Im}\{X(1)\} = \tfrac{1}{2}[X(7) - X(1)] = \tfrac{1}{2}\{-1.41 - 3.41\} = -2.41 \qquad (E.19)$$

where the left-hand side of (E.18) is associated with the FFT and the right-hand side with the FHT.

Exercise E.2: Sixteen-Point Fast Hartley Transform

Let the rectangular sequence $x(n)$ be represented by $x(0) = x(1) = \ldots = x(7) = 1$, and $x(8) = x(9) = \ldots = x(15) = 0$. A 16-point FHT flow graph can be arrived at, building on the 8-point FHT. The permutation of the input sequence before the first

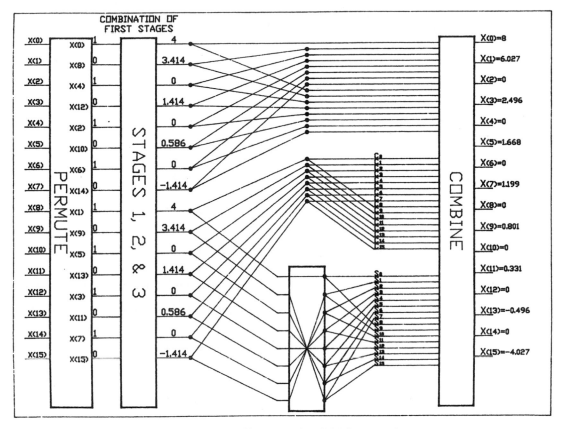

FIGURE E.2. Sixteen-point FHT flow graph.

stage is as follows for the first (upper) eight-point FHT: $x(0)$, $x(8)$, $x(4)$, $x(12)$, $x(2)$, $x(10)$, $x(6)$, $x(14)$ and for the second (lower) eight-point FHT: $x(1)$, $x(9)$, $x(5)$, $x(13)$, $x(3)$, $x(11)$, $x(7)$, $x(15)$. After the third stage, the intermediate output results for the upper and the lower eight-point FHTs are as obtained in the previous eight-point FHT example. Figure E.2 shows the flow graph of the fourth stage for the 16-point FHT. The intermediate output results from the third stage become the input to the fourth stage in Figure E.2. The output sequence $X(0)$, $X(1)$, ..., $X(15)$ from Figure E.2 can be verified using the results obtained with the 16-point FFT in Exercise 6.2. For example, using

$$Cn = \cos\frac{2\pi n}{N} = \cos\frac{\pi n}{8}$$

$$Sn = \sin\frac{2\pi n}{N} = \sin\frac{\pi n}{8}$$

with $N = 16$, $X(1)$ can be obtained from Figure E.2:

$$X(1) = 3.414 + 3.414C1 - 1.414S1 = 3.414 + 3.154 - 0.541 = 6.027$$

Equation (E.18) can be used to verify $X(1) = 1 - j5.028$, as obtained using the FFT in Exercise 6.2. Note that, for example,

$$X(15) = -1.414 + (-1.414C15) + (3.414S15)$$
$$= -1.414 - 1.306 - 1.306$$
$$= -4.0269$$

as shown in Figure E.2.

REFERENCES

1. R. N. Bracewell, The fast Hartley transform, *Proceedings of the IEEE*, Vol. 72, pp. 1010–1018, Aug. 1984.

2. R. N. Bracewell, Assessing the Hartley transform, *IEEE Transactions on Acoustics, Speech, and Signal Processing*, Vol. ASSP-38, pp. 2174–2176, 1990.

3. R. N. Bracewell, *The Hartley Transform*, Oxford University Press, New York, 1986.

4. R. N. Bracewell, *The Fourier Transform and its Applications*, McGraw Hill, New York, 2000.

5. H. V. Sorensen, D. L. Jones, M. T. Heidman, and C. S. Burrus, Real-valued fast Fourier transform algorithms, *IEEE Transactions on Acoustics, Speech, and Signal Processing*, Vol. ASSP-35, pp. 849–863, 1987.

6. H. S. Hou, The fast Hartley transform algorithm, *IEEE Transactions on Computers*, Vol. C-36, pp. 147–156, Feb. 1987.

7. H. S. Hou, Correction to "The fast Hartley transform algorithm," *IEEE Transactions on Computers*, Vol. C-36, pp. 1135–1136, Sept. 1987.

8. A. Zakhor and A. V. Oppenheim, Quantization errors in the computation of the discrete Hartley transform, *IEEE Transactions on Acoustics, Speech, and Signal Processing*, Vol. ASSP-35, pp. 1592–1601, Oct. 1987.

F

Goertzel Algorithm

Goertzel's algorithm performs a DFT using an IIR filter calculation. Compared to a direct N-point DFT calculation, this algorithm uses half the number of real multiplications, the same number of real additions, and requires approximately $1/N$ the number of trigonometric evaluations. The biggest advantage of the Goertzel algorithm over the direct DFT is the reduction of the trigonometric evaluations. Both the direct method and the Goertzel method are more efficient than the FFT when a "small" number of spectrum points is required rather than the entire spectrum. However, for the entire spectrum, the Goertzel algorithm is an N^2 effort, just as is the direct DFT.

F.1 DESIGN CONSIDERATIONS

Both the first order and the second order Goertzel algorithms are explained in several books [1–3] and in Ref. 4. A discussion of them follows. Since

$$W_N^{-kN} = e^{j2\pi k} = 1$$

both sides of the DFT in (6.1) can be multiplied by it, giving

$$X(k) = W_N^{-kN} \sum_{r=0}^{N-1} x(k) W_N^{+kr} \tag{F.1}$$

Digital Signal Processing and Applications with the TMS320C6713 and TMS320C6416 DSK,
Second Edition   By Rulph Chassaing and Donald Reay
Copyright © 2008 John Wiley & Sons, Inc.

which can be written

$$X(k) = \sum_{r=0}^{N-1} x(r) W_N^{-k(N-r)} \tag{F.2}$$

Define a discrete-time function as

$$y_k(n) = \sum_{r=0}^{N-1} x(r) W_N^{-k(n-r)} \tag{F.3}$$

The discrete transform is then

$$X(k) = y_k(n)|_{n=N} \tag{F.4}$$

Equation (F.3) is a discrete convolution of a finite-duration input sequence $x(n)$, $0 < n < N - 1$, with the infinite sequence W_N^{-kn}. The infinite impulse response is therefore

$$h(n) = W_N^{-kn} \tag{F.5}$$

The z-transform of $h(n)$ in (F.5) is

$$H(z) = \sum_{n=0}^{\infty} h(n) z^{-n} \tag{F.6}$$

Substituting (F.5) into (F.6) gives

$$H(z) = \sum_{n=0}^{\infty} W_N^{-kn} z^{-n} = 1 + W_N^{-k} z^{-1} + W_N^{-2k} z^{-2} + \cdots = \frac{1}{1 - W_N^{-2k} z^{-1}} \tag{F.7}$$

Thus, equation (F.7) represents the transfer function of the convolution sum in equation (F.3). Its flow graph represents the first order Goertzel algorithm and is shown in Figure F.1. The DFT of the kth frequency component is calculated by

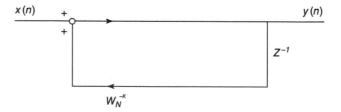

FIGURE F.1. First order Goertzel algorithm.

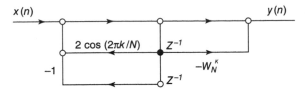

FIGURE F.2. Second order Goertzel algorithm.

starting with the initial condition $y_k(-1) = 0$ and running through N iterations to obtain the solution $X(k) = y_k(N)$. The $x(n)$'s are processed in time order, and processing can start as soon as the first one comes in. This structure needs the same number of real multiplications and additions as the direct DFT but $1/N$ the number of trigonometric evaluations.

The second order Goertzel algorithm can be obtained by multiplying the numerator and denominator of (F.7) by $1 - W_N^{-kn} z^{-1}$ to give

$$H(z) = \frac{1 - W_N^{+k} z^{-1}}{1 - 2\cos(2\pi k/N)z^{-1} + z^{-2}} \tag{F.8}$$

The flow graph for this equation is shown in Figure F.2. Note that the left half of the graph contains feedback flows and the right half contains only feedforward terms. Therefore, only the left half of the flow graph must be evaluated each iteration. The feedforward terms need only be calculated once for $y_k(N)$. For real data, there is only one real multiplication in this graph and only one trigonometric evaluation for each frequency. Scaling is a problem for fixed-point arithmetic realizations of this filter structure; therefore, simulation is extremely useful.

The second order Goertzel algorithm is more efficient than the first order Goertzel algorithm. The first order Goertzel algorithm (assuming a real input function) requires approximately $4N$ real multiplications, $3N$ real additions, and two trigonometric evaluations per frequency component as opposed to N real multiplications, $2N$ real additions, and two trigonometric evaluations per frequency component for the second order Goertzel algorithm. The direct DFT requires approximately $2N$ real multiplications, $2N$ real additions, and $2N$ trigonometric evaluations per frequency component.

This Goertzel algorithm is useful in situations where only a few points in the spectrum are necessary, as opposed to the entire spectrum. Detection of several discrete frequency components is a good example. Since the algorithm processes samples in time order, it allows the calculation to begin when the first sample arrives. In contrast, the FFT must have the entire frame in order to start the calculation.

REFERENCES

1. G. Goertzel, An algorithm for the evaluation of finite trigonometric series, *American Mathematics Monthly*, vol. 65, Jan. 1958.

2. A. V. Oppenheim and R. Schafer, *Discrete-Time Signal Processing*, Prentice Hall, Upper Saddle River, NJ, 1989.

3. C. S. Burrus and T. W. Parks, *DFT/FFT and Convolution Algorithms: Theory and Implementation*, Wiley, Hoboken, NJ, 1988.

4. http://ptolemy.eecs.berkeley.edu/papers/96/dtmf_ict/www/node3.html.

G

TMS320C6416 DSK

G.1 TMS320C64x PROCESSOR

Another member of the C6000 family of processors is the C64x, which can operate at a much higher clock rate. The C6416 DSK operates at 1 GHz for a 1.00-ns instruction cycle time. Features of the C6416 architecture include: four 16×16-bit multipliers (each .M unit can perform two multiplies per cycle), sixty-four 32-bit general-purpose registers, more than 1 MB of internal memory consisting of 1 MB of L2 RAM/cache, and 16 kB of each L1P program cache and L1D data cache [1–7].

The C64x is based on the architecture VELOCITI.2, which is an extension of VELOCITI [2]. The extra registers allow for packed data types to support four 8-bit or two 16-bit operations associated with one 32-bit register, increasing parallelism [3]. For example, the instruction MPYU4 performs four 8-bit multiplications within a single instruction cycle time. Several special-purpose instructions have also been added to handle many operations encountered in wireless and digital imaging applications, where 8-bit data processing is common. In addition, the .M unit (for multiply operations) can also handle shift and rotate operations. Similarly, the .D unit (for data manipulation) can also handle logical operations. The C64x is a fixed-point processor. Existing instructions are available to more units. Double-word load (LDDW) and store (STDW) instructions can access 64 bits of data, with up to a two double-word load or store instructions per cycle (read or write 128 bits per cycle).

Digital Signal Processing and Applications with the TMS320C6713 and TMS320C6416 DSK,
Second Edition   By Rulph Chassaing and Donald Reay
Copyright © 2008 John Wiley & Sons, Inc.

A few instructions have been added for the C64x processor. For example, the instruction

```
BDEC LOOP,B0
```

decrements a counter B0 and performs a conditional branch to LOOP based on B0. The branch decision is before the decrement, with the branch decision based on a negative number (not on whether the number is zero). This multitask instruction resembles the syntax used in the C3x and C4x family of processors.

Furthermore, with the intrinsic C function _dotp2, it can perform two 16×16 multiplies and adds the products together to further reduce the number of cycles. This intrinsic function in C has the corresponding assembly function DOTP2. With two multiplier units, four 16×16 multiplies per cycle can be performed, double the rate of the C62x or C67x. At 720 MHz, this corresponds to 2.88 billion multiply operations per second, or 5.76 billion 8×8 multiplies per second.

G.2 PROGRAMMING EXAMPLES USING THE C6416 DSK

Nearly all of the program examples described in Chapters 1–9 of this book will run on the C6416 DSK provided that the appropriate support files are used. Files c6416dskinit.c and c6416dskinit.h must be used in place of files c6713dskinit.c and c6713dskinit.h and library files csl6416.lib, dsk6416bsl.lib, and rts6400.lib must be used in place of csl6713.lib, dsk6713bsl.lib, and rts6700.lib. Slightly different compiler and linker build options are also required by the C6416 DSK.

Assuming that a C6416 DSK is being used in place of the C6713 DSK and that Code Composer Studio for that DSK has been installed, these issues can be resolved by copying the files supplied on the CD in folder C6416 into folder c:\CCStudio_v3.1\MyProjects. Support files appropriate to the C6416 DSK are stored in folder c:\CCStudio_v3.1\MyProjects\support and the project (.pjt) files provided have been set up to use those support files and with the appropriate compiler and linker options.

Three examples of the use of programs described earlier in this book are presented here.

Example G.1: Sine Wave Generation with DIP Switch Control (sine8_LED)

This example is equivalent to Example 1.1. Figure G.1 shows a listing of program sine8_LED.c provided for the C6416 DSK. The essential differences between this file and that listed in Figure 1.2 concern the header file included (dsk6416_aic23.h), and the support library functions called (e.g., DSK6416_DIP_INIT()). Figure G.2 shows the *Preprocessor Compiler* and *Basic Linker* options for the project. Compare Figures G.2a and G.2b with Figures 1.7 and 1.8 and note, for example, that the

```
//sine8_LED.c  sine generation with DIP switch control

#include "dsk6416_aic23.h"              //codec support
Uint32 fs = DSK6416_AIC23_FREQ_8KHZ;   //set sampling rate
#define DSK6416_AIC23_INPUT_MIC 0x0015
#define DSK6416_AIC23_INPUT_LINE 0x0011
Uint16 inputsource=DSK6416_AIC23_INPUT_MIC; //select input
#define LOOPLENGTH 8
short loopindex = 0;                   //table index
short gain = 10;                       //gain factor
short sine_table[LOOPLENGTH]=
  {0,707,1000,707,0,-707,-1000,-707};  //sine values

void main()
{
  comm_poll();                         //init DSK,codec,McBSP
  DSK6416_LED_init();                  //init LED from BSL
  DSK6416_DIP_init();                  //init DIP from BSL
  while(1)                             //infinite loop
  {
    if(DSK6416_DIP_get(0)==0)          //if DIP #0 pressed
    {
      DSK6416_LED_on();                //turn LED #0 ON
      output_left_sample(sine_table[loopindex++]*gain); //output
      if (loopindex >= LOOPLENGTH) loopindex = 0; //reset index
    }
    else DSK6416_LED_off(0);           //else turn LED #0 OFF
  }                                    //end of while(1)
}                                      //end of main
```

FIGURE G.1. Listing of program sine8_LED.c.

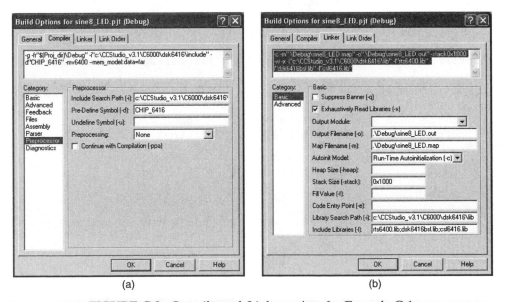

FIGURE G.2. *Compiler* and *Linker* options for Example G.1.

Include and *Library Search Paths*, the *Include Libraries*, and the *Pre-Define Symbol* options are different.

Project file sine8_LED.pjt has been provided so that in order to run the program sine8_LED.c it is necessary only to open that project, build, load, and run.

The functionality of the program, that is, a 1-kHz tone is output via LINE OUT and HEADPHONE sockets while DIP switch #0 is pressed down, is the same as that described in Example 1.1.

Example G.2: Loop Program Using the C6416 DSK (loop_intr)

Figure G.3 shows the C source file loop_intr.c that implements a loop program. Compare this program with that listed in Figure 2.4. Build the project as **loop_ intr** and verify that the results are similar to those described for Example 2.2.

Example G.3: Estimating Execution Times for DFT and FFT Functions

This example is similar to Example 6.2. Three different methods of computing the DFT of 128 sample values are implemented in programs dft.c, dftw.c, and fft.c. Using Code Composer's *Profile Clock*, an indication of the number of processor instruction cycles used for the computation can be obtained.

One of the main differences between the C6416 and C6713 processors is the absence of floating-point hardware in the case of the C6416. The same C programs,

```
//loop_intr.c loop program using interrupts

#include "DSK6416_AIC23.h"            //codec support
Uint32 fs=DSK6416_AIC23_FREQ_8KHZ;    //set sampling rate
#define DSK6416_AIC23_INPUT_MIC 0x0015
#define DSK6416_AIC23_INPUT_LINE 0x0011
Uint16 inputsource=DSK6416_AIC23_INPUT_MIC; //select input

interrupt void c_int11()              //interrupt service routine
{
  short sample_data;

  sample_data = input_left_sample(); //input data
  output_left_sample(sample_data);   //output data
  return;
}

void main()
{
  comm_intr();                       //init DSK, codec, McBSP
  while(1);                          //infinite loop
}
```

FIGURE G.3. Listing of program loop_intr.c.

using floating-point variables, can be compiled and run on the C6416 processor but it will use software routines in place of floating-point hardware in order to carry out floating-point arithmetic operations. In general, the C6416 will use more instruction cycles than the C6713 to carry out floating-point arithmetic. On the other hand, the processor on the C6416 DSK has a clock speed of 1 GHz whereas the C6713 DSK processor uses a 225-MHz clock.

As in the case of Example 6.2, edit the lines in programs dft.c and dftw.c that read

```
#define N 100
```

to read

```
#define N 128
```

Then:

1. Ensure that source file dft.c and not dftw.c is present in the project.
2. Select *Project → Build Options*. In the *Compiler* tab in the *Basic* category set the *Opt Level* to *Function(–o2)* and in the *Linker* tab set the *Output Filename* to .\Debug\dft.out.
3. *Build* the project and load dft.out.
4. *Open* source file dft.c by double-clicking on its name in the *Project View* window and set breakpoints at the lines dft(samples); and printf("done!\n");.
5. Select *Profile → Clock → Enable*.
6. Select *Profile → Clock View*.
7. Run the program. It should halt at the first breakpoint.
8. Reset the *Profile Clock* by double-clicking on its icon in the bottom right-hand corner of the CCS window.
9. Run the program. It should stop at the second breakpoint.

The number of instruction cycles counted by the *Profile Clock* (271,966,152) gives an indication of the computational expense of executing function dft(). On a 1-GHz C6416, 271,966,152 instruction cycles correspond to an execution time of 272 ms. Repeat the preceding experiment substituting file dftw.c for file dft.c. The modified DFT function using twiddle factors, dftw(), uses 6,256,266 instruction cycles, corresponding to 6.26 ms and representing a decrease in execution time by a factor of 43. At a sampling rate of 8 kHz, 6.26 ms corresponds to just over fifty sampling periods.

Finally, repeat the experiment using file fft.c (also stored in folder dft). This program computes the FFT using a function written in C and defined in the file fft.h. Function fft() takes 1,608,328 instruction cycles, or 1.61 ms (approximately

13 sampling periods at 8 kHz) to execute. The advantage, in terms of execution time, of the FFT over the DFT seen in Example 6.2 is repeated here. However, the floating-point computations take more than ten times longer on the C6416 processor.

REFERENCES

1. *TMS320C6416, TMS320C6415, TMS320C6416 Fixed-Pont Digital Signal Processors, SPRS146*, Texas Instruments, Dallas, TX, 2003.

2. *TMS320C6000 Programmer's Guide, SPRU198G*, Texas Instruments, Dallas, TX, 2002.

3. *TMS320C6000 CPU and Instruction Set, SPRU189F*, Texas Instruments, Dallas, TX, 2000.

4. *TMS320C64x Technical Overview, SPRU395*, Texas Instruments, Dallas, TX, 2003.

5. *How to Begin Development Today with the TMS320C6416, TMS320C6415, and TMS320C6416 DSPs Application Report, SPRA718*, Texas Instruments, Dallas, TX, 2003.

6. *TMS320C6000 Chip Support Library API User's Guide, SPRU401*, Texas Instruments, Dallas, TX, 2003.

7. *TMS320C6000 DSK Board Support Library API User's Guide, SPRU432*, Texas Instruments, Dallas, TX, 2001.

Index

Digital Signal Processing and Applications with the TMS320C6713 and TMS320C6416 DSK,
Second Edition By Rulph Chassaing and Donald Reay
Copyright © 2008 John Wiley & Sons, Inc.

CUSTOMER NOTE: IF THIS BOOK IS ACCOMPANIED BY SOFTWARE, PLEASE READ THE FOLLOWING BEFORE OPENING THE PACKAGE.

This software contains files to help you utilize the models described in the accompanying book. By opening the package, you are agreeing to be bound by the following agreement:

WILEY